Climate Variability and Extremes during the Past 100 Years

ADVANCES IN GLOBAL CHANGE RESEARCH

VOLUME 33

The titles published in this series are listed at the end of this volume.

Climate Variability and Extremes during the Past 100 Years

Stefan Brönnimann • Jürg Luterbacher
Tracy Ewen • Henry F. Diaz
Richard S. Stolarski • Urs Neu
Editors

Swiss Re

NCCR CLIMATE
Swiss Climate Research

Max und Elsa Beer-Brawand-Fonds

FNSNF
SWISS NATIONAL SCIENCE FOUNDATION

sc | nat

ProClim–
Forum for Climate and Global Change
Forum of the Swiss Academy of Sciences

Springer

Stefan Brönnimann
Institute for Atmospheric and Climate Science
ETH Zürich, Switzerland

Jürg Luterbacher
NCCR Climate and Oeschger Centre for Climate Change Research and Institute of Geography
University of Bern, Switzerland

Tracy Ewen
Institute for Atmospheric and Climate Science
ETH Zürich, Switzerland

Henry F. Diaz
NOAA-ESRL, Boulder CO
USA

Richard S. Stolarski
NASA Goddard Space Flight Center,
Greenbelt MD, USA

Urs Neu
ProClim, Bern
Switzerland

Cover: Deadvlei pan in the Namib Desert, Namibia (Photo: Tracy Ewen), flooding in Bern, Switzerland, 15 May 1999 (Photo: Stefan Brönnimann), satellite image of sea ice cover in the Barents and Kara sea region, 12 June 2001 (Jacques Descloitres, MODIS Land Rapid Response Team, NASA Visible Earth v1 ID 8126).

ISBN 978-1-4020-6765-5 e-ISBN 978-1-4020-6766-2

Library of Congress Control Number: 2007940817

Printed on acid-free paper.

9 8 7 6 5 4 3 2 1

springer.com

Preface

Large progress has been made in the past few years towards quantifying and understanding climate variability during past centuries. At the same time, present-day climate has been studied using state-of-the-art data sets and tools with respect to the physical and chemical mechanisms governing climate variability. Both the understanding of the past and the knowledge of the processes are important for assessing and attributing the anthropogenic effect on present and future climate. The most important time period in this context is the past approximately 100 years, which comprises large natural variations and extremes (such as long droughts) as well as anthropogenic influences, most pronounced in the past few decades.

Recent and ongoing research efforts steadily improve the observational record of the 20th century, while atmospheric circulation models are used to underpin the mechanisms behind large climatic variations. Atmospheric chemistry and composition are important for understanding climate variability and change, and considerable progress has been made in the past few years in this field. The evolving integration of these research areas in a more comprehensive analysis of recent climate variability was reflected in the organisation of a workshop "Climate variability and extremes in the past 100 years" in Gwatt near Thun (Switzerland), 24–26 July 2006. The aim of this workshop was to bring together scientists working on data issues together with statistical climatologists, modellers, and atmospheric chemists to discuss gaps in our understanding of climate variability during the past approximately 100 years. The peer-reviewed chapter contributions together provide an overview of this entire period, and incorporate both physical and chemical aspects of climate science, employing observational data and models. In summary, the book aims at providing an overview of the current state of research on 20th-century climate variability.

We would like to thank all of the contributors, who made this meeting an extraordinary event. Most importantly, we would like to thank the sponsors of the workshop, namely SwissRE/NCCR Climate, the Swiss National Science Foundation, ProClim, and the Max and Elsa Beer-Brawand-Fonds of the University of Bern.

Zürich, Switzerland — Stefan Brönnimann
Bern, Switzerland — Jürg Luterbacher
Zürich, Switzerland — Tracy Ewen
Boulder, Colorado, USA — Henry F. Diaz
Greenbelt, Maryland, USA — Richard S. Stolarski
Bern, Switzerland — Urs Neu

Contents

Preface v

Contributors xi

Introductory Paper

A Focus on Climate During the Past 100 Years 1
S. Brönnimann, T. Ewen, J. Luterbacher, H. F. Diaz,
R. S. Stolarski, and U. Neu

Section A The Observational Climate Record

From the Bottom to the Stratosphere – Arctic Climate Features as Seen from the First International Polar Year (1882–1883) Until the End of World War II 29
C. Lüdecke

Arctic Sea Ice Data Sets in the Context of Climate Change During the 20th Century 47
G. V. Alekseev, S. I. Kuzmina, A. P. Nagurny, and N. E. Ivanov

The Evolving SST Record from ICOADS 65
S. D. Woodruff, H. F. Diaz, E. C. Kent, R. W. Reynolds,
and S. J. Worley

Upper-air Temperature Trends: Current Problems and Some Recent Results 85
A. M. Sterin, V. M. Khan, and K. G. Rubinshtein

Atmospheric Reanalyses and Climate Variations 103
S. Uppala, A. Simmons, D. Dee, P. Kållberg,
and J.-N. Thépaut

Constructing Climate Quality Atmospheric Temperatures from Satellite Microwave Measurements 119
C. A. Mears

Total Ozone Observations During the Past 80 Years 129
S. Brönnimann, C. Vogler, J. Staehelin, R. Stolarski, and G. Hansen

Section B Climate Trends: Forcings and Processes

Arctic Sea Ice Variability During the Last Half Century 143
J. Stroeve and W. Maslowski

Decadal Changes in Surface Radiative Fluxes and Their Role in Global Climate Change 155
M. Wild

Observed Interdecadal Changes in Cloudiness: Real or Spurious? 169
J. R. Norris

The Role of Land–Atmosphere Interactions for Climate Variability in Europe 179
S. I. Seneviratne and R. Stöckli

Aerosol Effects on Precipitation Locally and Globally 195
U. Lohmann

Section C Climate Variability and Extremes

Human Activity and Climate Change in Africa 209
H. Paeth

Impacts of Climate Variability, Trends and NAO on 20th Century European Plant Phenology 221
A. Menzel, N. Estrella, and C. Schleip

Summer Heat Waves in Western Europe, Their Past Change and Future Projections. 235
P. M. Della-Marta and M. Beniston

Distribution Changes of Seasonal Mean Temperature in Observations and Climate Change Scenarios 251
S. C. Scherrer, M. A. Liniger, and C. Appenzeller

Section D Chemical Changes and the Variability of the Stratosphere

Long-term Tropospheric Ozone Trends: A Critical Review **271**
J. Staehelin and C. Schnadt Poberaj

Simulation of Long-term Evolution of Stratospheric Dynamics and Chemistry: Role of Natural and Anthropogenic Forcings **283**
M. Dameris and R. Deckert

Dynamical Changes in the Arctic and Antarctic Stratosphere During Spring **293**
U. Langematz and M. Kunze

Creating Knowledge from the Confrontation of Observations and Models: The Case of Stratospheric Ozone **303**
G. P. Brasseur

Response of the Earth's Atmosphere to the Solar Irradiance Variability **317**
E. Rozanov, T. Egorova, and W. Schmutz

Stratospheric Ozone Variations Caused by Solar Proton Events Between 1963 and 2005 .... **333**
C. H. Jackman and E. L. Fleming

Sunspots, the QBO, and the Stratosphere in the North Polar Region: An Update **347**
K. Labitzke, M. Kunze, and S. Brönnimann

Index **359**

Contributors

Genrikh Alekseev
Arctic and Antarctic Research Institute, 38 Beringa St., 199397 St.Petersburg, Russia, alexgv@aari.nw.ru

Christof Appenzeller
MeteoSwiss, Kräbühlstrasse 58, P.O. Box 514, CH-8044 Zurich, Switzerland, christof.appenzeller@meteoswiss.ch

Martin Beniston
Chair for Climate Research, University of Geneva, 7 route de Drize, CH-1227 Carouge, Switzerland, Martin.Beniston@unige.ch

Guy Brasseur
National Center for Atmospheric Research, 3450 Mitchell Lane, Boulder, CO 80305, USA, brasseur@ucar.edu

Stefan Brönnimann
Institute for Atmospheric and Climate Science, ETH Zürich, Universitätstr. 6, 8092 Zürich, Switzerland, stefan.bronnimann@env.ethz.ch

Martin Dameris
Institut für Physik der Atmosphäre, DLR Oberpfaffenhofen, 82230 Wessling, Germany, Martin.Dameris@dlr.de

Rudolf Deckert
Institut für Physik der Atmosphäre, DLR Oberpfaffenhofen, 82230 Wessling, Germany, Rudolf.Deckert@dlr.de

Dick Dee
European Centre for Medium Range Weather Forecasts, Shinfield Park, Reading, Berkshire, RG2 9AX, UK, Dick.Dee@ecmwf.int

Paul M. Della-Marta
MeteoSwiss, Kräbühlstrasse 58, P.O. Box 514, CH-8044 Zurich, Switzerland, Paul.Della-Marta@meteoswiss.ch

Henry F. Diaz
NOAA, Climate Diagnostics Centre, 325 Broadway, Boulder, CO 80305, henry.f.diaz@noaa.gov

Tatiana Egorova
PMOD/WRC, Dorfstrasse 33, CH-7260 Davos Dorf, Switzerland, t.egorova@pmodwrc.ch

Nicole Estrella
Department of Ecology, TU München, Am Hochanger 13, D-85354 Freising, Germany, estrella@met.forst.tu-muenchen.de

Tracy Ewen
Institute for Atmospheric and Climate Science, ETH Zürich, Universitätstr. 16, 8092 Zürich, Switzerland, tracy.ewen@env.ethz.ch

Eric L. Fleming
NASA Goddard Space Flight Center, Code 613.3, Greenbelt, MD 20771, USA, fleming@kahuna.nasa.gov

Georg H. Hansen
Norwegian Institute for Air Research (NILU), Polar Environmental Centre, N-9296 Tromsø, Norway, ghh@nilu.no

N. E. Ivanov
Arctic and Antarctic Research Institute, 38 Beringa st., 199397, St.Petersburg, Russia, neivanov@aari.nw.ru

Charles H. Jackman
NASA Goddard Space Flight Center, Code 613.3, Greenbelt, MD 20771, USA, Charles.H.Jackman@nasa.gov

Per Kållberg
The Swedish Meteorological and Hydrological Institute, SE-60176 Norrköping, Sweden, Per.Kallberg@smhi.se

Elizabeth C. Kent
National Oceanography Centre, European Way, Southampton, SO14 3ZH, UK, elizabeth.c.kent@noc.soton.ac.uk

Valentina Khan
Hydrometeorological Center of Russia, B. Predtechenskiy Lane, 9-11, 123242 Moscow, Russia, khan@mecom.ru

Markus Kunze
Institut für Meteorologie, Freie Universität Berlin, Carl-Heinrich-Becker-Weg 6-10, D-12165 Berlin, Germany, Markus.Kunze@met.fu-berlin.de

Svetlana Kuzmina
Nansen International Science Fund, 199034 St. Petersburg, Russia, svetak4@yandex.ru

Karin Labitzke
Institut für Meteorologie, Freie Universität Berlin, Carl-Heinrich-Becker-Weg 6-10, 12165 Berlin, Germany, Karin.Labitzke@mail.met.fu-berlin.de

Ulrike Langematz
Institut für Meteorologie, Freie Universität Berlin, Carl-Heinrich-Becker-Weg 6-10, D-12165 Berlin, Germany, Ulrike.Langematz@met.fu-berlin.de

Mark A. Liniger
MeteoSwiss, Krähbühlstrasse 58, Postfach 514, CH-8044, Zurich, Switzerland, mark.liniger@meteoswiss.ch

Ulrike Lohmann
Institute for Atmospheric and Climate Science, ETH Zurich, Universitätstr. 16, CH-8092 Zurich, Switzerland, Ulrike.lohmann@env.ethz.ch

Cornelia Lüdecke
AK Geschichte der Polarforschung, Valleystrasse 40, D-81371 München, C.Luedecke@lrz.uni-muenchen.de

Jürg Luterbacher
NCCR Climate and Oeschger Centre for Climate Change Research and Institute of Geography, University of Bern, Hallerstrasse 12, 3012 Bern, Switzerland, juerg@giub.unibe.ch

Wieslaw Maslowski
Department of Oceanography, Naval Postgraduate School, Monterey, CA 93943, USA, maslowsk@nps.edu

Carl Mears
Remote Sensing Systems, 438 First Street, Suite 200, Santa Rosa, CA 95401, USA, mears@remss.com

Annette Menzel
Department of Ecology, TU München, Am Hochanger 13, D-85354 Freising, Germany, menzel@forst.tu-muenchen.de

Andrey Nagurny
Arctic and Antarctic Research Institute, 38 Beringa st., 199397 St.Petersburg, Russia, nagurny@aari.nw.ru

Urs Neu
ProClim, Schwarztorstrasse 9, CH-3007 Bern, Switzerland, neu@scnat.ch

Joel Norris
Scripps Institution of Oceanography, University of California, San Diego, 9500 Gilman Drive DEPT 0224, La Jolla, CA 92093-0224, jnorris@ucsd.edu

Heiko Paeth
Institut für Geographie, Universität Würzburg, Am Hubland, 97074 Würzburg, heiko.paeth@mail.uni-wuerzburg.de

Richard W. Reynolds
NOAA National Climatic Data Center, Asheville, NC 28801, USA,
Richard.W.Reynolds@noaa.gov

Eugene Rozanov
PMOD/WRC, Dorfstrasse 33, CH-7260 Davos Dorf, Switzerland,
e.rozanov@pmodwrc.ch

Konstantin G. Rubinshtein
Hydrometeorological Center of Russia, B. Predtechenskiy Lane 9-11, 123242
Moscow, Russia, rubin@mecom.ru

Simon C. Scherrer
Climate and Global Dynamics Division, National Center for Atmospheric
Research, P.O. Box 3000, Boulder, CO 80307-3000, USA, scherrer@ucar.edu

Christoph Schleip
Department of Ecology, TU München, Am Hochanger 13 D-85354 Freising,
Germany, Christoph.Schleip@web.de

Werner Schmutz
PMOD/WRC, Dorfstrasse 33, CH-7260 Davos Dorf, Switzerland,
w.schmutz@pmodwrc.ch

Christina Schnadt Poberaj
Institute for Atmospheric and Climate Science, ETH Zurich, Universitätstr.
16, CH-8092 Zurich, Switzerland, Christina.schnadt@env.ethz.ch

Sonia I. Seneviratne
Institute for Atmospheric and Climate Science, ETH Zürich, Universitätstr.
16, CH-8092 Zürich, Switzerland, sonia.seneviratne@env.ethz.ch

Adrian Simmons
European Centre for Medium Range Weather Forecasts, Shinfield Park, Reading,
Berkshire, RG2 9AX, UK, Adrian.Simmons@ecmwf.int

Johannes Staehelin
Institute for Atmospheric and Climate Science, ETH Zurich, Universitätstr.
16, CH-8092 Zurich, Switzerland, johannes.staehelin@env.ethz.ch

Alexander M. Sterin
Russian Research Institute for Hydrometeorological Information, World Data
Center (RIHMI-WDC), 6, Korolev st., Obninsk, Kaluga, Region, 249035,
Russia, sterin@meteo.ru

Reto Stöckli
Institute for Atmospheric and Climate Science, ETH Zurich, Universitätstr.
16, CH-8092 Zurich, Switzerland, reto.stoeckli@env.ethz.ch

Richard S. Stolarski
Code 613.3, NASA Goddard Space Flight Center, Greenbelt, MD 20771, USA, Richard.S.Stolarski@nasa.gov

Julienne Stroeve
NSIDC, CIRES, Campus Box 449, University of Colorado, Boulder, CO 80309-0449, USA, stroeve@kodiak.colorado.edu

Jean-Noël Thépaut
European Centre for Medium Range Weather Forecasts, Shinfield Park, Reading, Berkshire, RG2 9AX, UK, Jean-Noel.Thepaut@ecmwf.int

Sakari Uppala
European Centre for Medium Range Weather Forecasts, Shinfield Park, Reading, Berkshire, RG2 9AX, UK, Sakari.Uppala@ecmwf.int

Christian Vogler
Institute for Atmospheric and Climate Science, ETH Zürich, Universitätstr. 16, CH-8092 Zürich, Switzerland, christian.vogler@alumni.ethz.ch

Martin Wild
Institute for Atmospheric and Climate Science, ETH Zurich, Universitätstr. 16, CH-8092 Zurich, Switzerland, wild@env.ethz.ch

Scott Woodruff
NOAA Earth System Research, Laboratory (R/PSD3), 325 Broadway, Boulder, CO 80305, USA, Scott.D.Woodruff@noaa.gov

Steven J. Worley
National Center for Atmospheric Research, P.O. Box 3000, Boulder, CO 80307-3000, USA, worley@ucar.edu

A Focus on Climate During the Past 100 Years

S. Brönnimann[1], T. Ewen[1], J. Luterbacher[2], H. F. Diaz[3], R. S. Stolarski[4], and U. Neu[5]

Abstract The past 100 years are a key period for understanding climate variability and climate change as it marks the changeover from a climate system dominated by natural influences to one significantly dominated by anthropogenic activities. This volume is a compilation of contributions to a workshop dealing with different aspects of climate change, variability, and extremes during the past 100 years. The individual contributions cover a broad range of topics, from the re-evaluation of historical marine data to the effect of solar variability on the stratosphere. In this introductory chapter we provide an overview of the book in the context of recent research.

1 Introduction

The latest report of Working Group I of the International Panel on Climate Change (IPCC 2007) concludes that, very likely (probability >90 %), human influence has contributed to a warming of the planet during the past 50 years and that it virtually certainly (>99 %) will lead to an accelerated warming in the future. In order to understand, quantify and predict anthropogenic impacts on climate, the study of past climate variability is of utmost importance. The current understanding of past climate and its variability on interannual-to-decadal time scales is still fragmentary owing to the complexity of the coupling between atmosphere, ocean, and land surface and

[1] Institute for Atmospheric and Climate Science, ETH Zürich, Switzerland
stefan.bronnimann@env.ethz.ch, tracy.ewen@env.ethz.ch

[2] NCCR Climate and Oeschger Centre for Climate Change Research and Institute of Geography, University of Bern, Switzerland
juerg@giub.unibe.ch

[3] NOAA-ESRL, Boulder CO, USA
henry.f.diaz@noaa.gov

[4] NASA Goddard Space Flight Center, Greenbelt MD, USA
Richard.S.Stolarski@nasa.gov

[5] ProClim, Bern, Switzerland
neu@scnat.ch

S. Brönnimann et al. (eds.), *Climate Variability and Extremes during the Past 100 Years.*

complex dynamics within the atmosphere itself. Open questions also concern the chemical and dynamical variability of the stratosphere, which arguably plays an important role in climate variability.

Probably the most important time period for assessing climate variability and understanding the underlying processes is the past 100 years. First, with respect to climate trends, this period marks the changeover of a climate system driven by natural forcings to a climate system in which anthropogenic forcings play a major role. Second, with respect to climate variability, the past 100 years are the closest precedent to the present and future. Understanding the processes governing interannual climate variability during the recent past may eventually contribute to improved seasonal forecasts. Third, with respect to extremes, the period is important as it comprises several volcanic eruptions, severe droughts and floods, heat waves and other climate extremes. Some of these extremes are expected to become more frequent in the future and therefore past extremes may serve as an analogue. Finally, the past 100 years are accessible, at least to some extent, for both direct observational analyses and modelling studies.

The aim of this book is to give a broad overview of climate variability and extremes during the past 100 years through a compilation of selected contributions to a workshop held in Gwatt, Switzerland in July 2006. Some of the contributions present new research results on a specific topic, while others have a review character and provide an overview with a broader scope. However, the book is by no means complete. This introductory chapter gives an overview of this volume in the context of recent research, highlighting some of the key findings and concepts.

The chapter is structured as follows. In Sect. 2 we use the example of Arctic warming to demonstrate why the focus on the past 100 years is important and how science might progress in the future. The remaining sections then follow the structure of the book, starting with the observational record (Sect. 3), climate trends and processes (Sect. 4), climate variability and extremes (Sect. 5) as well as chemical changes and the variability of the stratosphere (Sect. 6). Conclusions are drawn in Sect. 7.

2 The Focus on the Past 100 Years: An Example

An example that illustrates the importance of a focus on the past 100 years as well as the challenges in understanding the complex interactions in the climate system is Arctic warming (see also Alekseev et al. and Stroeve and Maslowski, this volume). During the past 100 years the Arctic experienced two pronounced warming periods (Fig. 1). Between 1915 and 1945, annual mean temperatures increased by about 1.8°C. This period was followed by a cooling and a more recent warming, which started around 1970 and is still ongoing. Stroeve and Maslowski (this volume) report accelerated melting of sea ice during the past decades, particularly since the late 1990s. In fact, the two most recent winters (2005/2006 and 2006/2007) were extremely warm and were accompanied by record low sea ice extent (see also Serreze et al. 2007). The change of Arctic climate is a major research focus (reflected in the Arctic Climate Impact Assessment and the International Polar Year 2007/2008) because of the sensitive nature of Arctic systems.

The recent warming appears to be related to global warming, which is expected to be amplified in the Arctic due to feedback effects (e.g., ice-albedo feedback or cloud feedback). Due to this "polar amplification", most climate models produce an accelerated future Arctic warming of three to four times the global average (Holland and Bitz 2003). As a consequence, Arctic sea ice might change abruptly in the future (Holland et al. 2006). How sure are we about the mechanisms underlying the current and future Arctic warming in the light of the early 20th century warming? Global temperatures (land and sea) also increased in the first part of the 20th century (Fig. 1), but anthropogenic influences were much smaller at that time, and the Arctic warming could also just reflect an extreme realisation of internal variability in the climate system (Zhang et al. 2007). Some authors postulate a low-frequency oscillation (Polyakov et al. 2003) in the Arctic ocean–sea ice system which underlies any long-term trend, producing phases of accelerated warming. Bengtsson et al. (2004) suggest that the early 20th century warming was caused by increased transport of warm Atlantic water into the Arctic due to stronger westerly winds between Norway and Spitsbergen, reinforced by ocean–atmosphere–sea ice feedbacks in the Barents Sea (see also Fig. 2). For the recent warming, Stroeve and Maslowski (this volume) found that atmospheric circulation does not contribute considerably. Figure 1 shows that Arctic temperature is correlated with both global temperature and with the sea-level pressure (SLP) gradient between Norway and Spitsbergen (an approximation of the strength of westerly winds, bottom curve in Fig. 1). Given the similarity of response in Arctic temperatures during the early and late 20th century warming, the question remains: To what extent is Arctic temperature controlled by global warming, by the regional atmospheric circulation, or by lower frequency oceanic processes?

Our understanding of the climate mechanisms operating in the Arctic on different timescales is still limited. From the data side, it would be important to have better

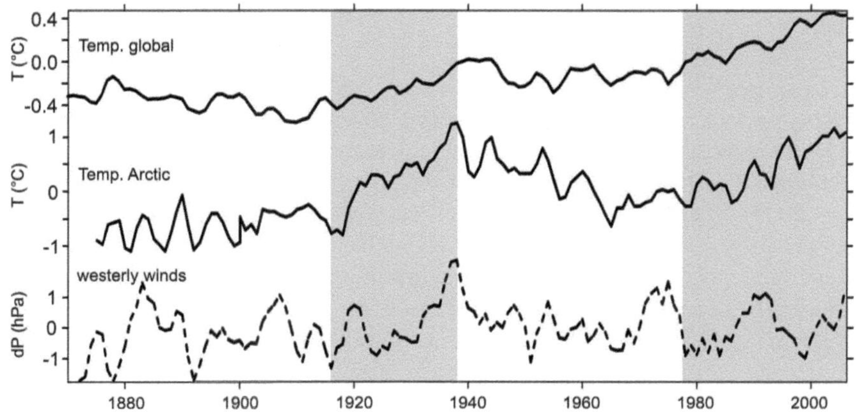

Fig. 1 Annual surface air temperature for the globe (top, HadCRUT2v, Jones and Moberg 2003) and north of 62°N (middle, Polyakov et al. 2003). The bottom curve shows the annual difference in sea-level pressure between Tromsø and Spitsbergen (HadSLP2, Allan and Ansell 2006). All series are anomalies with respect to 1961–1990, smoothed with a 3-year moving average. Grey shadings denote the two phases of pronounced Arctic warming.

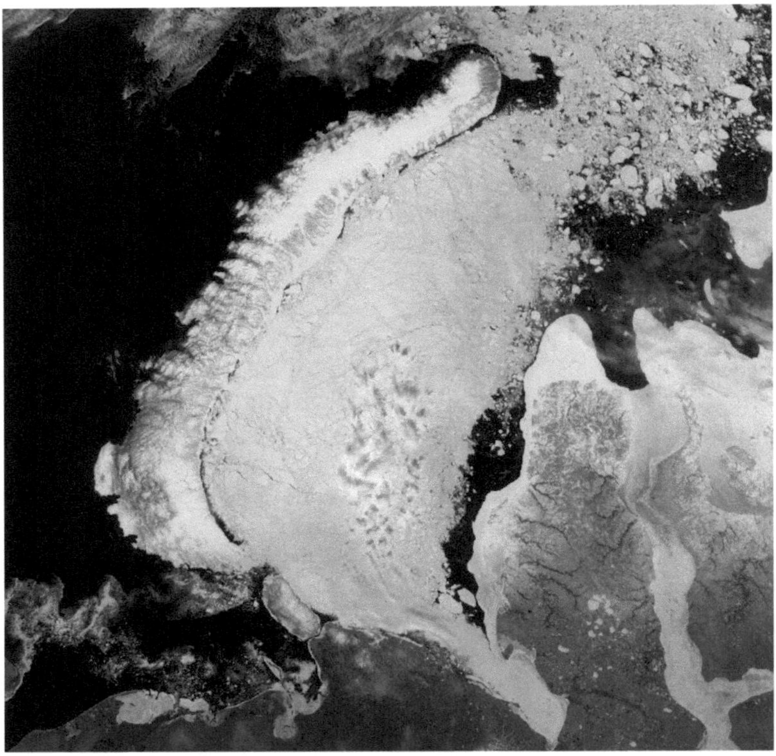

Fig. 2 Sea ice cover in the Barents sea region plays an important role in Arctic climate variability (Bengtsson et al. 2004). True colour image (Terra/MODIS) of the Novaya Zemlya islands and the coast of Russia on 12 June 2001. The Barents sea (left) is ice free, whereas the Kara sea (right) is still ice covered (Jacques Descloitres, MODIS Land Rapid Response Team, NASA Visible Earth v1 ID 8126).

surface air temperature data as well as sea ice coverage and upper-air data, for both warming periods and the intermittent cooling phase. The data then need to be analysed with respect to seasonal to interannual variability of Arctic climate, including extremes, and with respect to the large-scale atmospheric circulation (Overland and Wang 2005). Land-surface feedback might play a particularly important role (Chapin et al. 2005). From the model side, one would wish to have models capable of reproducing both the early 20th century warming and the recent trend. However, there are still strong limitations in this respect (e.g., Wang et al. 2007).

Perhaps Arctic warming can only be understood by taking atmospheric chemistry into account. Pollutants emitted in the midlatitudes can be transported to the Arctic, where they lead to the phenomenon of "Arctic Haze". Its effects on clouds and long-wave radiation in the Arctic are suggested to be significant contributors to the recent warming (e.g., Lubin and Vogelmann 2006; Garrett and Zhao 2006; Law and Stohl 2007).

Arctic weather and climate are also affected by the stratosphere and the stratospheric ozone layer. Through dynamical coupling of the stratosphere and troposphere, stratospheric disturbances can propagate downward and affect weather and climate at the ground (e.g., Baldwin and Dunkerton 2001).

This list of research directions – data work, analyses of trends, variability, and extremes in observations and models, analyses of coupling processes and feedbacks, analyses of chemical climate and aerosols, analyses of stratospheric processes – not only holds for the problem of Arctic warming, but also for many other open questions in climate research. Therefore, we have structured the volume along these lines.

3 The Observational Climate Record

In this section we attempt to give an overview of the current efforts and remaining problems related to climate data over the past 100 years. The examples in this book cover the most important observation platforms and data types, though they are by no means exhaustive. Figure 3 shows the platforms and measurement systems used to study the Earth's atmosphere and climate since the late 19th century. Many of the widely used global climate data sets for the Earth's surface reach back to the second half of the 19th century (Fig. 3). These data sets, comprising sea-surface and surface-air temperature, pressure, precipitation, and some other variables, are the most important climate and weather data prior to the more recent period (1948–present), which has been reanalysed using data-assimilation techniques. There are, however, changes within these data sets that affect the quality of the analyses. National meteorological networks, the Brussels Maritime Conference of 1853 and, after

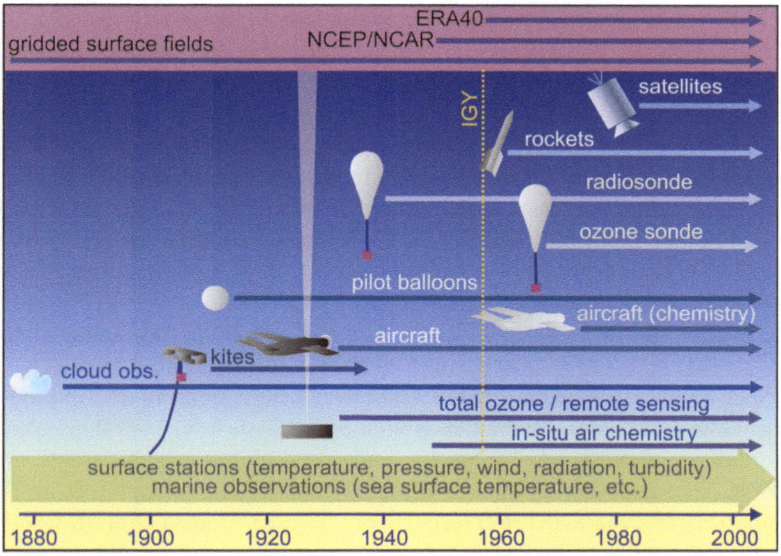

Fig. 3 Measurement platforms for 20th century climate data (adapted from Brönnimann et al. 2005).

1873, the International (now World) Meteorological Organization established standards for weather observations worldwide. The data quality has increased ever since, in some cases gradually, sometimes in steps – as has generally data coverage over both the terrestrial and marine domains.

The observational data sets are constantly being reanalysed and improved, but they could also be complemented and extended back in time. A large fraction of the meteorological data has still not been digitised. Concentrated efforts are needed (and are partly underway) to digitise and re-evaluate these data. However, changes in instrumentation, measurement practices and reporting, location, station environments, and lack of meta information inevitably lead to heterogeneous data sets. It is a challenging task to homogenize the data in order to assess climate trends reliably. The contributions by Woodruff et al. and Alekseev et al. (this volume) summarize these efforts for the examples of sea-surface temperature (SST) and sea ice. The situation is very similar for land (station) temperature and pressure data (e.g., Klein Tank et al. 2002; Allan and Ansell 2006; Ansell et al. 2006; Nicholls et al. 2006; Moberg et al. 2006; Della-Marta and Wanner 2006). Della-Marta and Beniston (this volume) report the additional efforts needed for homogenizing daily temperature data for the analysis of temperature extremes. With daily climate data it is necessary to homogenize the entire distribution function, not just the inhomogeneities in the mean.

Climate and weather data near the Earth's surface only provide a very limited view of the large-scale atmospheric circulation. Upper-air measurements started in the late 19th century (Fig. 3). During the first decades these measurements were in the early stages of development. Lüdecke (this volume) describes this pioneering work for the case of the Arctic, where kites, tethered and free-flying balloons, and airships were used as platforms. In many countries, operational upper-air networks using aircraft, kites, and pilot balloons gradually established during the 1900s–1920s, while radiosonde networks started only in the late 1930s and 1940s (Fig. 3). An important event in the history of climate observations was the International Geophysical Year (IGY) 1957/58 (Fig. 3), which not only led to the establishment of meteorological stations in Antarctica, a ground-based global total ozone network, and to measurements of CO_2 in the atmosphere, but also to an improved, global radiosonde network. Still, as these measurements were predominantly obtained for operational weather forecast, achieving trend quality remains a major problem (Free and Seidel 2005; Sherwood et al. 2005). In the contribution by Sterin et al. (this volume) issues related to the quality of upper-air observations and differences between data sets are discussed.

The availability of global radiosonde data also marks the beginning of current reanalysis data sets, which have become invaluable for atmospheric research. Using numerical weather prediction models combined with a filtering and interpolation scheme, radiosonde data, observations from the Earth's surface, from ships, satellites, and aircraft are assimilated. The most widely used global reanalysis data sets are NCEP/NCAR reanalysis (Kistler et al. 2001) and ERA40 (Uppala et al. 2005), reaching back to 1948 and 1957, respectively (Fig. 3). Uppala et al. (this volume) assess the value of reanalysis data products for climate analysis. While caution is still necessary when using the data for trend analysis (see Sterin et al. this volume), they are invaluable for the analysis of global atmospheric circulation variability and offer opportunities to reconcile information from independent platforms simultaneously. Efforts are currently underway to extend three-dimensional atmospheric data sets back to the 19th century based on surface data and historical upper-air data using statistical reconstruction techniques and data assimilation (Brönnimann et al. 2005; Compo et al. 2006).

Information from a reanalysis can also help in the process of homogenizing radiosonde temperature records. Instrument-dependent radiation errors as well as breakpoints in groups of stations have been identified and corrected by these methods (Andrae et al. 2004; Haimberger 2007; see also Sterin et al., this volume).

Another important step in the history of climate observations was the advent of space-borne Earth observation in the 1970s (Fig. 3). Satellites potentially provide a near-global coverage of numerous climate variables. For instance, the contributions by Alekseev et al., Mears, Norris, and Stroeve and Maslowski (this volume) address products based on satellite observations such as sea ice coverage (see also Fig. 2), cloud cover, and the vertical temperature structure. At least as important as for surface data, quality remains a fundamental issue for satellite data. In particular, the overlap between different sensors is often too short to obtain reliable transfer functions and the necessary corrections have been a matter of debate (Mears and Wentz 2005; Mears, this volume).

Not only was the physical state of the atmosphere measured during the past 100 years, but also its chemical composition. Spectral measurements in the UV wavelength range that provide information on the amount of ozone present locally in the Earth's atmosphere (i.e., total ozone) were performed since the 1920s (see Brönnimann et al., this volume). Starting around the 1950s to 1970s, in situ chemical measurements were performed at the ground, in the troposphere (e.g., using aircraft) and in the stratosphere (using balloons, see Staehelin et al., this volume). Since the 1970s, space-borne sensors have been delivering a wide range of chemical information, including ozone concentrations. Moreover, satellites measure aerosol properties and land-surface information that is extremely important for climate studies (e.g., MODIS on NASA's Terra Aqua satellites, see Fig. 2). Monitoring the atmosphere from space not only allows the climate system to be studied from a chemical and physical perspective, but also allows external (e.g., volcanic, solar) climate forcings to be quantified. Examples are found in the contributions by Jackman and Fleming and by Rozanov (this volume).

Large efforts have been devoted to obtaining data sets for the past 100 years that are suitable for climate applications. Reprocessing and homogenizing the available data in order to obtain a sufficient level of quality to enable the analysis of climate trends is an important, ongoing task. At the same time, constant efforts are also needed to ensure the quality of the present and future monitoring systems. In particular, space-borne observations offer new possibilities, but also new challenges with respect to long-term data quality.

While quality is one problem, data quantity is another, particularly in the early decades. Much could still be learned from analysing the past 100 years and concentrated data rescue efforts are needed to extend and supplement the early part of the record.

4 Climate Trends: Forcings and Processes

During the past 100 years global climate has experienced major changes. Several contributions in this volume deal with trends in the climate system and the underlying physical processes. In this section we provide a background for these chapters by briefly discussing climate forcings and temperature trends.

As shown in Fig. 1, two phases of distinct global (and Arctic) warming can be addressed. However, these two warming periods had a different spatial signature, as is shown in Fig. 4 for the boreal winter (January–March) period. The 1910–1945

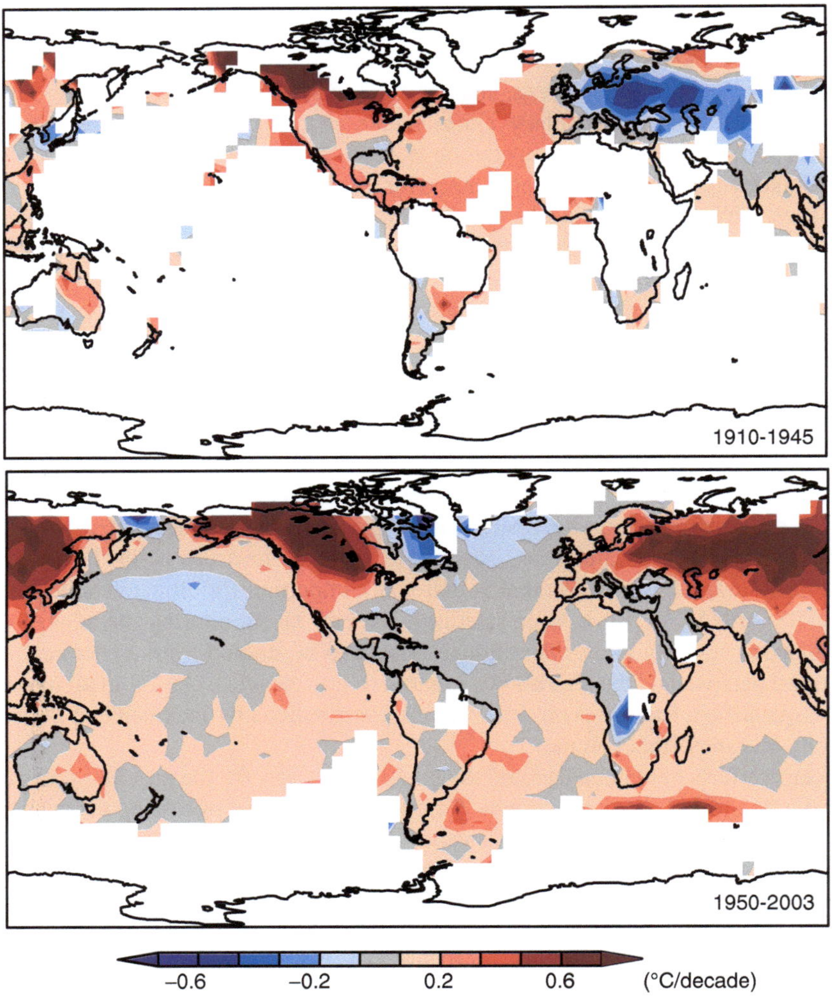

Fig. 4 Linear trend of boreal winter (January–March) surface air temperature for the period 1910–1945 (top) and 1950–2003 (bottom) using HadCRUT2v data (Jones and Moberg 2003). At least 20 winter mean values must be available in order to calculate a trend.

trend was most pronounced over the Atlantic and North America, while Europe experienced a winter cooling. In contrast, the 1950–2003 period exhibited a strong winter warming of the northern hemispheric land masses (see also Jones et al. 1999; Trenberth et al. 2007). What may have caused this difference in climatic response?

Multidecadal climate trends can be understood as the product of external forces (due to both natural and anthropogenic factors) and internal variability of the climate system. The development of the external factors (expressed as radiative forcing relative to 1870–1880) is shown in Fig. 5. Despite remaining uncertainties, which mainly concern the solar and aerosol effects, it is clear that the past 100 years mark the transition of a climate system where changes are dominated by natural forcings to one where changes are dominated by anthropogenic forcings. Was the 1910–1945 trend a result of "natural variability" and the 1950–2003 trend an "anthropogenic" warming?

Detection and attribution studies use climate models to predict the effect of the forcings and to estimate internal variability and then use these results for statistical analyses of the observed climate record (Allen et al. 2006; Hegerl et al. 2006; 2007). These studies often find that the early 20th century warming was caused by both increasing solar irradiance and large natural variability, while anthropogenic effects could have also contributed. The "large natural variability" could be related to the so-called Atlantic Multidecadal Oscillation (AMO; Delworth and Mann 2000) – the basin-wide average of North Atlantic SSTs, which is believed to be related to the strength of the thermohaline oceanic circulation (Knight et al. 2005; Zhang et al. 2007).

The global temperature trend during the past 30 years is dominated by the anthropogenic climate forcing (e.g., Allen et al. 2006). The anthropogenic influence is also detectable in SLP, ocean heat content, and tropopause altitude (e.g., Santer et al. 2003). The rate of change observed during the past 30 years was particularly large and affected not only surface air temperatures but also troposheric and stratospheric temperatures (Mears, Sterin et al., this volume; for an overview on

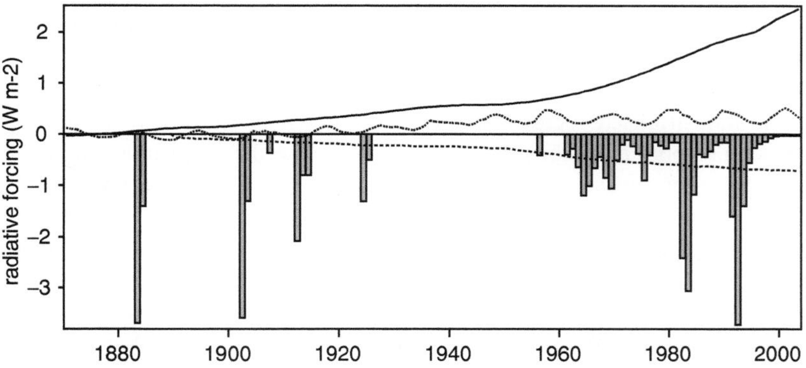

Fig. 5 Radiative forcing of climate since 1870. Solid: Greenhouse gases, dotted: solar, dashed: tropospheric aerosols, bars: volcanic forcings (updated from Crowley (2000) and Lean (2004)). All series are shown as differences with respect to 1870–1880.

stratospheric temperature trends see Ramaswamy et al. 2001, 2006), Arctic sea ice coverage (Stroeve and Maslowski, this volume) and thickness (Alekseev et al., this volume) and many other variables. Menzel et al. (this volume) demonstrate that particularly rapid changes have also occurred in several phenological time series. Trends in cloud cover and solar radiation are discussed in Norris (this volume) and Wild (this volume), respectively.

In terms of radiative forcing, natural factors dominate only immediately after strong tropical volcanic eruptions (Fig. 5). At least five major tropical eruptions occurred during the past 125 years, which allows volcanic effects on climate to be studied and better understood. Figure 6 (top) shows surface air temperature and SLP anomalies averaged over the first post-eruption winters (January–March). Volcanic eruptions illustrate that the climate response to a radiative forcing is more complex than a uniform temperature change. Rather, there are pronounced seasonal and regional patterns, and short and long-term effects need to be distinguished. While over the oceans, volcanic eruptions lead to cooling related to decreased shortwave radiation (with long-term effects, see Church et al. 2005), they cause a winter warming of the northern extratropical land masses (see Robock 2000; Fischer et al. 2007). This is an indirect dynamical effect that might proceed via the stratosphere (see further discussion in Sect. 6). Depending on the time period, this pattern might contribute to winter temperature trends (e.g., from 1950 to 2003; Fig. 4 bottom).

Among the anthropogenic forcings, greenhouse gases are the most important and arguably the best understood. However, other influences such as tropospheric aerosols and land-use change also play a role. These influences are addressed in the contributions by Lohmann, Wild, and Seneviratne and Stöckli (this volume). Figure 7 shows schematically the fluxes of energy, matter, and momentum in the troposphere and how they are affected by anthropogenic activities. Greenhouse gases primarily affect the long-wave radiation balance, while the effect of aerosols on the energy and water balance is more complex. Lohmann (this volume) provides an excellent overview.

The aerosol effect is often addressed in terms of direct and indirect effects. A consequence of the direct effect is a reduction of total solar radiation reaching the Earth's surface (Fig. 7), which is known as "global dimming" (Ohmura and Lang 1989). Wild (this volume) shows that for the last 20 years, this trend has reversed over the industrialised regions (in fact, over much of the globe) in line with a decrease in sulphur emissions (Wild et al. 2005). In order to disentangle the effects of increasing long-wave radiation (due to greenhouse gases) and changing short-wave radiation (due to aerosols, affecting only daytime temperature), the diurnal temperature range is often analysed. Wild (this volume) shows that its change agrees with a trend reversal from "global dimming" to "global brightening".

The indirect aerosol effect (see Lohmann, this volume) comprises the effects on the cloud optical and microphysical properties as well as on the cloud amount and lifetime (see Fig. 7). Its magnitude is one of the largest remaining uncertainties in the assessment of anthropogenic climate influences (Forster et al 2007). Addressing trends in cloud cover (which aerosols and other factors also influence) is much more difficult than

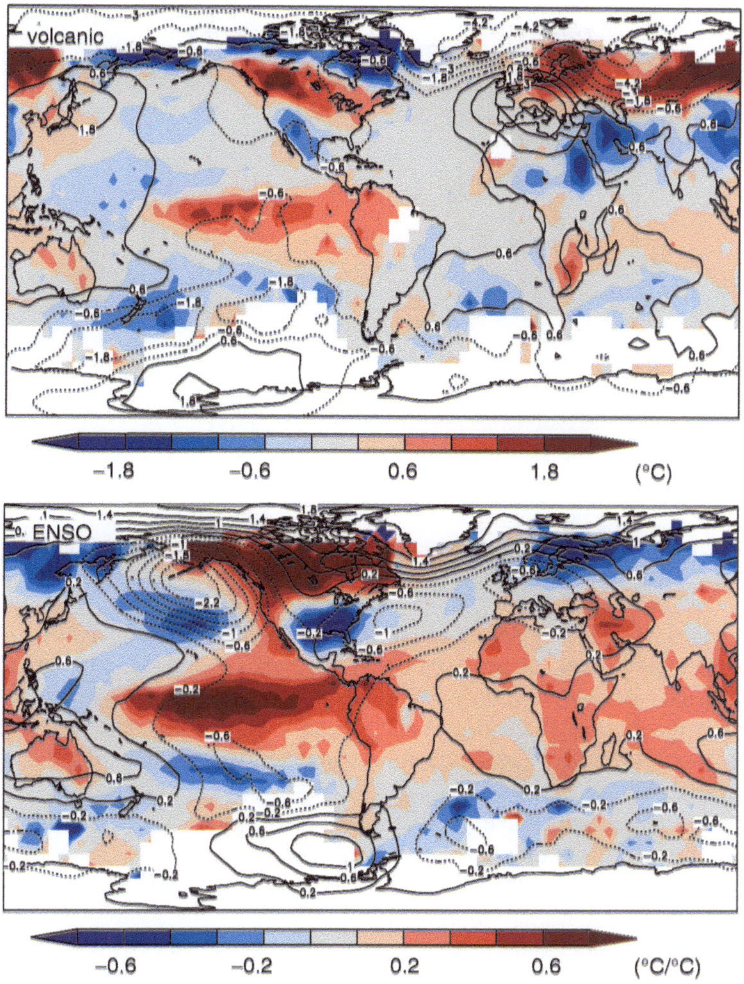

Fig. 6 Effect of tropical volcanic eruptions (top) and ENSO (bottom) on boreal winter (January–March) surface air temperature and SLP since 1880. Temperature and SLP data were detrended. The top panel shows a composite of the first winters after five tropical eruptions (Krakatoa, Santa Maria, Mt. Agung, El Chichón, Pinatubo). The bottom panel shows regression coefficients using a NINO3.4 index (September–February average) after removing two winters after each major volcanic eruption (same as above plus Katmai, see Brönnimann 2007).

addressing temperature trends. Norris (this volume) reports that upper-level cloud cover has decreased over the past decades, while the data show large discrepancies and inconsistencies when it comes to low clouds (see also Evan et al. 2007).

Another important player in the climate system is the land surface. It affects climate through the storage of water and CO_2 in soils and vegetation, through controlling

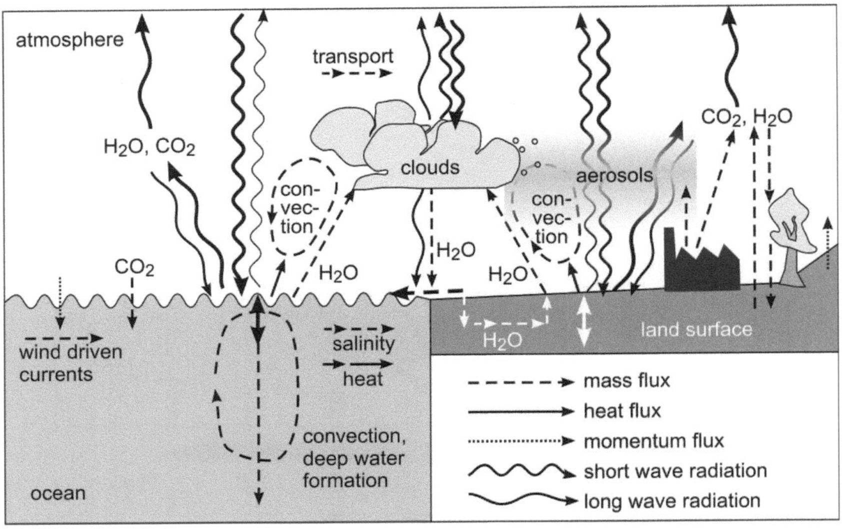

Fig. 7 Schematic depiction of the fluxes of matter, energy, and momentum in the climate system.

their exchange with the atmosphere and through modifying the energy balance (Fig. 7). Land-surface processes are not static but themselves depend on climate (e.g., albedo). Moreover, human induced land-use changes can affect climate via land–atmosphere interactions. Seneviratne and Stöckli (this volume) give an overview of the processes relevant for land–atmosphere interactions and their implications for climate variability.

Marked climate trends have been observed in the global climate system during the past 100 years, most importantly during the past 30 years, when the anthropogenic influence was strongest (IPCC 2007). These trends are increasingly well documented; examples are found throughout this volume. While some of the processes are relatively well understood today, this is not the case for others, in particular when coupling processes and feedbacks come into play. Focusing on the past 100 years, which comprise periods with strong and weak trends in anthropogenic forcing, might help to narrow down some of the remaining uncertainties.

5 Climate Variability and Extremes

The processes depicted in Fig. 7 show the direct, "static" effect of climate forcings on the budgets of energy, matter, and momentum. As mentioned for the case of volcanic eruptions, there are important indirect, "dynamic" effects as these processes also change the large-scale atmospheric circulation and its variability. For instance, they might affect the strength of the Hadley cell (see Diaz and Bradley 2004 for an overview) or the extratropical circulation. In order to properly assess future climate scenarios it is therefore important to study the interannual climate variability that was observed or modelled during the past 100 years. Moreover, a better understanding of interannual climate variability helps to improve seasonal forecasts. The interest is particularly strong for extreme climate events that have large economical impacts such as large-scale droughts and heat waves. In fact, several contributions in this volume address heat waves and droughts.

One main line of research on interannual climate variability and extremes concerns the concept of variability modes and their impact on climate, predictability, and change in a future climate. Arguably the most important mode is El Niño/Southern Oscillation (ENSO). El Niño is an episodic warming of the eastern and central tropical Pacific, associated with an eastward shift of convective activity from the western tropical Pacific (Philander 1989; Diaz and Markgraf 2000; McPhaden et al. 2006). Via changes in the Hadley circulation and interaction with the extratropical circulation, ENSO affects temperature and SLP globally as is shown for the boreal winter period in Fig. 6 (bottom, regression coefficients using the September–February average of the NINO3.4 index, see Brönnimann 2007). El Niño often brings drought to different parts of the world (e.g., Diaz et al. 2001). Apart from affecting the Indian monsoon, it also influences monsoonal rainfall over southern West Africa (Paeth, this volume). Because ENSO is to some extent predictable (e.g., Chen et al. 2004), studying the relationship between the tropical Pacific and climate in more remote regions is an important area of research.

The dominant mode of interannual variability in the North Atlantic-European sector is the North Atlantic Oscillation (NAO, see Wanner et al. 2001; Hurrell et al. 2003). It is defined as the SLP difference between the Azores high and the Icelandic low (the two dominant centres of action over the Atlantic) and affects climate in Europe, North Africa, and eastern North America. The NAO is an important concept for assessing impacts of climate variability. For instance, Menzel et al. (this volume) show that the phase of the NAO affects spring leaf unfolding in Germany. Large efforts are currently devoted towards a seasonal forecasting of the NAO, even though its predictability seems to be limited.

Climate extremes are often related to extremes in the above-mentioned variability modes, although in general, different factors act together to produce extremes. This is shown in Fig. 8 for the example of droughts. The key factors identified here are oceanic forcing, aerosol effects, and land–atmosphere interactions. As a consequence, global change can affect the frequency of extremes in various ways.

Oceanic forcing of the atmosphere can alter the large-scale moisture flux, including the monsoon systems and other circulation systems that normally transport humidity to the drought-prone areas. Several major droughts were observed during the past 100 years and were analysed in relation to oceanic forcing. The importance of large-scale SST anomalies in forcing atmospheric circulation patterns that are conducive to drought has been demonstrated for the Great Plains (Schubert et al. 2004), the Sahel region (Giannini et al. 2003; see also Paeth, this volume), and central Europe (Sutton and Hodson 2005; Black and Sutton 2007; see also Hoerling and Kumar 2003). For instance, large-scale SST anomalies related to a positive phase of the AMO cause dry conditions across the USA, while a cold tropical Pacific (e.g., La Niña) often leads to droughts in the south-west of the USA and the Great Plains (McCabe et al. 2004). Both effects acted together in the 1930s and set

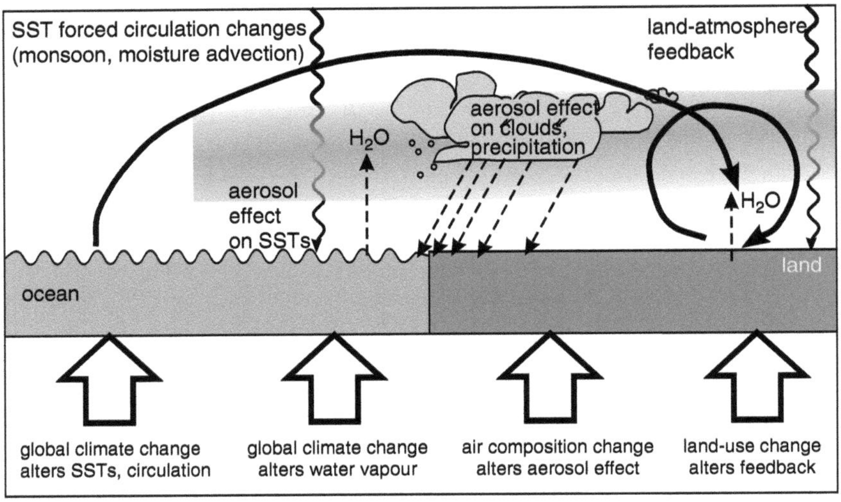

Fig. 8 Processes involved in the development of droughts and pathways through which they can be affected by climate change.

the stage for "Dust Bowl" droughts in the USA (Schubert et al. 2004). However, other processes such as land–atmosphere feedbacks also played an important role (Schubert et al. 2004; see also Seneviratne and Stöckli, this volume). The "Dust Bowl" is not the only region where land–atmosphere coupling is important. In a future climate, this might also be the case for central Europe (Seneviratne et al. 2006).

Paeth (this volume) gives an overview of drought conditions in the Sahel region. While large-scale SST anomalies and changes in atmospheric circulation are an immediate cause, the long-term trends can also be explained by other causes, including soil properties, vegetation cover, and aerosols (see also Rotstayn and Lohmann 2002). Lohmann (this volume) discusses the effect of aerosols on precipitation and points to droughts in Southern China that may have been linked to a decreased land–sea temperature gradient (due to the presence of aerosols) and a weak monsoon circulation (see also Lau et al. 2006). However, the suggested mechanisms differ between the studies and the results are not yet conclusive, as current climate models do not account for all relevant microphysical properties.

Droughts are also experienced in Europe. The Mediterranean region is particularly vulnerable (e.g., Xoplaki et al. 2004), but other regions are also affected by droughts. The most prominent recent example is the heat wave in central Europe in 2003 (Schär et al. 2004; Beniston 2004). Have droughts and heat waves in Europe become more frequent? Della-Marta and Beniston (this volume) analyse changes in the occurrence of summer heat waves in Europe during the past 126 years. Apart from important data obstacles they had to overcome, the authors report that heat waves have increased both in number and in persistence.

The topic of climate extremes became apparent to the participants of the meeting from which this book results, as the event took place during a heat wave. In many parts of Switzerland, July 2006 was the hottest July on record (Fig. 9 top left). Fall 2006, winter 2006/2007 and spring 2007 were equally extreme (shown for the series from Bern, Switzerland, in Fig. 9 bottom, see also Luterbacher et al. 2007). Had the meeting taken place a few months earlier, the participants would have witnessed another extreme: In August 2005 the conference venue was flooded (Fig. 9 top right).

Is this evidence that climate has become more variable? Several authors have suggested that increasing temperature variability contributes to the increased frequency in heat waves (Schär et al. 2004). For the occurrence of extremes, changes in variability are as important as changes in the mean (Katz and Brown 1992).

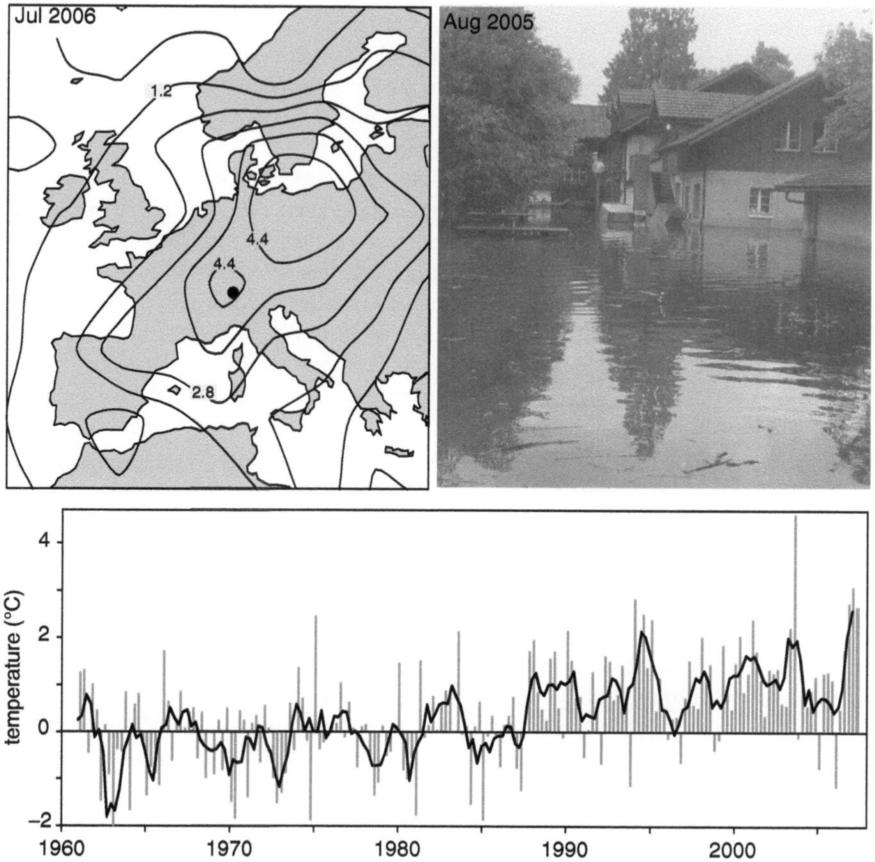

Fig. 9 Surface air temperature anomalies in July 2006 (top left, in degree Celsius with respect to the 1961–1990 average), Jones and Moberg 2003. The conference site in Gwatt, Switzerland (dot in left panel) in August 2005 (top right) following an extended period of much above normal precipitation. Lower panel displays seasonal temperature anomalies (with respect to the 1961–1990 average) in Bern, Switzerland from 1961 to 2007 (winter 1963 is off the scale), data provided by MeteoSwiss. The solid line shows a running average over four seasons.

Scherrer et al. (this volume) analysed climate data as well as climate scenario runs for the 21st century with both respects, focussing on central European temperature. For the summer season, they found an increase in the mean during the past 40 years and in the variability. Analyses based on daily summer temperature from Della-Marta and Beniston (this volume) show that summer temperature variability has increased significantly over the last 126 years in central western Europe. Relatively small changes in the variability of daily summer temperature (+11%) have contributed approximately 40% to the increase in extremely hot days. For the 21st century, Scherrer et al. (this volume) report large relative increases in mean temperature for all seasons with maxima in summer, accompanied by an increase in variability. In Central and Eastern Europe, this appears linked with a mean decrease of soil moisture and a consequent strengthening of land–atmosphere coupling as concentrations in greenhouse gas increase (Seneviratne et al. 2006; see also Seneviratne and Stöckli, this volume), which could also explain part of the recent trend during the 20th century.

The contributions in this volume show that progress has been made in addressing and documenting changes in climate extremes. Such progress is important, as extremes are rare events that are occur infrequently enough for statistical analyses, and are difficult to model numerically. Concerning the underlying processes, the contributions show that several factors act together and therefore must be considered together, similar as outlined in Sect. 2 for the problem of Arctic warming. For droughts, for instance, the large-scale oceanic forcing of atmospheric circulation must be considered as well as land-surface processes and aerosols. A lot still has to be learned in order to better assess extremes and their possible change in a future climate and in order to obtain better forecasts. In this effort, the past 100 years have been and still are an important learning ground.

6 Chemical Changes and the Variability of the Stratosphere

In the past 100 years, climate has undergone changes not just near the Earth's surface but also in the middle and upper atmosphere. Also, changes have been observed in the chemical properties of the atmosphere, some of which have affected the climate near the ground. The link between chemical and physical aspects of climate is a comparably new and rapidly evolving topic of climate research that is covered only to a limited extent by the contributions in this book. A key role is played by ozone and by stratospheric processes – the focus of this section.

Trace gases naturally occur in the atmosphere due to emissions by living organisms, geogenic emissions, lightning, or forest fires. Humans have started to influence the composition of the atmosphere substantially and on a global scale at least since the advent of the industrialisation (see Staehelin and Schnadt Poberaj, this volume). In addition to the greenhouse gases CO_2 and methane (CH_4), reactive gases are being emitted. Combustion processes produce nitrogen oxides (NO_x), carbon monoxide (CO) and, depending on the fuel, sulphur dioxide and a range of organic compounds. Industrial processes and households have emitted ozone depleting substances, and man-induced biomass burning produces large amounts of aerosols. These constituents have profound effects on both air quality and climate.

A particularly important tropospheric constituent is ozone. It is an irritant to humans, a greenhouse gas, and one of the main determinants of the oxidizing capacity of the troposphere (and hence the lifetime of other greenhouse gases). It is formed in the stratosphere via oxygen chemistry in the presence of UV radiation. In the troposphere, it can be formed from the oxidation of CO, CH_4, or higher hydrocarbons in the presence of NO_x. How much of the present tropospheric ozone burden is a consequence of the anthropogenic influence and how much is natural (e.g., influx from the stratosphere)? Staehelin and Schnadt Poberaj (this volume) assess the changes in tropospheric ozone during the past decades. They find that tropospheric ozone levels over Europe have doubled since World War II due to anthropogenic emissions.

Ozone plays a particularly important role in the stratosphere, where it forms the ozone layer that shields life on Earth from harmful UV radiation. Ozone depleting substances have caused a considerable depletion of stratospheric ozone since the 1970s and have led to the phenomenon of the Antarctic ozone hole (Solomon 1999; Staehelin et al. 2001; WMO 2007; see also Brasseur and Brönnimann et al., this volume). Figure 10 shows the trend of the ozone column in February between 1979

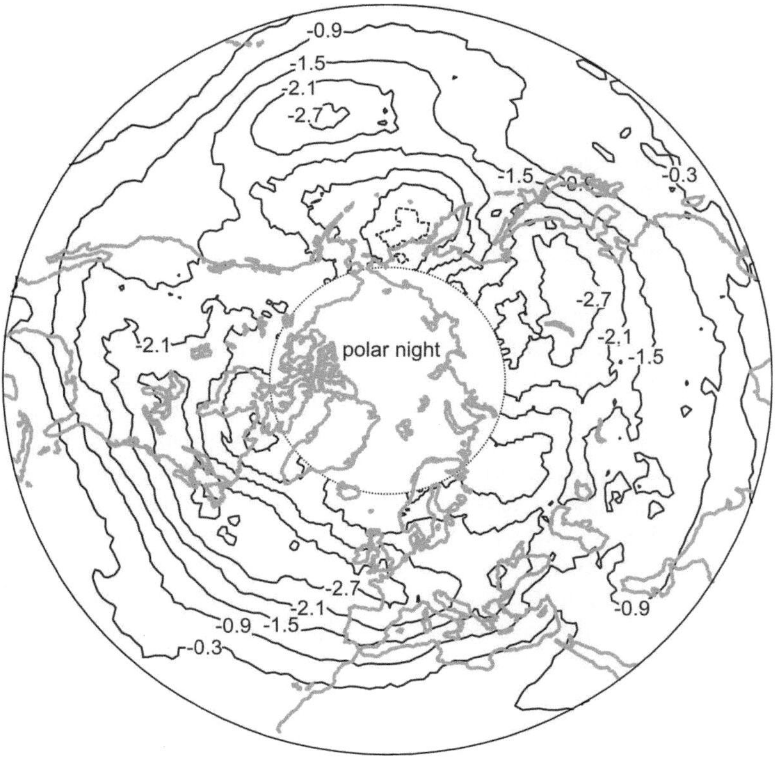

Fig. 10 Linear trend of the ozone column (DU/Year) in February from 1979 to 2001 based on Total Ozone Mapping Spectrometer data.

and 2000. Apart from the ozone depletion signal, the trend has a clear spatial pattern which can not be explained by chemical depletion but rather points to dynamical processes. Hence, understanding the ozone layer forces us to combine chemical and physical perspectives and to use different methods and concepts in order to understand the atmosphere. Using stratospheric ozone as an example, Brasseur (this volume) shows that scientific understanding can only develop due to the interplay of observations, models, and laboratory measurements. All three methods are necessary to confirm, contradict, and constrain each other – until new observations (such as the discovery of the ozone hole) reveal an inadequacy in our understanding, leading to a new learning cycle.

How good are the chemical and dynamical stratospheric processes represented in current models? Dameris and Deckert (this volume) integrated a model over a relatively long period of 40 years. They show that their model is able to capture the main features of temporal variability of stratospheric ozone on different time scales. However, in addition to ozone depleting substances, other important forcing factors (greenhouse gases, SSTs and sea ice cover, Quasi-Biennial Oscillation, solar variability, volcanic aerosols) also need to be included. Currently, large efforts are being devoted to a process-oriented validation of chemistry climate models (Eyring et al. 2006). This allows the models to be used for simulating future changes in the stratospheric ozone layer with sufficient confidence.

Apart from affecting UV radiation at the Earth's surface, changes in stratospheric ozone might also change climate at the ground (Thompson and Solomon 2002; Gillett and Thompson 2003). Of particular interest in this context are the stratospheric polar vortices which form in winter. On the one hand, the polar vortices both affect and are affected by ozone chemistry and are a location of chemical-dynamical interaction. On the other hand, the polar vortices also take part in the dynamical coupling between the stratosphere and the troposphere (Holton et al. 1995). Langematz and Kunze (this volume) discuss trends in the intensity and persistence of the vortices at both poles during the past decades. They find a high variability on multiannual timescales, which makes trend detection very difficult. For instance, a series of very cold winters in the Arctic stratosphere during the 1990s was then followed by a trend to a more disturbed stratosphere in recent years. Longer time series are required to assess trends in the dynamics of the polar vortices.

An important contribution to the variability of the northern polar vortex is that of "Major Midwinter Warmings" (see Labitzke et al., this volume). In these events, the vortex collapses completely in conjunction with an abrupt warming of the middle stratosphere by several tens of degrees within a few days. Such events are mostly triggered by anomalies in planetary wave activity propagating upward from the troposphere (see Limpasuvan et al. 2004). Once the vortex is significantly weakened in the middle stratosphere, the signal often propagates downward (Baldwin and Dunkerton 2001). Figure 11 shows the time–altitude evolution of the daily Northern Annular Mode index (measuring the strength of the polar vortex) in 2006. Day 0 marks the onset of a major warming in the middle stratosphere (30 km), which propagates downward within a week or two and affects the troposphere repeatedly until the end of February.

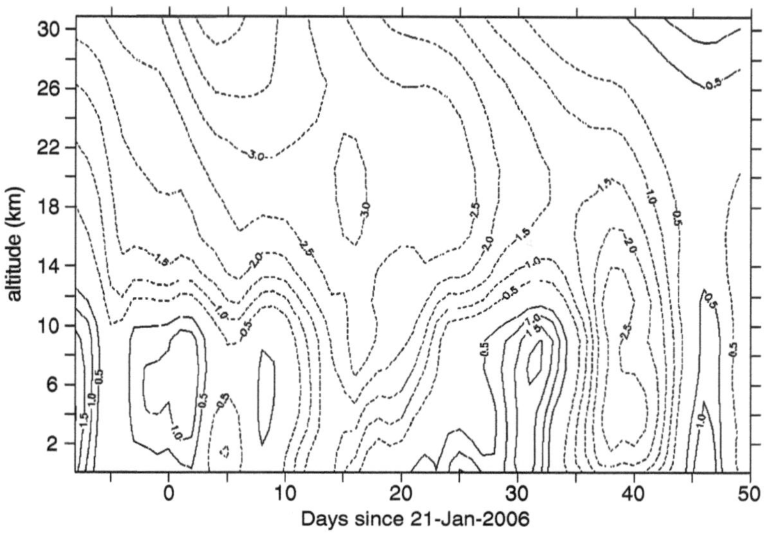

Fig. 11 Development of the standardized daily index of the Northern Annular Mode in January and February 2006 (data were provided by Mark Baldwin). Day 0 is defined as the first day when the index exceeds three standard deviations at 10 hPa (ca, 30 km).

Downward propagation offers a mechanism by which forcings that strongly act on the stratosphere, such as solar or volcanic effects, can have an influence on climate at the ground. As schematically depicted in Fig. 12, strong tropical volcanic eruptions not only cool the Earth's surface by blocking short-wave radiation, but also heat up the stratosphere considerably. During the winter season this can lead to a differential heating between the sunlit areas and the polar night, which in turn may strengthen the polar vortex and, through downward propagation, weather and climate at the ground (see also Stenchikov et al. 2004). The winter warming following tropical eruptions (Fig. 5 top) could be related to this mechanism.

A combination of direct thermal effects (White et al. 1997; Meehl et al. 2002) and indirect dynamical effects via the stratosphere could also be responsible for solar effects on climate (see Calisesi et al. 2007 for a recent compilation of papers on Sun–climate links). Rozanov et al. (this volume) give an overview of the effects via the stratosphere (see also Shindell et al. 1999; Kodera and Kuroda 2002). The variability of solar radiation over an 11-year sunspot cycle is much larger in the UV wavelength range than total solar irradiance. Correspondingly, clear solar effects appear in upper-stratospheric ozone and temperature, which might lead to changes in stratospheric circulation and, through downward propagation, climate at the Earth's surface (see Rozanov et al., this volume). Changes in the Sun might also affect climate on Earth via energetic particles. Jackman and Fleming (this volume) provide an overview of the effect of solar proton events on the mesosphere and stratosphere. However, much about the solar influence still remains to be learned. For instance, Labitzke et al. (this volume) show that solar variability does affect the

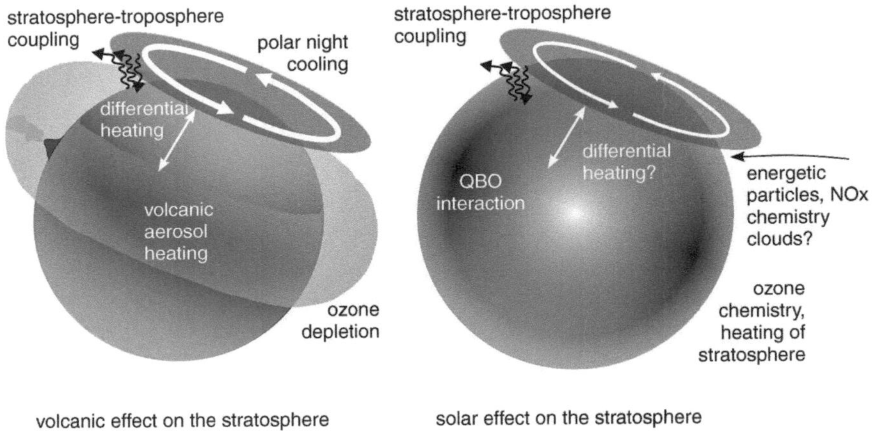

Fig. 12 Schematic depiction of the effects of volcanic eruptions and solar irradiance changes on the stratosphere.

northern polar vortex, but the effect depends on the phase of the Quasi-Biennial Oscillation in the tropical stratosphere – a demonstrated effect that is not well understood.

Chemical climate variability and stratospheric processes are today considered important areas of climate research. This is reflected, for instance, in the IGAC (International Global Atmospheric Chemistry) project of the International Geosphere-Biosphere Programme and the SPARC (Stratospheric Processes and their Role in Climate) project of the World Climate Research Programme. Recent research results have further demonstrated the importance of stratospheric processes (Baldwin et al. 2007). In particular, the concept of downward propagation provides hope that some of the long-standing issues in climate research such as the effect of solar variability can finally be addressed.

7 Conclusions

While major progress has been made in the past few years towards quantifying and understanding climate variability, trends, and extremes, there are still many open questions. Answering these questions will be important for our understanding of the climate system, in particular for assessing future climate change and for improving seasonal forecasts. A key period in this context are the past 100 years, which mark the changeover of a climate system dominated by natural forcings to a climate system dominated by anthropogenic influences.

The contributions in this volume provide a broad overview of the current state of the research on climate variability and extremes during the past 100 years. In this introductory chapter we embed the contributions in a scientific context and summarise some important findings and key mechanisms that are currently discussed. At the same time, we show that our knowledge is still incomplete. We argue that the complexity of many current problems in climate research calls for a three-dimensional, physical-chemical view of the climate system.

Acknowledgements SB, TE, and JL were funded by the Swiss National Science Foundation; HFD was supported by a grant from the US Department of Energy. We would like to acknowledge NCCR Climate, Swiss RE, the Max and Elsa Beer-Brawand Foundation, the Swiss National Science Foundation, and ProClim for sponsoring the meeting and the Gwatt Centre (Switzerland) for hosting the workshop and providing the photo in Fig. 9. Mark Baldwin provided the NAM data (Fig. 11). We also thank the Climatic Research Unit (Univ. East Anglia, UK), the UK Met Office, MeteoSwiss, and the NASA TOMS team for providing data. The paper was sent to all contributors of this volume, and we thankfully included all of their comments.

References

Allan, R. and T. Ansell, 2006: A new globally complete monthly historical gridded mean sea level pressure data set (HadSLP2): 1850–2003. *J. Climate*, **19**, 5816–5842.

Allen, M. R. et al., 2006: Quantifying anthropogenic influence on recent near-surface temperature change. *Surv. Geophys.*, **27**, 491–544.

Andrae, U., N. Sokka, and K. Onogi, 2004: The radiosonde temperature bias correction in ERA-40. ECMWF ERA-40 Project Rep. Series 15, 34 pp.

Ansell, T. J. et al., 2006: Daily mean sea level pressure reconstruction for the European-North Atlantic region for the period 1850–2003. *J. Climate*, **19**, 2717–2742.

Baldwin, M. and T. J. Dunkerton, 2001: Stratospheric harbingers of anomalous weather regimes. *Science*, **294**, 581–584.

Baldwin, M., M. Dameris, and T. G. Shepherd, 2007: How will the stratosphere affect climate change? *Science*, **316**, 576–577.

Bengtsson, L., V. A. Semenov, and O. Johanessen, 2004: The early century warming in the Arctic – a possible mechanism. *J. Climate*, **17**, 4045–4057.

Beniston, M., 2004: The 2003 heat wave in Europe: a shape of things to come? *Geohpys. Res. Lett.*, **31**, 2022–2026.

Black, E. and R. Sutton, 2007: The influence of oceanic conditions on the hot European summer of 2003. *Clim. Dynam.*, **28**, 53–66.

Brasseur, G. P., 2007: Creating knowledge from the confrontation of observations and models: the case of stratospheric ozone (this volume).

Brönnimann, S., 2007: The impact of El Niño/Southern Oscillation on European climate. *Rev. Geophys*, **45**, RG3003, doi:10.1029/2006 RG000199.

Brönnimann, S., G. P. Compo, P. D. Sardeshmukh, R. Jenne, and A. Sterin, 2005: New approaches for extending the 20th century climate record. *Eos*, **86**, 2–7.

Brönnimann, S., C. Vogler, J. Staehelin, R. Stolarski, and G. Hansen, 2007: Total ozone observations during the past 80 years (this volume).

Calisesi, Y., R. M. Bonnet, L. Gray, J. Langen, and M. Lockwood, eds., 2007: Solar variability and planetary climates. *Space Sciences Series of ISSI*, vol. 23, Springer, New York.

Chapin III, F. S. et al. 2005: Role of land-surface changes in Arctic summer warming. *Science*, **310**, 627–628.

Chen, D., M. A. Cane, A. Kaplan, S. E. Zebiak, and D. J. Huang, 2004: Predictability of El Niño over the past 148 years. *Nature*, **428**, 733–736.

Church, J. A., N. J. White, J. M. Arblaster, 2005: Significant decadal-scale impact of volcanic eruptions on sea level and ocean heat content. *Nature*, **438**, 74–77.

Compo, G. P., J. S. Whitaker, and P. D. Sardeshmukh, 2006: Feasibility of a 100-year reanalysis using only surface pressure data. *Bull. Am. Meteorol Soc.*, **87**, 175–190.
Crowley, T. J., 2000: Causes of climate change over the past 1,000 years. IGBP PAGES/World Data Center for Paleoclimatology Data Contribution Series #2000–045. NOAA/NGDC Paleoclimatology Program, Boulder, CO.
Dameris, M. and R. Deckert, 2007: Simulation of long-term evolution of stratospheric dynamics and chemistry – role of natural and anthropogenic forcings (this volume).
Della-Marta, P. M. and H. Wanner, 2006: A method of homogenizing the extremes and mean of daily temperature measurements. *J. Climate*, **19**, 4179–4197.
Delworth, T. L. and M. E. Mann, 2000: Observed and simulated multidecadal variability in the northern hemisphere. *Clim. Dynam.*, **16**, 661–676.
Diaz, H. F. and R. S. Bradley, eds., 2004: The Hadley Circulation: present, past and future. *Advances in Global Change Research 21.*
Diaz, H. and V. Markgraf, eds., 2000: *El Niño and the Southern Oscillation: Multiscale Variability and Global and Regional Impacts*. Cambridge University Press, Cambridge.
Diaz, H. F., M. P. Hoerling, and J. K. Eischeid, 2001: ENSO variability, teleconnections, and climate change. *Int. J. Climatol.*, **21**, 1845–1862.
Evan, A. T., A. K. Heidinger, and D. J. Vimont, 2007: Arguments against a physical long-term trend in global ISCCP cloud amounts. *Geophys. Res. Lett.*, **34**, L04701, doi:10.1029/2006 GL028083.
Eyring, V. et al. (2006) Assessment of temperature, trace species, and ozone in chemistry-climate model simulations of the recent past. *J. Geophys. Res.*, **111**, D22308, doi:10.1029/2006 JD007327.
Fischer, E., J. Luterbacher, E. Zorita, S. F. B. Tett, C. Casty, and H. Wanner, 2007: European climate response to major volcanic eruptions for the last half millennium. *Geophys. Res. Lett.*, **34**, L06707, doi:10.1029/2006GL029068.
Free, M. and Seidel DJ (2005) Causes of differing temperature trends in radiosonde upper air data sets. *J. Geophys. Res.*, **110**, D07101, doi:10.1029/2004JD005481.
Forster, P. et al., 2007: Changes in atmospheric constituents and in radiative forcing. In: *Climate Change 2007: The Physical Science Basis*. Contribution of Working Group I to the Fourth Assessment Report of the Intergovernmental Panel on Climate Change, Cambridge University Press, Cambridge.
Garrett, T. J. and C. Zhao, 2006: Increased Arctic cloud longwave emissivity associated with pollution from mid-latitudes. *Nature*, **440**, 787–789.
Giannini, A., R. Saravanan, and P. Chang, 2003: Oceanic forcing of Sahel rainfall on interannual to interdecadal time scales. *Science*, **302**, 1027–1030.
Gillett, N. P. and D. W. J. Thompson, 2003: Simulation of recent southern hemisphere climate change. *Science*, **302**, 273–275.
Haimberger, L., 2007: Homogenization of radiosonde temperature time series using innovation statistics. *J. Climate.*, **20**, 1377–1403.
Hegerl, G. C. et al., 2006: Climate change detection and attribution: beyond mean temperature signals. *J. Climate.*, **19**, 5058–5077.
Hegerl, G. C., F. W. Zwiers, P. Braconnot, N. P. Gillett, Y. Luo, J. A. Marengo, J. A. Orsini, N. Nicholls, J. E. Penner, and P. A. Stott, 2007: Understanding and attributing climate Change. In: *Climate Change 2007: The Physical Science Basis*. Contribution of Working Group I to the Fourth Assessment Report of the Intergovernmental Panel on Climate Change. Cambridge University Press, Cambridge.
Hoerling, M. and A. Kumar, 2003: The perfect ocean for drought. *Science*, **299**, 691–694.
Holland, M. M. and C. M. Bitz, 2003: Polar amplification of climate change in coupled models. *Clim. Dynam.*, **21**, 221–232.
Holland, M. M., C. M. Bitz, and B. Tremblay, 2006: Future abrupt transitions in the summer Arctic sea ice. *Geophys. Res. Lett.*, **33**, L23503, doi:10.1029/2006GL028024.
Holton, J. R., P. H. Haynes, M. E. McIntyre, A. R. Douglass, R. B. Rood, and L. Pfister, 1995: Stratosphere-troposphere exchange. *Rev. Geophys.*, **33**, 403–439.

Hurrell, J., Y. Kushnir, G. Ottersen, and M. Visbeck, eds., 2003: The North Atlantic Oscillation: climatic significance and environmental impact. AGU, Geophysical Monograph Series 134. Washington, DC.

IPCC, 2007: *Summary for Policymakers*. Contribution of Working Group I to the Fourth Assessment Report of the Intergovernmental Panel on Climate Change. Geneva.

Jones, P. D. and A. Moberg, 2003: Hemispheric and large-scale surface air temperature variations: an extensive revision and update to 2001. *J. Climate*, **16**, 206–223.

Jones, P. D., M. New, D. E. Parker, S. Martin, and I. G. Rigor, 1999: Surface air temperature and its change over the past 150 years. *Rev. Geophys.*, **37**, 173–200.

Katz, R. W. and B. G. Brown, 1992: Extreme events in a changing climate: variability is more important than averages. *Clim. Change*, **21**, 289–302.

Kistler, R. et al., 2001: The NCEP-NCAR 50-year reanalysis: monthly means CD-ROM and documentation. *Bull. Am Meteorol Soc.*, **82**, 247–268.

Klein Tank, A. et al., 2002: Daily dataset of 20th-century surface air temperature and precipitation series for the European Climate Assessment. *Int. J. Climatol.*, **22**, 1441–1453.

Knight, J. R., R. J. Allan, C. Folland, M. Vellinga, and M. E. Mann, 2005: A signature of persistent natural thermohaline circulation cycles in observed climate. *Geophys. Res. Lett.*, **32**, L20708, doi:10.1029/2005GL024233.

Kodera, K. and Y. Kuroda 2002: Dynamical response to the solar cycle. *J. Geophys. Res.*, **107**, 4749, doi:10.1029/2002JD002224.

Lau, K. M., M. K. Kim, and K. M. Kim, 2006: Asian monsoon anomalies induced by aerosol direct effects. *Clim. Dynam.*, **26**, 855–864.

Law, K. S. and A. Stohl, 2007: Arctic air pollution: origins and impacts. *Science*, **16**, 1537–1540.

Lean, J., 2004: Solar irradiance reconstruction. IGBP PAGES/World Data Center for Paleoclimatology, Data Contribution Series # 2004–035. NOAA/NGDC Paleoclimatology Program, Boulder, CO.

Limpasuvan, V., D. W. J. Thompson, and D. L. Hartmann, 2004: On the life cycle of northern hemisphere stratospheric sudden warming. *J. Climate*, **17**, 2584–2596.

Lubin, D. and A. M. Vogelmann, 2006: A climatologically significant aerosol longwave indirect effect in the Arctic. *Nature*, **439**, 453–456.

Luterbacher, J., M. A. Liniger, A. Menzel, N. Estrella, P. M. Della-Marta, C. Pfister, T. Rutishauser, and E. Xoplaki, 2007: The exceptional European warmth of Autumn 2006 and Winter 2007: historical context, the underlying dynamics and its phenological impacts. *Geophys. Res. Lett.*, **34**, L12704 doi:10.1029/2007GL029951.

McCabe, G. J., M. A. Palecki, and J. L. Betancourt, 2004: Pacific and Atlantic Ocean influences on multidecadal drought frequency in the United States. *P Natl. Acad. Sci. USA*, **101**, 4136–4141.

McPhaden, M. J., S. E. Zebiak, and M. H. Glantz, 2006: ENSO as an intergrating concept in Earth science. *Science*, **314**, 1740–1745.

Mears, C. A. and F. J. Wentz, 2005: The effect of diurnal correction on satellite-derived lower tropospheric temperature. *Science*, **309**, 1548–1551.

Meehl, C. A., W. M. Washington, T. M. L. Wigley, J. M. Arblaster, and A. Dai, 2002: Solar and greenhouse gas forcing and climate response in the twentieth century. *J. Climate*, **16**, 426–444.

Moberg, A. et al., 2006: Indices for daily temperature and precipitation extremes in Europe analysed for the period 1901–2000. *J. Geophys. Res.*, **111**, D22106.

Nicholls, N., D. Collins, B. Trewin, and P. Hope, 2006: Historical instrumental climate data for Australia – quality and utility for palaeoclimatic studies. *J. Quat. Sci.*, **21**, 681–688.

Ohmura, A. and H. Lang, 1989: Secular variation of global radiationover Europe. In: *Current Problems in Atmospheric Radiation*, J. Lenoble and J. F. Geleyn, ed., A. Deepak, Hampton, VA, 98–301.

Overland, J. E. and M. Wang, 2005: The third Arctic climate pattern: 1930s and early 2000s. *Geophys. Res. Lett.*, **32**, L23808, doi:10.1029/2005GL024254.

Philander, S. G., 1989: *El Niño, La Nina, and the Southern Oscillation*, Academic Press, San Diego, CA.

Polyakov, I. V., R. V. Bekryaev, G. V. Alekseev, U. Bhatt, R. L. Colony, M. A. Johnson, A. P. Makshtas, and D. Walsh, 2003: Variability and trends of air temperature and pressure in the maritime Arctic, 1875–2000. *J. Climate*, **16**, 2067–2077.

Robock, A., 2000: Volcanic eruptions and climate. *Rev. Geophys.*, **38**, 191–219.

Ramaswamy, V. et al., 2001: *Rev. Geophys.*, **39**, 71–122.

Ramaswamy, V., M. D. Schwarzkopf, W. J. Randel, B. D. Santer, B. J. Soden, and G. L. Stenchikov, 2006: Anthropogenic and natural influences in the evolution of lower stratospheric cooling. *Science*, **311**, 1138–1141.

Rotstayn, L. D. and U. Lohmann, 2002: Tropical rainfall trends and the indirect aerosol effect. *J. Climate*, **15**, 2103–2116.

Santer, B. et al., 2003: Contributions of anthropogenic and natural forcing to recent tropopause height changes. *Science*, **301**, 479–483.

Schär, C., P. L. Vidale, D. Lüthi, C. Frei, C. Häberli, M. Liniger, and C. Appenzeller, 2004: The role of increasing temperature variability for European summer heat waves. *Nature*, **427**, 332–336.

Schubert, S. D., M. J. Suarez, P. J. Pegion, R. D. Koster, and J. T. Bacmeister, 2004: On the Cause of the 1930s Dust Bowl. *Science*, **303**, 1855–1859.

Seneviratne, S. I., D. Lüthi, M. Litschi, and C. Schär, 2006: Land-atmosphere coupling and climate change in Europe. *Nature*, **443**, 205–209.

Serreze, M. C., M. M. Holland, and J. Stroeve, 2007: Perspectives on the Arctic's shrinking sea-ice cover. *Science*, **16**, 1533–1536.

Sherwood, S. C., J. R. Lanzante, and C. L. Meyer, 2005: Radiosonde daytime biases and late-20th century warming. *Science*, **309**, 1556–1559.

Shindell, D., D. Rind, N. Balachandran, J. Lean, and P. Lonergran, 1999: Solar cycle variability, ozone, and climate. *Science*, **284**, 305–308.

Solomon, S., 1999: Stratospheric ozone depletion: a review of concepts and history. *Rev. Geophys.*, **37**, 275–316.

Staehelin, J., N. R. P. Harris, C. Appenzeller, and J. Eberhard, 2001: Ozone trends: a review. *Rev. Geophys.*, **39**, 231–290.

Stenchikov, G., K. Hamilton, A. Robock, V. Ramaswamy, and M. D. Schwarzkopf, 2004: Arctic Oscillation response to the 1991 Pinatubo Eruption in the SKYHI GCM with a realistic Quasi-Biennial Oscillation. *J. Geophys. Res.*, **109**, D03112, doi:10.1029/2003JD003699.

Sutton, R. T. and L. R. Hodson, 2005: Atlantic forcing of North American and European summer climate. *Science*, **309**, 115–118.

Thompson, D. W. J. and S. Solomon, 2002: Interpretation of recent southern hemisphere climate change. *Science*, **296**, 895–899.

Trenberth, K. E., P. D. Jones, P. Ambenje, R. Bojariu, D. Easterling, A. Klein Tank, D. Parker, F. Rahimzadeh, J. A. Renwick, M. Rusticucci, B. Soden, and P. Zhai, 2007: Observations: surface and atmospheric climate change. In: *Climate Change 2007: The Physical Science Basis*. Contribution of Working Group I to the Fourth Assessment Report of the Intergovernmental Panel on Climate Change. Cambridge University Press, Cambridge.

Uppala, S. M. et al., 2005: The ERA-40 reanalysis. *Q. J. Roy. Meteorol Soc.*, **131**, 2961–3012.

Wang, M., J. Overland, V. Kattsov, J. Walsh, X. Zhang, and T. Pavlova, 2007: Intrinsic versus forced variation in coupled climate model simulations over the Arctic during the 20th century. *J. Climate*, **20**, 1084–1098.

Wanner, H., S. Brönnimann, C. Casty, D. Gyalistras, J. Luterbacher, C. Schmutz, D. Stephenson, and E. Xoplaki, 2001: North Atlantic Oscillation – concept and studies. *Surv. Geophys.*, **22**, 321–381.

White, W. B., J. Lean, D. R. Cayan, and M. Dettinger, 1997: A response of global upper ocean temperature to changing solar irradiance. *J. Geophys. Res.*, **102**, 3255–3266.

Wild, M. et al., 2005: From dimming to brightening: decadal changes in solar radiation at the Earth's surface. *Science*, **308**, 847–850.

WMO, 2007: Scientific Assessment of Ozone Depletion: 2006. WMO Global Ozone Research and Monitoring Project – Report No. 50.

Xoplaki, E., J. F. Gonzalez-Rouco, J. Luterbacher, and H. Wanner, 2004: Wet season Mediterranean precipitation variability: influence of large-scale dynamics and trends. *Clim. Dynam.*, **23**, 63–78.

Zhang, R., T. L. Delworth, and I. M. Held, 2007: Can the Atlantic Ocean drive the observed multidecadal variability in northern hemisphere mean temperature? *Geophys. Res. Lett.*, **34**, L02709, doi:10.1029/2006GL028683.

Section A
The Observational Climate Record

From the Bottom to the Stratosphere

Arctic Climate Features as Seen from the First International Polar Year (1882–1883) Until the End of World War II

C. Lüdecke

Abstract After the establishment of national weather services and the development of early weather forecasts towards the end of the 19th century, the Arctic region was seen as the home of cyclones, which very often resulted in violent storms in northern latitudes and influenced the weather of middle latitudes. Between 1882 and 1945 Arctic climate features were observed starting from ground-based measurements of a temporary network of stations in the framework of the First International Polar Year (1882–1883) and expanding to the third dimension with continuous observations of the upper atmosphere at Greenland and Svalbard (Spitsbergen), and ending with monitoring of the atmosphere around the Arctic in a dense network of stations during the Second International Polar Year (1932–1933). In addition, the airship "Graf Zeppelin" presented a platform for various kinds of instruments to measure meteorological parameters below, along, and above the flight track. During World War II, secret German weather stations were maintained in the Arctic, which sent their data to Norway to enter daily weather forecasts. Today these historical data series provide very important insight into climate features of the Arctic, which are mostly observed only during expeditions or through continuous observations during certain periods. They can be used to develop or test theories.

1 Introduction

Everybody knows that the Arctic influences the weather of northern and even middle latitudes. Winterstorms destroy logistic connections including transport systems and electricity, sometimes resulting in a longer stay at home with cold meals over candlelight. During the first years of the 21st century, we have experienced various kinds of winters – very cold ones and much too warm ones. Can these climate variations be explained by human activities or greenhouse effects? The Arctic plays an important role within the climate system and presents very impressive examples of climate change. Early 20th-century warming can already be seen by the permanent retreat of glacier

Institute for History of Natural Sciences, Mathematics and Technology, Valleystrasse 40, D-81371 Munich, Germany
C.Luedecke@lrz.uni-muenchen.de

S. Brönnimann et al. (eds.), *Climate Variability and Extremes during the Past 100 Years.*

tongues in West Greenland, for example, by the most impressive retreat of the Jakobshavns Isbrae in Disko Bay. Increasing average temperatures by 3–5° in some parts of the Arctic result in a reduction in ice thickness of the Arctic Ocean. Melting permafrost at high latitudes leads to the destruction of historical buildings along the Arctic coastline and polar heritage sites, like the remains of the 17th-century whaling station at Jan Mayen, which are in danger. Shall they be removed and built up or reconstructed in safer places? Discussions have just started within the International Polar Heritage Committee.

If you want to investigate climate impacts you have to look at Arctic climate features. The measurement of ground-based meteorological data and of upper air data is a special challenge in a windy and cold environment, especially during the polar night when it is dark for several months of the year. Nevertheless, scientists knew about the importance of their data and invested much effort in taking them for further analysis.

Since the end of the 19th century, instruments were improved and new devices were developed. In the 1930s, meteorologists started to investigate the third (vertical) dimension. Before it became routine to measure meteorological parameters from satellites, there were only a few locations in the Arctic where data was collected. In addition, these data were taken only during certain periods. The paper will follow this development and give examples of measurements at different times, from the First International Polar Year until the end of World War II.

2 The First International Polar Year (1882–1883)

When the first national weather services were established and early weather forecasts developed in the second half of the 19th century, the Arctic region was seen as the home of cyclones, which very often resulted in violent storms in the inhabited countries of northern latitudes. During various expeditions, meteorological data were taken only occasionally, but nonetheless described a hostile climate which seemed to influence middle latitudes. The idea to coordinate the investigation of the weather conditions of the Arctic in the framework of an International Polar Year goes back to Georg von Neumayer (1826–1909), director of the German Maritime Observatory (Deutsche Seewarte) in Hamburg, and the German Lieutenant of the Austro-Hungarian Navy Carl Weyprecht (1838–1881). Weyprecht led the Austro-Hungarian expedition (1872–1874) to the Russian Arctic and discovered Franz Josef Land but lost his ship. Both independently had the idea of establishing a belt of scientific stations around the Arctic for simultaneous meteorological, magnetic and other geophysical measurements over at least 1 year (Baker 1982; Corby 1982). During the 2nd International Meteorological Congress in Rome 1879, Weyprecht's proposal was discussed and a list of stations regularly distributed around the Arctic adopted. In addition, Neumayer suggested that stations in the southern hemisphere also be included.

An International Polar Commission was founded under Neumayer's presidency, which prepared the then so called International Polar Year (Lüdecke 2004). After a first meeting in Hamburg (October 1879), the measurements had to be postponed for a year during the second meeting in Bern (August 1880) because only four nations were willing to contribute to the programme. Furthermore, due to a lack of financial support from the German Empire (Deutsches Reich), Neumayer resigned from the presidency of the commission. Under his successor Heinrich von Wild (1833–1902), director of the Central Observatory at St. Petersburg, the Polar Year was fixed during the 3rd meeting at St. Petersburg in August 1881, and a meteorological and magnetic programme was decided upon, including contributions of navy and mercantile ships. The Polar Year was the first international meteorological experiment organised as a cooperation of 11 nations, which established 12 observing stations around the Arctic (Fig. 1) and two stations in the southern hemisphere, collecting data from 1 August 1882 until 1 September 1883 (Table 1) (Barr 1985).

After 1 year of ground-based measurements, the first meteorological data set describing the climate of the Arctic was available and entered the textbooks. However, spacing of the network had been too wide for a detailed analysis of the relationship between the higher and lower latitudes. Only a German Ph.D. thesis published synoptic charts of monthly mean values of pressure and temperature for January and July (Ehrhart 1902). Circumpolar isotherms for January 1883 showed the influence of the cold current in the Davis Strait and the warm Gulf Stream east of Spitsbergen as never seen before (Fig. 2).

Fig. 1 Circum Arctic stations established for the 1st International Polar Year (1882–1883) (see plate 2 in Neumayer 1901).

Table 1 Stations of the 1st International Polar Year (1882–1883) (after Corby 1982)

Station	Location	Country
	Northern hemisphere	
Bossekop	North Norway	Norway
Cape Thordsen	Spitsbergen	Sweden
New Holland	Drifting station in Kara Sea	The Netherlands
Fort Rae	Great Slave Lake, Canada	Great Britain
Godthaab	Western Greenland	Denmark
Jan Mayen	North Atlantic	Austro-Hungary
Karmakuly	Novaya Zemlya	Russia
Kingua Fjord	Baffin Island (North-east Canada)	Germany
Lady Franklin Bay	Ellesmere Island, North-east Canada	USA
Point Barrow	Alaska	USA
Sagastyr	Lena Delta, Siberia	USSR
Sodankylä	Northern Finland	Finland
	Southern hemisphere	
Cape Horn	Terra del Fuego	France
South Georgia	South Atlantic	Germany

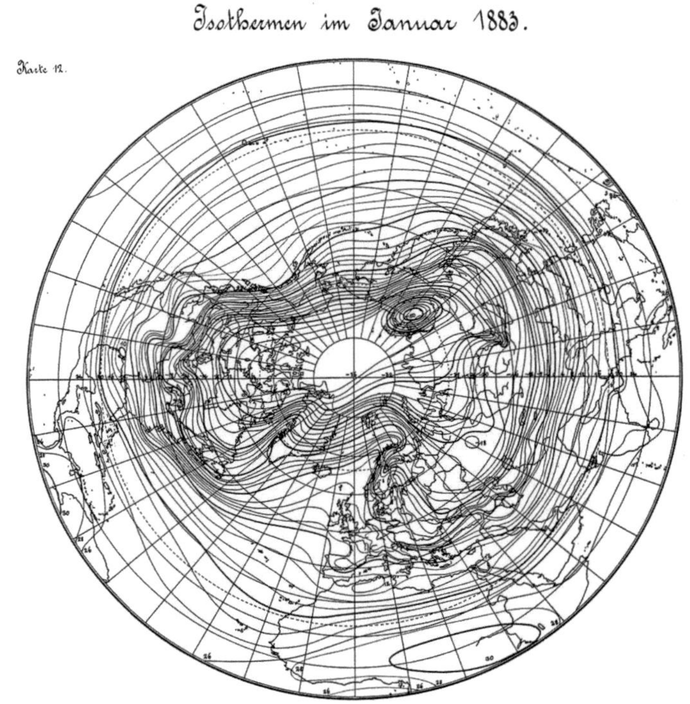

Fig. 2 Circumpolar chart of isothermes of January 1883 constructed from 799 stations in the northern hemisphere and 101 data points in the Atlantic Ocean (see Ehrhart 1902, Meteorological Institute, University of Munich, Germany, unpublished data).

Regarding measuring techniques, the first Polar Year was 50 years ahead of time but it nevertheless paved the way for further international cooperation in science. The data were used to construct daily circumpolar synoptic charts of 1–5 March, 17–20 April, 7–10 May, and 4–7 June 1883. These charts were needed to plan an expedition with the airship "Graf Zeppelin" in 1931 (see Sect. 4.2).

3 Aerological Cross Sections in Time and Space

Since the development of meteorological measuring techniques with manned balloons, unmanned captive balloons, pilot balloons, and kites equipped with registering instruments, coordinated investigation of the upper atmosphere started at the end of the 19th century. In 1896 an International Commission for Aeronautics was established under the chairmanship of Hugo Hergesell (1859–1938). Independent balloon ascents in Trappes (France) and Berlin (Germany) led to the discovery of the tropopause in 1902 (Hoinka 1997). With the establishment of the kite station in Großborstel close to Hamburg (1903) and the aeronautical observatory in Lindenberg east of Berlin (1905), aerology was institutionalised in Germany. At these stations daily ascents were carried out and the meteorological data gathered, delivered timely cross sections of the weather conditions in upper levels.

3.1 Introduction of Aerology to Polar Research (1906–1908)

Following his brother Kurt Wegener (1878–1964), who worked in Lindenberg, Alfred Wegener (1880–1930) soon became an expert in aerological soundings. When he applied to participate in the Danish expedition to Greenland (1906–1908) led by Ludwig Mylius-Erichsen (1872–1907), he had the opportunity to introduce aerology, using kites and rubber balloons, into polar research (A Wegener 1909). The main task of the expedition was to explore the uninhabited coastline of north-east Greenland between 77 and 83°N. At Danmarks Havn, Wegener established a meteorological station. Aerological equipment included steel bottles filled with hydrogen and a car used as a winch. After the first year of experience his kite reached a maximum height of 3,110 m on 3 August 1907. In the same month he set up a second manned weather station for the detailed study of foehn and other phenomena (Brand and Wegener 1912). By analysing the vertical profiles of his measurements, Wegener showed aerology in high latitudes to be very promising. In Fig. 3 he summarised the deviation of temperature, wind speed, wind direction and relative humidity from the mean values of these parameters between the surface and 1,500 m level of 125 ascents (99 with kites, 26 with balloons) made at Danmarks Havn. He found a pronounced layer of discontinuity at 1,200 m height, which corresponded to similar conditions at ground level. These results were comparable with the conditions he knew from Lindenberg.

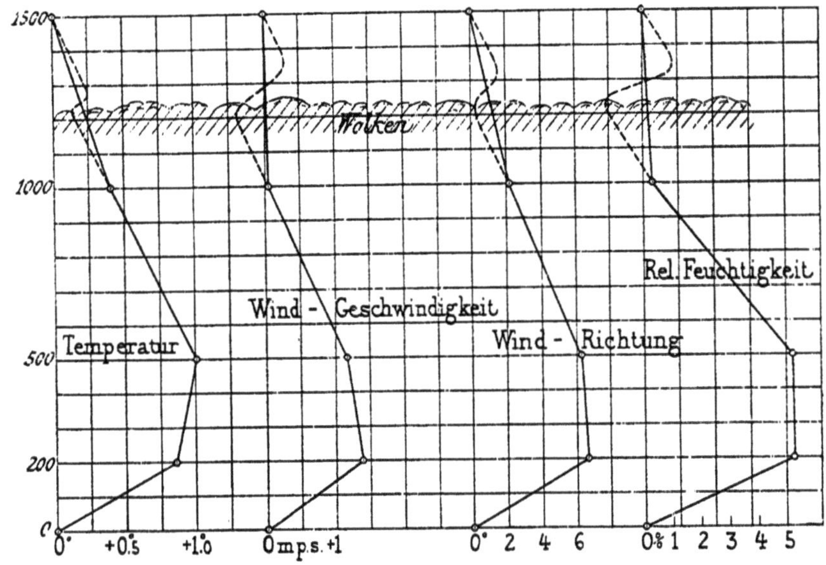

Fig. 3 Deviation of temperature, wind speed, wind direction and relative humidity from the mean values of these parameters between bottom and 1,500 m level of 125 ascents made at Danmarks Havn, East Greenland (1906–1908) (see p. 61 in A. Wegener 1909).

3.2 *German Geophysical Station in Spitsbergen (1911–1914)*

Aerological measurements in the Arctic were continued during the study trip of the German Arctic Zeppelin Expedition to Spitsbergen in 1910. Graf Ferdinand von Zeppelin (1838–1917) wanted to investigate if meteorological conditions would allow long distance journeys with airships to explore the still vast unknown regions of the Arctic (Miethe and Hergesell 1911). Hergesell was in charge of the aerological measurements during the trip, which detected strong winds off land restricted to some hundred metres above ground level. He had already encountered such strong local winds off lands during an earlier expedition to Spitsbergen in 1906 (Hergesell 1906–1908). Above these orographic winds, wind direction is dominated by the general circulation. The results of 1910 were not sufficient however to make a clear decision in favour of secure airship travel. As a consequence he proposed to install an observatory in Spitsbergen to provide a continuous record of aerological data from the upper atmosphere. The German Geophysical Observatory was installed in an existing hut at Hotelnesset, southwest of Advent Bay (the present day airport of Longyearbyen) in 1911 (Hergesell 1914). During the first year, 78 ascents were made, 17 with kites (maximum height 3,310 m) and 61 with balloons (maximum height 2,480 m) (p. 144 in Kopp 1935). Figure 4 shows a picture of a captive balloon ascent in the snow covered area at Advent Bay in May 1912.

Mai 1912

Fig. 4 Captive balloon ascent at the German Geophysical Station at Hotelnesset (Advent Bay, Spitsbergen) on May 1912 (see plate 2 in Rempp and Wegener 1914).

The results of five representative ascents from 8 March to 2 May 1912 (Fig. 5) show several inversion layers at different levels. This information was used for planning the previously mentioned airship expedition in 1931 (see Sect. 4.2).

During the next summer, the first station was abandoned and a new station was built at Ebeltofthamna at the northern entrance of Cross Bay. Kurt Wegener became station leader during the 1912–1913 overwintering. His group accomplished 114 ascents; 15 with kites up to 1,590 m and 99 with balloons up to a maximum height of 5,400 m. Daily measurements of meteorological parameters at ground level are given in Fig. 6. During winter, rapid temperature variations of more than 20–30° were observed several times and resulted in temperatures above the freezing point due to changing ice conditions of the bay. At the beginning of World War I the station was closed.

Unfortunately the third group could not publish their results from the 1913–1914 overwintering. However, a comprehensive list of publications of the German Geophysical Station at Spitsbergen is given in Dege (1962).

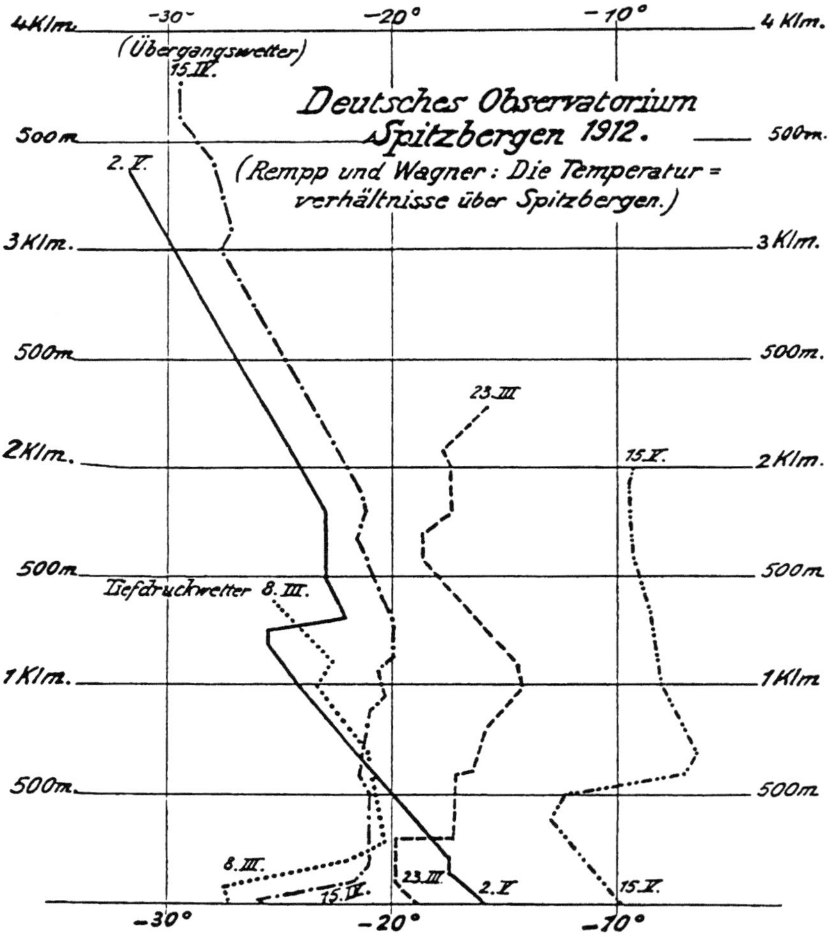

Fig. 5 Temperature conditions at Spitsbergen up to 4,000 m between 8 March and 15 May 1912 (see plate 9 in Internationale Studiengesellschaft 1924).

3.3 *Crossing of Greenland (1912–1913)*

During the Danish inland-ice expedition of 1912–1913, under the leadership of his former comrade lieutenant Johan Peter Koch (1870–1928), Alfred Wegener could investigate the meteorological conditions on the 3,000 m high ice cap during the crossing of Greenland from Danmarks Havn to the west coast (Koch and Wegener 1930). Besides glaciological and meteorological work at the wintering station close to the inland-ice, Wegener was interested in the cold region at the centre of Greenland, which was interpreted as the "glacial anticyclone" (glaziale Anticyklone) – a

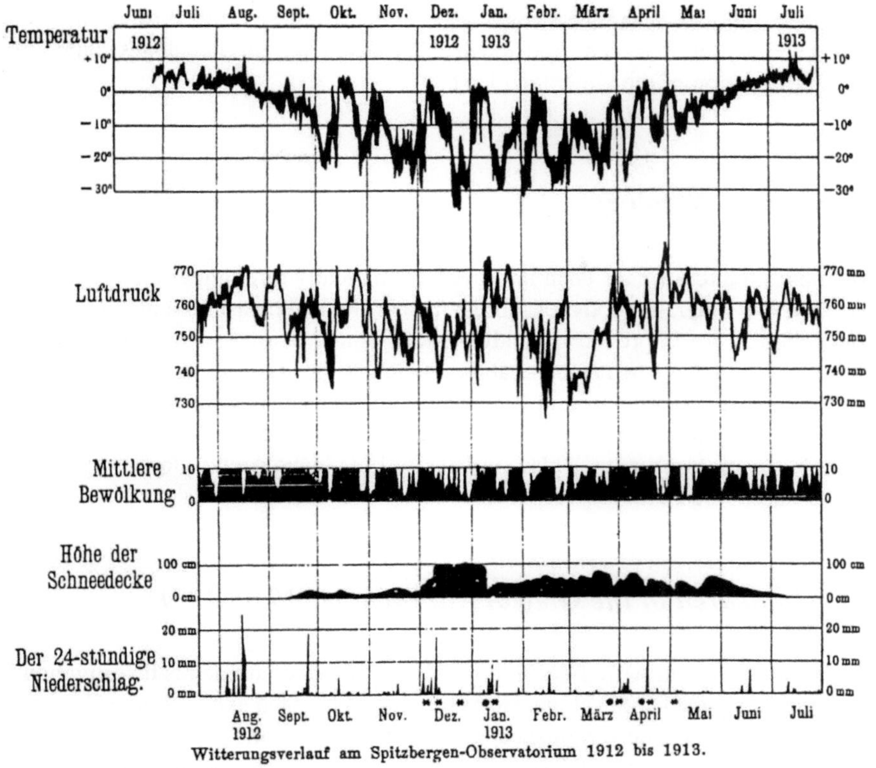

Fig. 6 Daily temperature, pressure, mean cloudiness, height of snow cover, and daily precipitation at ground level at the German Geophysical Station (Cross Bay, Spitsbergen) from June 1912 until July 1913 (see Fig. 2 in Wegener 1916).

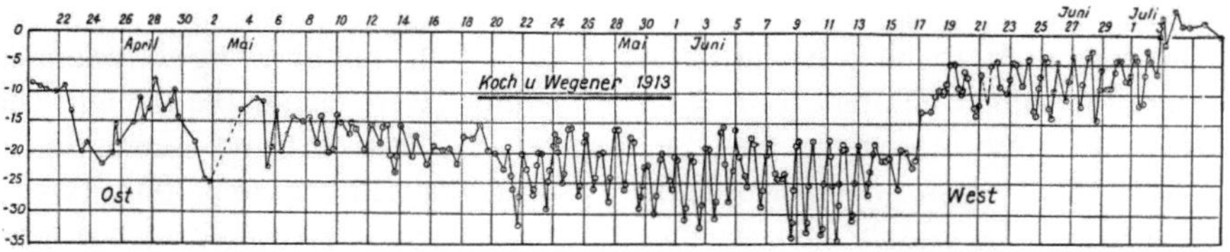

Fig. 7 Daily temperatures of the crossing of northern Greenland from east to west (21 April–4 July 1913) showing the central cold region (see Fig. 286 in Koch and Wegener 1930).

permanent high caused by the cold glacier below. The change of wind direction during the crossing, with clear weather and stable temperatures, confirmed the idea of a central cold region (Fig. 7). But Wegener did not know its behaviour in wintertime or if the conditions at the west coast were similar to those at the east coast, which he was already familiar with. And above all, he asked himself if cyclones could cross the central region of the Greenland ice cap.

4 Modern Aerology

Airships and aircraft have been improved dramatically throughout their use during World War I. Later, methods for remote meteorological measurements in the upper atmosphere were improved by adding radio transmission to send the data from the instruments down to a receiver at ground level. This resulted in the development of the radiosonde system around 1930, which offered new possibilities to investigate the upper air by means of a throwaway instrument. A successful test of radiosounding up to the stratosphere took place in January 1931 in the Arctic, close to Murmansk (Moltchanoff 1931).

4.1 In Search of the Glacial Anticyclone (1930–1931)

Not before 1930, did Alfred Wegener have a chance to continue his investigation of the supposed glacial anticyclone. He did so during his own expedition to Greenland (1930–1931), when he established three aerological stations on a cross section along 71°N over a period of 1 year. Two of the stations were set up at the west and east coast respectively. The most challenging task was the establishment of the central station "Eismitte" on the ice cap 400 km east of West station. Unfortunately aerological measurements could not be coordinated (K. Wegener 1939). Only once did all three stations carry out balloon ascents at the same time; if one considers "at the same day" to be simultaneous (Fig. 8a). Synchronous measurements at two stations were taken only on 11 days (Fig. 8b).

At "Eismitte" ground level was already 3,000 m high, thus allowing considerable heights with simple equipment, if weather conditions were favourable enough. The best results were received at East Station with 155 (mostly) kite ascents up to 4,700 m, while a lesser number of balloon ascents reached only to 3,700 m (see p. 144 in Kopp 1935). In the end, the data was insufficient to provide evidence for a stationary glacial anticyclone. In addition, meteorological ground level data were taken at each station, describing the differences in local climate during all seasons. Surprisingly, on 24 March (12:00) 1931 a low pressure zone was passing "Eismitte", which had already been observed at West Station on 23 March (16:00) and which was also measured at East Station on 25 March (2:00) (Kopp and Holzapfel 1939). This indicated that cyclones did cross Greenland and that the supposed glacial Anticyclone can not operate as a wall, preventing such crossings. Tracks of low pressure zones derived from weather charts of the German Maritime Observatory also indicated the low crossing Greenland between 24 and 25 March 1931.

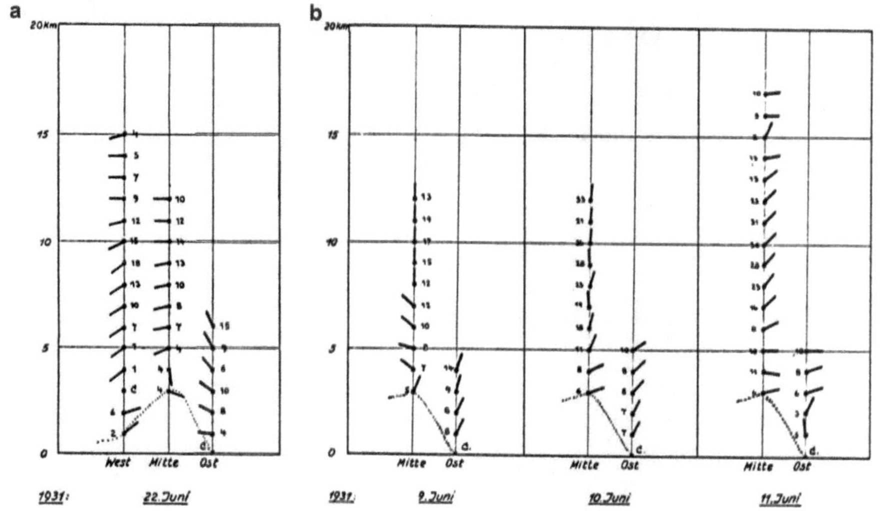

Fig. 8 (a) Simultaneous pilot balloon ascents at West Station, "Eismitte", and East Station showing a cross section of wind direction in upper levels above Greenland on 22 June 1931 (see Fig. 8 in K. Wegener 1939). (b) Continuous simultaneous pilot balloon ascents at "Eismitte" and East Station showing a cross section of wind direction in upper levels above Greenland from 9 to11 June 1931 (see Fig. 7 in Wegener 1939).

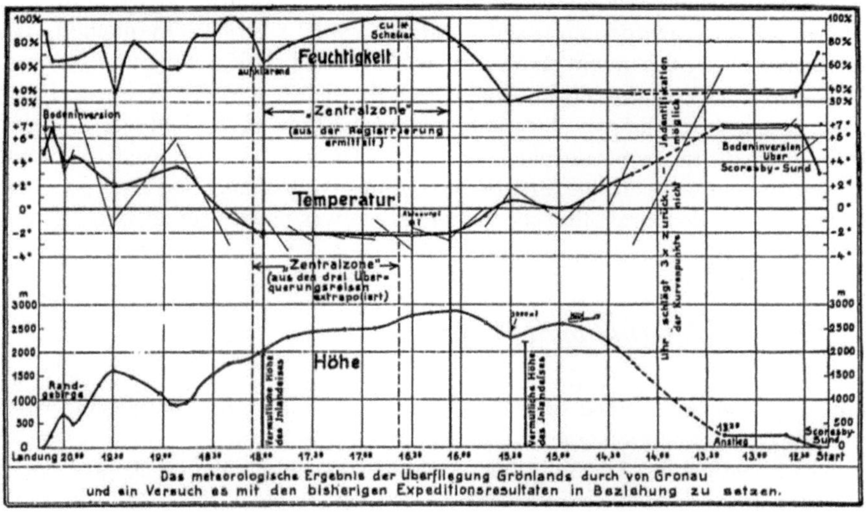

Fig. 9 Humidity and temperature measured during the flight over Greenland from east to west on 15 August 1931 (see Fig. 5 in Becker and Baumann 1933).

It was a pity that there was no cooperation with the British Air Route Expedition (1930–1931), which also maintained a wintering station on the ice cap about 550 km south of "Eismitte" and made a crossing of Greenland in the following summer.

Nor was there cooperation with the Michigan Expedition (1930–1931) which maintained two stations on the west coast.

It should be added that Wolfgang von Gronau (1893–1977) crossed Greenland with his Dornier Wal aircraft, flying from Germany on 15 August 1931. Temperature and humidity measured during the flight not only showed the cold region above the ice cap, but also served to describe the weather conditions of a flight track over Greenland (Becker and Baumann 1933) (Fig. 9).

4.2 The Airship as a Measuring Platform (1931)

At the end of the 1920s, the plan of trans-Arctic air routes were highly discussed. Resuming Graf Zeppelin's idea of exploring the Arctic with airships, an expedition with the airship LZ 127 "Graf Zeppelin" was planned in the Russian Arctic. The potential of radiosondes for the expedition in 1931 was already recognised at a very early stage of preparation. The best time period for the expedition was defined using the data of the 1st International Polar Year and the German Geophysical Observatory at Spitsbergen (see Sect. 2 and Sect. 3.2) (Internationale Studiengesellschaft 1924). During the six-day flight (26–30 July 1931) the airship was used as a platform for different measurements (Weickmann 1933). In addition to meteorological data continuously taken from instruments on board while in-flight, four radiosondes were released during the flight to observe pressure, temperature, and humidity, using a very tricky method. The balloon with the instrument fixed to it was released with a weight, so it went downward first. After a certain time, when the balloon would be about 100 m above ground level, the weight was blown away allowing the balloon to ascend up to 17,000 m in the stratosphere (Fig. 10). The results were used for online weather forecasts aboard the airship. As airships were abandoned after the crash of LZ 129 "Hindenburg", this special technique was never used again.

4.3 Second International Polar Year (1932–1933)

Almost 50 years had passed since the International Polar Year (1882–1883), when meteorology was on the verge of a new era, with the introduction of radiosondes for daily investigation of the upper air, at aerological stations all over the world. The polar front theory was already available for weather forecasting, but the relationship between the Arctic and middle latitudes was not well enough understood. A region of very strong winds close to the polar front (jet stream) had just been detected on the northern hemisphere. Now meteorologists wanted to organise another Polar Year to investigate the upper atmosphere with a dense aerological network in the Arctic. Due to the worldwide financial crisis, temporary stations in Antarctica could

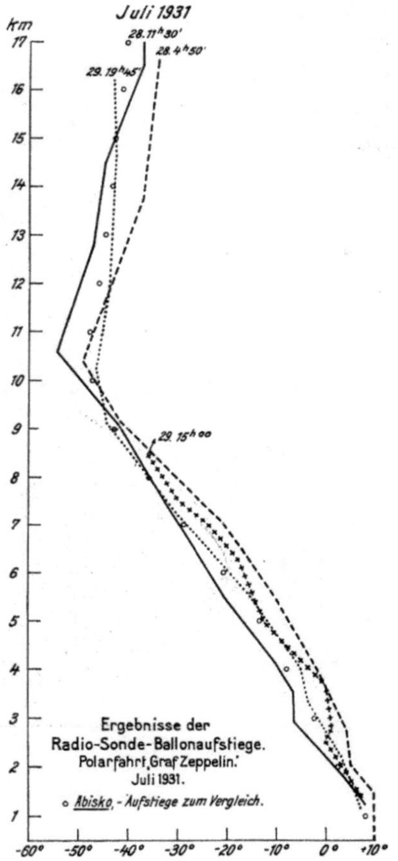

Fig. 10 Radiosondes ascents from the airship LZ 127 "Graf Zeppelin" during 28–29 July 1931. o: Ascent at Abisko (Norway) for comparison (see Fig. 8 of plate 18 in Weickmann 1933).

not be included in the plans (Laursen 1982). In the end, 44 nations participated in the 2nd International Polar Year, lasting from 1 August 1932 until 31 August 1933. They installed a network of 27 circum Arctic stations to explore the three-dimensional weather conditions, while Reykjavik became the centre for ascents with aircraft and radiosondes (Fig. 11).

Other countries, which could not finance expeditions, intensified their measurements at home. Continuously probing of meteorological parameters in time and space was now possible for the first time. The results were published in daily synoptic weather charts, constructed by the Deutsche Seewarte at Hamburg and finished in 1950. Only data from the last 15 days were destroyed during World War II.

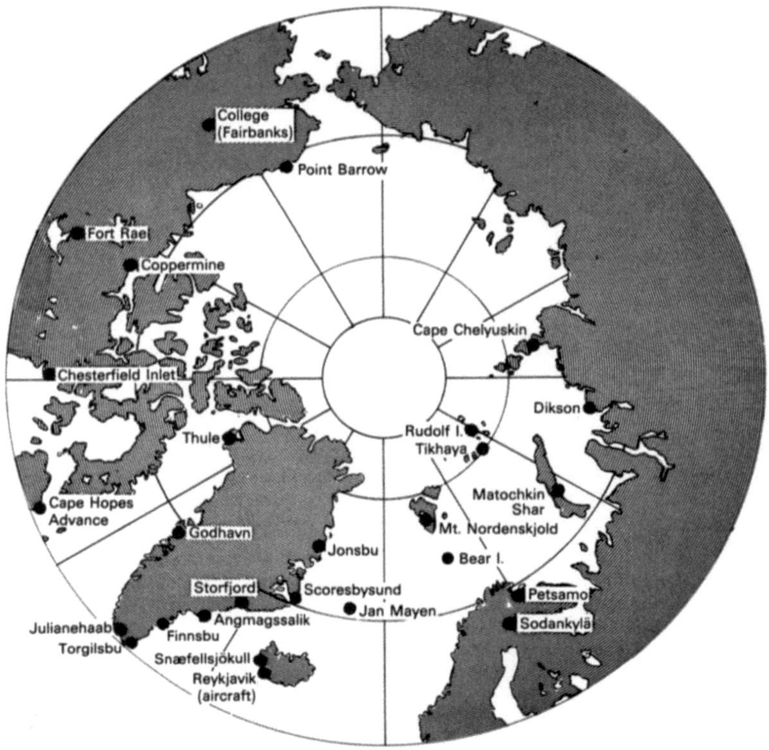

Fig. 11 Circum Arctic stations established for the 2nd International Polar Year (1932–1933) (see p. 215 in Laursen 1982).

5 Secret German Weather Stations in the Arctic During World War II (1941–1945)

After the 2nd International Polar Year, radiosondes were used for upper air measurements all over the world (Trifonov 1980). When Germany no longer received weather data from the North Atlantic region during World War II, automatic instruments connected to radio transmission, were invented to be used at unmanned meteorological stations, independently set up in the Arctic by the German Airforce and Navy (Fig. 12) (Kington and Selinger 2006).

For operations in the Northern Sea, the upper air was surveyed from secretly manned weather stations at Spitsbergen, Greenland, and Franz Josef Land since 1941 (Selinger 2001). The last weather troop stationed at Nordaustlandet (eastern Spitsbergen) was the first one to perform daily radiosonde ascents up to the stratosphere during the winter of 1944/45, after having overcome the problems of the previous years (Dege 1954, 1960). After surrender of Germany on 8 May 1945, this station continued its work until 9 September 1945, when it was closed by Norway. Due to this special circumstance,

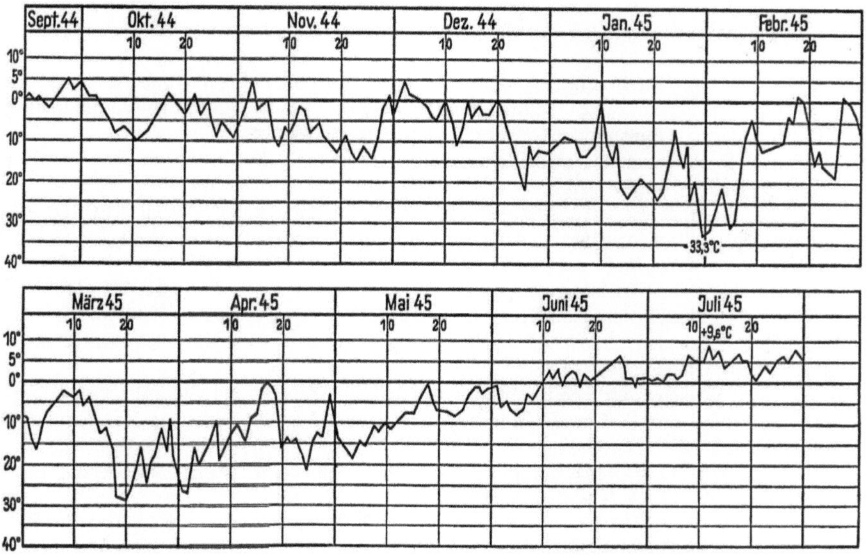

Fig. 12 Daily temperature in °C of station "Haudegen" (Nordaustlandet, Spitsbergen) (see p. 145 in Dege 1954).

observations did not disappear in secret daily weather forecasts for military purposes, but were published later. Altogether 222 ascents were made, 140 by radiosondes up to 18,000 m and 86 by pilot balloons up to over 20 km.

6 Outlook

After World War II, historical Arctic data were used to support new ideas and theories. One of them dealt with an expanded theory concerning foehn in Spitsbergen and Greenland (Rossman 1950). The author referred to data taken at the Swedish station in Spitsbergen during the 1st International Polar Year (1882–1883) and at the German Geophysical Observatory in Advent Bay (1911–1912), as well as data taken during Erich von Drygalski's expedition (1892–1893) to Greenland and the Wegener's Danmark expedition (1906–1908).

An overview of the state of aerology in polar regions after World War II is given by Flohn (1952), who discussed measurements at Jakusk (Russia) from 1940–1941 to 1948–1951 and the development of the temperature inversion above the troposphere over Russia and North America as well as over Antarctica during winter.

In comparison with recent long time series of Arctic climate data, historical observations of weather features become very important, especially if you want to investigate climate change. Due to this, historical data should no longer be hidden in old publications or archives.

References

Baker, F. W. G., 1982: The first international polar year, 1882–83. *Polar Record*, **21**, 275–285.

Barr, W., 1985: The expeditions of the first International Polar Year, 1882–83. The Arctic Institute of North America, University of Calgary, Technical Paper 29.

Becker, R. and G. H. Baumann, 1933: Beiträge zur Meteorologie des Luftweges über Grönland. *Archiv der Deutschen Seewarte*, **52**.

Brand, W. and A. Wegener, 1912: Meteorologische Beobachtungen der Station Pustervig. – In: Danmark-Ekspeditionen til Grønlands Nordøstkyst 1906–1908. Bind II, Nr. 6, Meddelelser om Grønland, vol. 42, København, 447–562.

Corby, G. A., 1982: The first International Polar Year (1882/83). *WMO Bull.*, **31**, 197–214.

Dege, W., 1954: Wettertrupp Haudegen. Eine deutsche Arktisexpedition 1944/45. Eberhard Brockhaus, Wiesbaden. Translated and edited by W. Barr, war north of 80°. University of Calgary Press, Calgary, 2003.

Dege, W., 1960: Wissenschaftliche Beobachtungen auf dem Nordostland von Spitzbergen 1944–1945. Selbstverlag des Deutschen Wetterdienstes, Offenbach, Berichte des Deutschen Wetterdienstes, **72**.

Dege, W., 1962: Deutsches Observatorium Ebeltofthafen – Spitzbergen. Zur 50. Wiederkehr der 1. Überwinterung 1912/13. *Polarforschung*, **32**, 136–140.

Ehrhart, S. B., 1902: Die Verteilung der Temperatur und des Luftdruckes auf der Erdoberfläche im Polarjahre 1882/1883. Inauguraldissertation. Stuttgart.

Flohn, H., 1952: Zur Aerologie der Polargebiete. *Meteorologische Rundschau*, **5**, 81–87, 121–128.

Hergesell, H., 1906–1908: Die Erforschung der freien Atmosphäre über dem Polarmeer. *Beiträge zur Physik der freien Atmosphäre*, **2**, 96–98.

Hergesell, H. ed., 1914: Das Deutsche Observatorium in Spitzbergen. Beobachtungen und Ergebnisse I, *Schriften der Wissenschaftlichen Gesellschaft in Straßburg*, **21**.

Hoinka, K. P., 1997: The tropopause: discovery, definition and demarcation. *Meteorologische Zeitschrift*, Neue Folge, **6**, 281–303.

Internationale Studiengesellschaft ed., 1924: Das Luftschiff als Forschungsmittel in der Arktis. Eine Denkschrift. Internationale Studiengesellschaft zur Erforschung der Arktis mit dem Luftschiff, Berlin.

Kington, J. A. and F. Selinger, 2006: The development and use of weather buoys 1940–2005. *Weather*, **61**(6), 164–166.

Koch, J. P. and A. Wegener, 1930: Wissenschaftliche Ergebnisse der Dänischen Expedition nach Dronning Louises-Land und quer über das Inlandeis von Nordgrönland 1912–13. *Meddelelser om Grønland*, **75**, Bd. 2.

Kopp, W., 1935: Die Oststation. In: *Wissenschaftliche Ergebnisse der Deutschen Grönland-Expedition Alfred Wegener 1929 und 1930/31*, K. Wegener, ed., Meteorologie, F.A. Brockhaus, Leipzig, Bd. IV, 1, 1–190.

Kopp, W. and R. Holzapfel, 1939: Beiträge zum Mechanismus des Witterungsverlaufs über Grönland. In: *Wissenschaftliche Ergebnisse der Deutschen Grönland-Expedition Alfred Wegener 1929 und 1930/31*, K. Wegener, ed., Meteorologie, F.A. Brockhaus, Leipzig, Bd. IV, 2, 274:325.

Laursen, V., 1982: The second International Polar Year (1932/33). *WMO Bull.*, **31**, 214–222.

Lüdecke, C., 2004: The first International Polar Year (1882–83): a big science experiment with small science equipment. *History Meteorol.*, **1.1**, 54–63. http://www.meteohistory.org/2004proceedings1.1/

Miethe, A. H. and H. Hergesell, eds., 1911: Mit Zeppelin nach Spitzbergen. Deutsches Verlagshaus Bong, Berlin.

Moltchanoff, P., 1931: Die Methode der Radiosonde und ein Versuch ihrer Anwendung bei der Erforschung der höheren Atmosphärenschichten in den Polargebieten. *Gerlands Beiträge zur Geophysik*, **34**, 36–56.

Neumayer, G. V., 1901: Auf zum Südpol! 45 Jahre Wirkens zur Förderung der Erforschung der Südpolarregion 1855–1900. Vita Deutsches Verlagshaus, Berlin.

Rempp, G. and A. Wagner, 1914: Die Station in der Adventbai 1911/1912. In: H. Hergesell ed., Das Deutsche Observatorium in Spitzbergen, Beobachtungen und Ergebnisse I, *Schriften der Wissenschaflichen Gesellschaft in Straßburg*, **21**, 6–20.

Rossman, F., 1950: Über den Foehn auf Spitzbergen und Grönland. *Polarforschung*, **20**(1/2), 347–353.

Selinger, F., 2001: Von »Nanok« bis »Eismitte«: Meteorologische Unternehmungen in der Arktis 1940–1945. Schriften des Deutschen Schiffahrtsmuseums, Bd. 53, Convent Verlag, Hamburg.

Trifonov, G. P., 1980: Fifty years' radiosounding of the atmosphere. *WMO Bull.*, **29**, 177–180.

Wegener, A., 1909: Drachen- und Fesselballonaufstiege. In: Danmark-Ekspeditionen til Grønlands Nordøstkyst 1906–1908. Bind II, Nr. 1, Meddelelser om Grønland, vol. 42, København, 1–75.

Wegener, K., 1916: Die meteorologischen Aufzeichnungen am Spitzbergen Observatorium 1912 bis 1913. *Meteorologische Zeitschrift*, **33**, 59–64.

Wegener, K., 1939: Vorläufige Zusammenfassung der meteorologischen Ergebnisse. In: *Wissenschaftliche Ergebnisse der Deutschen Grönland-Expedition Alfred Wegener 1929 und 1930/31*, K. Wegener, ed., F.A. Brockhaus, Leipzig, vol. Bd. IV, 2, 363–380.

Weickmann, L., 1933: Die meteorologischen Aufgaben bei der 1. Polarfahrt des Graf Zeppelin Juli 1931. *Petermanns Geographische Mitteilungen, Ergänzungsheft*, **216**, 48–59.

Arctic Sea Ice Data Sets in the Context of Climate Change During the 20th Century

G. V. Alekseev[1], S. I. Kuzmina[2], A. P. Nagurny[1], and N. E. Ivanov[1]

Abstract Available estimates of sea ice extent in the northern hemisphere cover the period from the early part of the 20th century to present day. We analyze changes in ice extent and thickness in the Arctic and its relation to surface air temperature over this period. Time series obtained from different data sets demonstrate better agreement after the 1950s and especially since 1979 with the onset of regular remote sensing observations from satellites. Statistics of time series show minima ice extent in August–September. Mean square deviations reach maxima in July–August. The distributions of trend coefficients show a more significant decrease of summer ice extent. Statistics of monthly ice extent in the Siberian Arctic seas show a similar distribution. September ice extent in the majority of the Siberian Arctic seas and in the Barents Sea reveal rapid shrinking during Arctic warmings in the 1920–1940s and 1990s. Significant correlation between surface air temperature and ice extent occurs in summer months with maximum in June under the influence of June maximum solar irradiation, and amplified by heat advection in the atmosphere and ice extent anomalies in the previous months. The relationship between variations of winter air temperature and ice extent is weaker because winter ice extent anomalies depend on air temperature anomalies as well as on the area occupied by a freshened upper layer. Good agreement between variations of the sum of summer air temperature in the marine Arctic and sea ice extent in September is found (correlation coefficient is 0.85). It confirms that summer melting plays the most important role in the sea ice volume decrease. The renewed observations in 2004–2005 at the Russian "North Pole" drifting stations revealed that the area-averaged perennial ice in the Arctic Basin decreased by 110 cm relative to the 1990 value. But the land-fast ice thickness in the Kara and Laptev Seas show an insignificant positive linear trend for 1934–2005 in agreement with the sum of winter air temperature. The negative trend of land-fast ice thickness becomes apparent starting from the 1970s.

[1] Arctic and Antarctic Research Institute, St. Petersburg, Russia
alexgv@aari.nw.ru, nagurny@aari.nw.ru, neivanov@aari.nw.ru

[2] Nansen International Environmental and Remote Sensing Center, St. Petersburg, Russia
svetak4@yandex.ru

S. Brönnimann et al. (eds.), *Climate Variability and Extremes during the Past 100 Years.*

1 Introduction

Sea ice cover in the Arctic is a clear indicator of climate change and an important player for the interaction between the Arctic and the globe. As seen from experiments with global climate models (Johannessen et al. 2004), ice cover decrease is strengthened by the warming. On the other hand, freshening of the upper layer during climate warming is favorable for sea ice spread during winter, leading to a decrease in the warming effect of the ocean to the atmosphere. Available estimates of sea ice extent (SIE) in the northern hemisphere (NH) cover the period from the 19th century for ice in the Nordic seas and adjacent regions (Vinje 2001; Brinck Løyning et al. 2003). Since 1900 ice information is available for the NH (Chapman and Walsh 1993) and since 1924 for the Siberian Arctic seas and Barents Sea (Zakharov 2003). Instrumental measurements of sea ice from satellites started in 1973 but as regular monitoring are provided from 1978. Since the 1950s, the observations of ice extent in the regions within the navigation route have been conducted mostly visually from airplanes. Based on these observations, data sets on sea ice extent in the NH were prepared and supplemented by reconstructions back to 1900. In this paper we analyze changes in ice extent and thickness in the Arctic and its relations to surface air temperature (SAT).

2 Monthly Sea Ice Data in the Northern Hemisphere

2.1 Walsh and Chapman Data Set

The Walsh and Chapman (WCH) data set covers the period from 1901 to 1998. This set is a compilation of data from several sources integrated into a single gridded ice concentration product by John Walsh and Bill Chapman, University of Illinois (Chapman and Walsh 1993). Later, it was supplemented by new data based on the remote sensing measurements (NSIDC 2005). SIE is calculated from this data based on grid cells which have an ice concentration greater than 15%. WCH data are available from http://www.nsidc.colorado.edu.

2.2 Met Office Hadley Centre's Data Set

The HADISST1 data set covers the period from 1870 to present and is described by Rayner et al. (2003). This data set has global coverage with a resolution of 1° latitude × 1° longitude, and incorporates a variety of sea ice data sources. In the Arctic, this data set is based primarily on compilations by John Walsh for 1901–1990 and on satellite data for the 1990s (http://hadobs.metoffice.com/hadisst/).

2.3 Satellite Passive Microwave Sensor Data Set

More than two decades of data have now been obtained from the Scanning Multichannel Microwave Radiometer (SMMR) and Special Sensor Microwave/Imager (SSM/I). We will refer to both types of these data as remote sensing (RMS) data. These data are processed into space–time series of sea ice concentration. Digital ice concentration charts were collected for 1979–1999 on CD (Johannessen et al. 2000) and are supplemented to present day in the National Ice and Snow Data Center (NISDC) Boulder, Colorado, USA. (nsidc.org/data/). Ice extent and ice area are retrieved from satellite microwave sensor measurements as the area within the ice-ocean margin defined as 15% ice concentration and extent minus the open water area, respectively (Fig. 1). Sets of sea ice parameters retrieved from satellite measurements are among the longest satellite-retrieved geophysical records. Satellite data have shown (Johannessen et al. 2004) that the winter maximum ice area at the northern hemisphere is typically about 14×10^6 km^2 while the summer minimum ice area is about 7×10^6 km^2.

2.4 Sea Ice Data Set Created within the WMO Project "Global Digital Sea Ice Data Bank"

This data set was prepared by V. Smolyanitsky at the Arctic and Antarctic Research Institute (AARI) for 1950–1998. It is based on sea ice concentration charts archived within the WMO project "Global Digital Sea Ice Data Bank." SIE is calculated

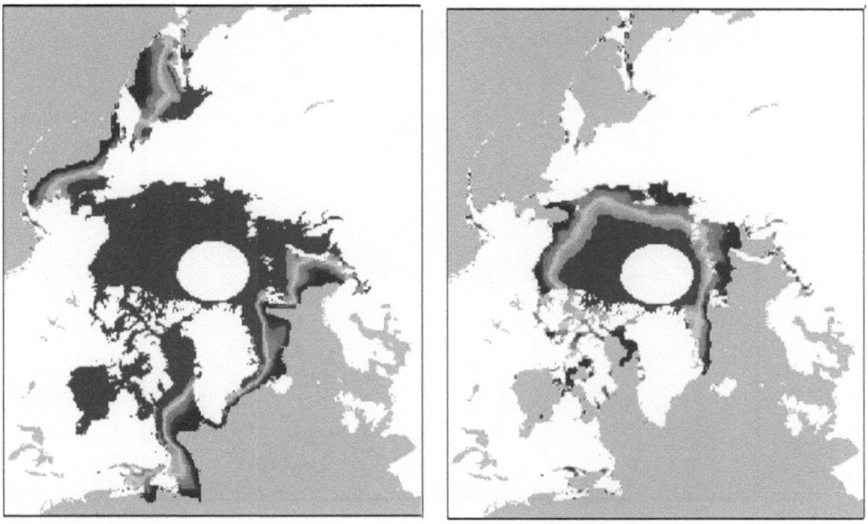

Fig. 1 Mean satellite-retrieved sea ice concentration for (a) March and (b) September in the northern hemisphere, 1978–2002.

from this data based on the grid cells which have an ice concentration greater than 15%. This data set (hereafter referred to as "Smo") is now continually supplemented by new ice concentration charts and available from http://www.aari.nw.ru/gdsid.

2.5 *Monthly Sea Ice Extent in the Nordic Seas and Adjacent Areas*

This data set was compiled by T. Vinje and presented in his paper (Vinje 2001). SIE is determined for the area of 30°W–70°E to south of 80°N for April (1864–1998) and August (1920–1998). A more detailed description of the data and its origin is given in Loyning et al. (2003) and Shapiro et al. (2003). This set is based on historical sea ice charts covering the Greenland, Iceland, Norwegian, Barents and Kara Seas, extending from 30°W to 70°E (Fig. 2). The earliest chart comes from 1553 and the

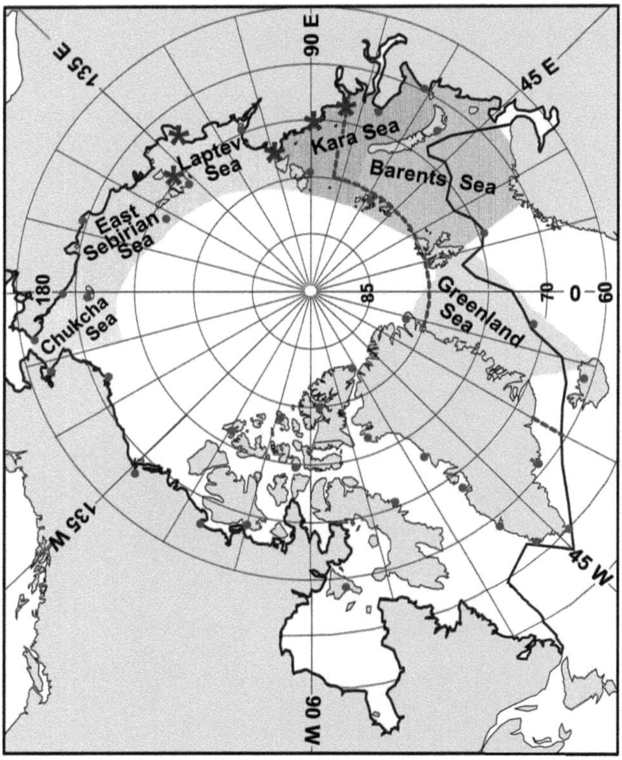

Fig. 2 Map of the Arctic Seas. Locations of meteorological stations are shown by circles. Line on water area denotes winter ice edge. Stations where ice thickness data were collected are indicated by stars. The dotted line shows the border of Vinje ice extent region.

set contains 6,007 maps in total for the period up to 2002. The outer ice edge has been used as the limit for the ice extent. The error in the position for estimating the outer ice edge is estimated as 30 km along the whole ice border in the Nordic seas.

2.6 *Monthly Sea Ice Extent in the Siberian Arctic Seas and Barents Sea*

SIE data sets for the Siberian seas (Kara, Laptev, East Siberian, and western part of Cukchi Sea) and the Barents Sea (Fig. 2) have been created by V. Zakharov (Zakharov 2003). The origin of the Zakharov data set is the same as data used in Polyakov et al. (2003) and Johannessen et al. (2004). All data sets are based on the sea ice chart collection stored in the AARI and are based on shipboard, airplane and RMS observations. Systematic airplane and ship observations of the sea ice from the Kara Sea to the Chukchi Sea began in 1932 with the start of the Northern Sea Route exploitation. The gaps during World War II (1942–1945) have been reconstructed using regression models relating the sea level pressure gradient and surface air temperature to ice extent (Polyakov et al. 2003). After 1979 a combination of satellite and aircraft summer ice-edge observations was used. Ice edge was defined where ice concentration is less than 10%. Since 1990, ice extent observations have been satellite based. These data sets were supplemented by V. Zakharov up to 2005 and now cover the period from 1924 to 2005. The period from 1924 to 1932 is covered by historical data collected by V. Vize (1944). Ice extent data for the Barents Sea is based on sea ice charts collected in the Danish Meteorological Institute for 1900–1956 and in the Arctic and Antarctic Research Institute for 1946–2005.

2.7 *Arctic Sea Ice Thickness Data*

The amount of data on sea ice thickness, an important sea ice parameter, is much less. This data is primarily based on observations of ice draft from submarines using the upward-looking sonars. Observations were made between 1958 and 1999 and served as a basis for estimates of climatic ice thickness distribution in the Arctic Basin and changes over this period (Wadhams 1997; Rothrock et al. 1999; Wadhams and Davis 2000). Another source of instrumental data on the sea ice draft in the Arctic is stationary upward-looking sonars, deployed under ice in different regions of the Arctic Ocean. Direct ice thickness measurements, by means of drilling from the surface, were mainly undertaken on land-fast ice during depth measurements or other special studies. In particular, ice thickness measurements were conducted in the Canadian Archipelago straits in 1971–1980 (Melling 2002). At Russian Arctic hydrometeorological stations, land-fast ice thickness measurements have been carried out from the 1930s up to now (Polyakov et al. 2003). In

1970–1991 the Arctic and Antarctic Research Institute carried out regular observations of sea ice cover oscillations in the Arctic Basin at the North Pole drifting stations. These observations were used by Nagurny et al. (1994) to calculate an integral sea ice thickness in the Arctic Basin. Estimated monthly ice thicknesses for 1970–1991 supplemented by data from 2004–2005 observations will be considered in Section 3.4.

3 Interannual Variability of Sea Ice Extent and Thickness from Different Data Sets

3.1 Sea Ice Extent in the Northern Hemisphere

Estimates obtained from different data sets show significant disagreements between variability of sea ice extent in the northern hemisphere for certain months of the year, and the best match for the annual average (Fig. 3). Estimates after 1979 for sea ice extent for the whole northern hemisphere calculated from different data sets are close to the RMS estimates, as all the data sets are based on the satellite data. The reconstructed series from 1900 to 1958 are of poor quality, thus for statistical comparison of interannual variability, data from 1958 on are used. The Smo time series of SIE is excluded as it has lower values for 1958–1978 in comparison with the other series. After 1979, this series shows good agreement with RMS series.

Mean values of monthly SIE in the northern hemisphere show the minima between August and September for all series (Fig. 4). Interannual variability of SIE from this series, based on mean square deviations and excluding RMS, reached the maxima in July–August and minima in January–February. The same distributions show negative

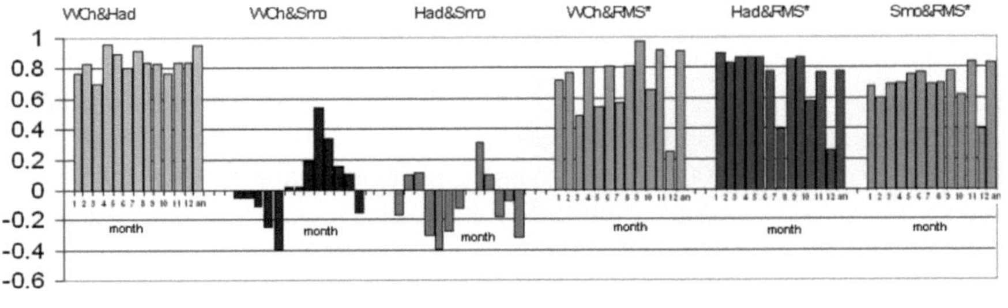

Fig. 3 Correlation coefficients between different series of mean monthly and annual sea ice extent in the northern hemisphere for 1958–1998 (95% significance is 0.31) * – both for 1979–1998 (95% significance is 0.44) WCH – Walsh and Chapmen data set, Had-Met Office Hadley Centre data set, RMS – remote sensing data set, Smo-data set was produced by V. Smolyanitsky (AARI).

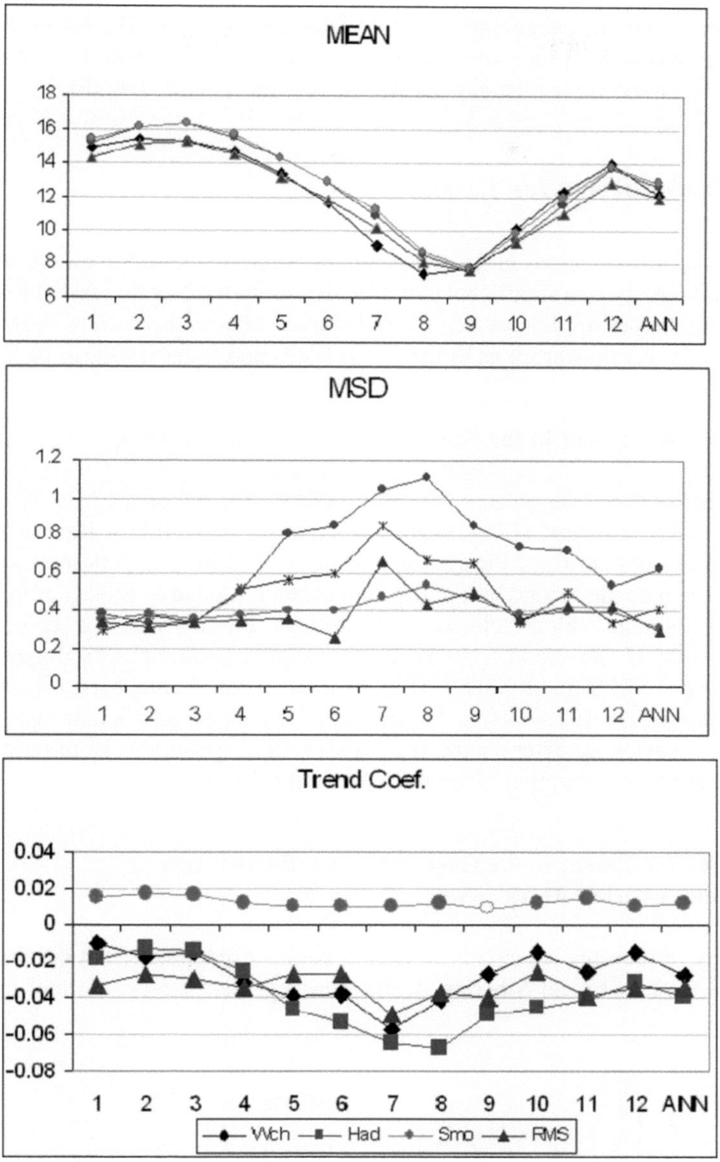

Fig. 4 Statistics (mean, $\times 10^6$ km^2; mean square deviations (MSD), $\times 10^6$ km^2 and trend coefficient, $\times 10^6$ km^2/year) of monthly and annual sea ice extent at the northern hemisphere from different data sets (WCH, 1958–1998; Had, 1958–2003; Smo, 1958–1998; RMS, 1979–2004). The symbol which is not filled indicates insignificant at the 95% level trend.

trend coefficients, revealing significant decreases in summer SIE. An exception to this is the Smo series which shows a positive trend for 1958–1998. For 1979–1998, however, the trend is significantly negative during the summer months.

3.2 *Regional Sea Ice Extent*

There are only two regions in the marine Arctic where sea ice observations cover the entire past century or a significant part of it. These are the Nordic seas with adjacent areas, and Siberian Arctic seas (Fig. 2). Data series collected here, allow the evolution of sea ice for Arctic warmings in the 1920–1940s and in the 1990s to be traced.

3.2.1 Sea Ice Extent in the Nordic Seas and Adjacent Areas

SIE variability was estimated by T. Vinje (2001) for the area of 30°W–70°E to south of 80°N for April (1864–1998) and August (1920–1998). Recently Divine and Dick (2006) confirmed the conclusions of Vinje (2001), of persistent ice retreat in this area since the second half of the 19th century. Zakharov (2003) estimated ice extent for the area which includes the Greenland, Barents seas and ice belt along the east coast of Greenland from 1900 for the spring–summer (April–September) months. April SIE from both series (Fig. 5) shows a continuous decrease with some acceleration in the 1920–1940s. August series from Zakharov's data (gray curve) reveals decreasing ice extent in the 1920–1940s more apparently. Both series do not demonstrate an acceleration of ice decrease in the 1990s.

3.2.2 Sea Ice Extent in the Greenland Sea, Barents Sea and Siberian Arctic Seas

Variability of monthly ice extent for each Siberian Arctic Sea (from the Kara up to Chukcha Sea) as for all Siberian Arctic seas together, shows minimal mean ice

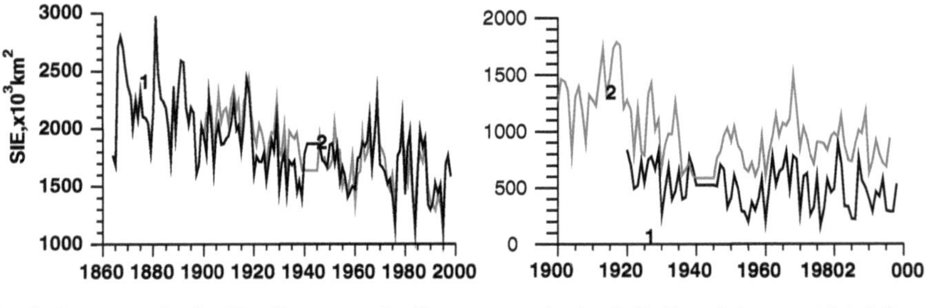

Fig. 5 Ice extent in the Nordic seas and adjacent areas in April (left) and August (right) from Vinje (1 – black curve) and Zakharov (2 – gray curve) data sets.

extent in September, maximal mean square deviations (MSD) in August–September as well as the strongest negative trend. Ice extent series for the Greenland and Barents seas demonstrate minimal SIE in September, but the maximum MSD falls on March and May respectively. This dissimilarity from other Arctic seas is due to variations of winter ice transport to both seas from the Arctic Basin (Kwok et al. 2005). The strongest negative trend of ice extent is revealed in September for the Greenland Sea and in April for the Barents Sea (Fig. 6).

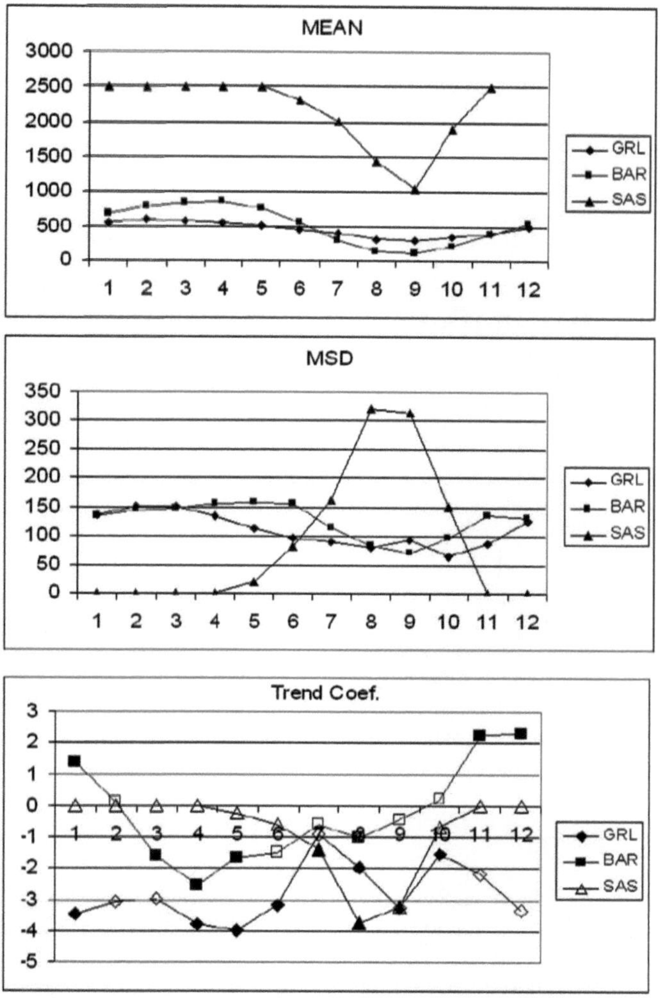

Fig. 6 Statistics (mean, $\times 10^3$ km^2; mean square deviations (MSD), $\times 10^3$ km^2 and trend coefficient, $\times 10^3$ m^2/year) of monthly sea ice extent in the Arctic seas from Zakharov data sets. (GRL-Greenland Sea, 1958–2004; BAR-Barents Sea, 1928–2005; SAS-Siberian Arctic seas, 1924–2005). The symbol which is not filled indicates insignificant at the 95% level trend.

3.2.3 Changes of Ice Extent in the Barents Sea and Siberian Arctic Seas for Seasonal Minimum

Rapid shrinking of SIE during Arctic warmings in the 1920–1940s and 1990s can be seen from time series of September SIE in the majority of the Siberian Arctic seas and in the Barents Sea (Fig. 7). September SIE in the Arctic seas in Fig. 7 reveals a negative trend for the entire period of observations up to 2005 (trend-line 1). But 95% significant trends are found in the East Siberian, Chukcha and all Siberian seas taken together. The trend from 1979 to 2005 (trend-line 3) is significant for all seas excluding the Kara Sea where a significant negative trend was found for 1924–1950 (trend-line 2).

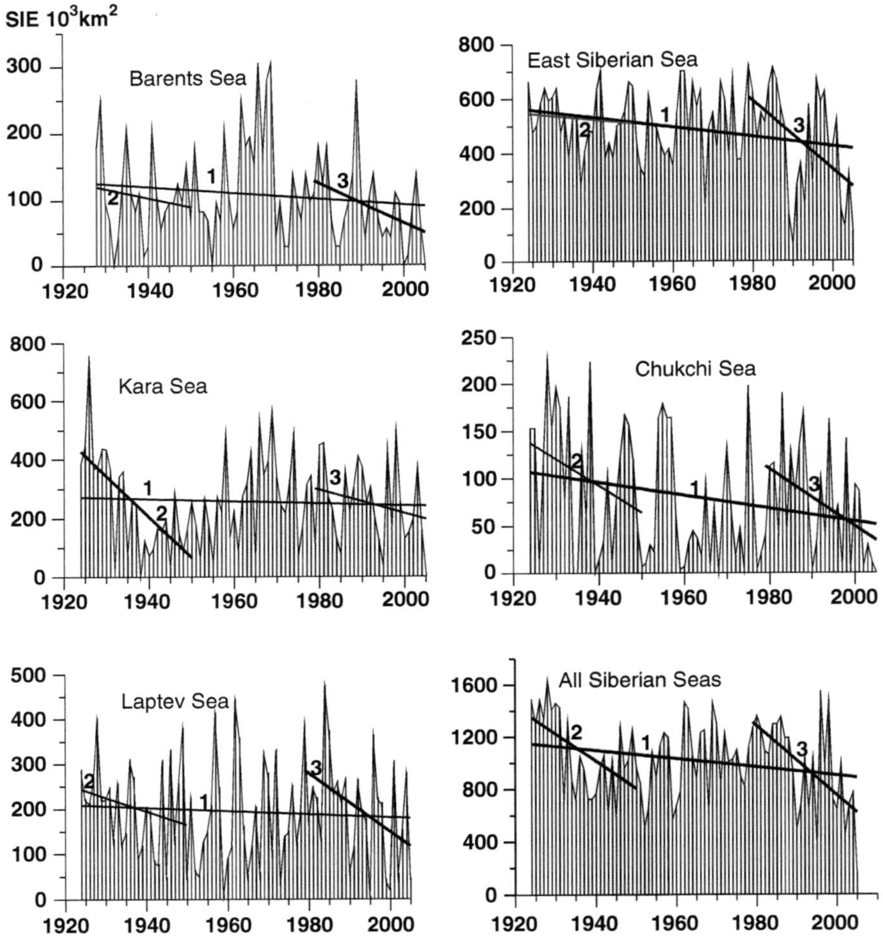

Fig. 7 Ice extent in the Siberian Arctic seas in September. Lines 1, 2, 3 denote linear trends for entire period, for 1924(1928)–1950, for 1979–2005. Thick line indicates 95% significant trend.

3.3 Change of Ice Thickness in the Arctic Ocean for 1970–2005

On the basis of a linear theory of free vibrations of sea ice, an average thickness of ice can be determined from the frequency of its resonant vibrations (Nagurny et al. 1994). The thickness value retrieved is the modal thickness of undeformed multiyear ice averaged over a great circle path from the ice edge to the measuring site. Regular observations of sea ice vibrations using seismometers and tiltmeters were carried out at the Russian "North Pole" drifting stations during 1970–1990 (Nagurny 2003; Johannessen et al. 2004). Records of vibrations which included waves with 30–40 s period had 1–2 h duration with 0.1 s sampling time. 30–60 such records from different wave directions per month have been used to determine the monthly ice thickness and to provide a basin-wide mean thickness estimate. The calculations revealed an 8–10 cm decrease in the mean ice thickness over a period of observations from April as well as from August data. This decrease is about 3% of the mean value (around 3 m), and is much less than the 1950s/1970s to 1990s submarine sonar data analysis (Rothrock et al. 1999), but close to trend estimation for perennial ice in the Beaufort Sea (0.07 m/decade for 1991–2003, Melling et al. 2005). The estimated thickness varied through a seasonal range of about 40 cm. The annual mean thickness appeared to follow an oscillation amplitude of 20 cm and a period of approximately 6–8 years. The observations were interrupted for the period 1991–2003. The renewal of observations in 2004–2005 revealed that the area-averaged perennial ice pack in the Arctic Basin decreased by 110 cm (Nagurny et al. 2007) (Fig. 8).

The precision of these observations was ±40 cm (95% confidential interval) as compared with ±5 cm for the 1970–1990 observations. The method was the same but the number of instruments and consequently number of records were fewer.

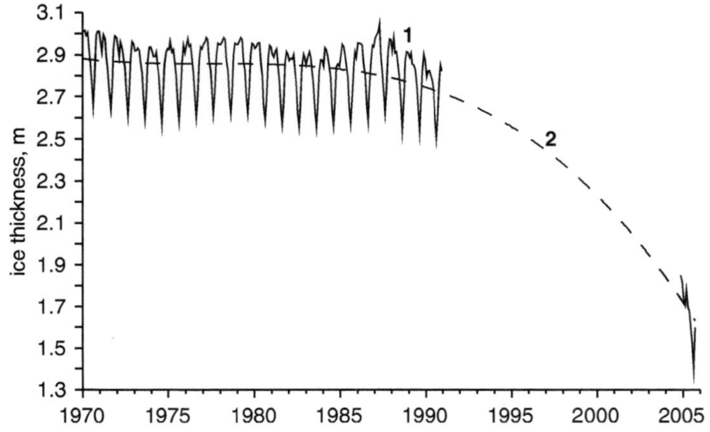

Fig. 8 Area-average mean monthly ice thickness in the Arctic Basin (1 – solid line) calculated from elastic-gravity vibrations of sea ice for 1970–1990 and 2004–2005 (2 – dashed line) – polynomial of degree three approximation.

4 Relation Between Surface Air Temperature and Sea Ice at the Northern Hemisphere

Analysis of the sea ice extent in the northern hemisphere and the Arctic regions revealed a significant decrease in the later decades of the last century. The reduction was found to vary during different seasons and was most pronounced in summer for the whole northern hemisphere (see Figs. 4 and 6). This conclusion is in agreement with analysis by Francis et al. (2005) and Comiso (2006). Opposite tendencies of sea ice extent and surface air temperature changes are manifested in negative correlations between them, both for monthly means and for annual values (Alekseev et al. 2005). Correlation between annual values is stronger due to the removal of uncorrelated monthly variations in the course of SAT and SIE annual averaging. After trend removal, a stronger negative correlation falls on summer months with a maximum in June (Table 1). Significant correlation between SAT and SIE in summer is explained by a similarity in the heat fluxes which determines the heat budget of the near surface air layer on surface sea ice. Besides, air temperature influences sensible and latent heat fluxes towards the ice. Correlation in winter is weaker as a consequence of a deceleration of both ice thickness and extent together with ice growth (negative feedback). Again there is a limit in the winter spreading of ice, which is determined by the shoreline in the Arctic seas and limited area of a low salinity layer where ice can be formed in the North Atlantic, Nordic seas and Pacific (Fig. 2). Correlation maximum in June presumably results from anomalies of insolation, advective heat fluxes and ice extent during the previous months,

Table 1 Correlation coefficients between surface air temperature (from NCEP Reanalysis data) and sea ice extent in the NH (WCH data) for both monthly mean and annual values for 1958–1999. Bold figures denote maximal values. 95% significant level is 0.30

	Correlation between SAT and SIE		Correlation between detrended SAT and SIE	
Month	65–90°N SAT	The NH SAT	65–90°N SAT	The NH SAT
1	−0.28	−0.29	−0.22	−0.11
2	−0.39	−0.44	−0.22	−0.32
3	−0.45	−0.4	−0.25	−0.11
4	−0.49	−0.49	−0.22	−0.01
5	−0.54	**−0.67**	−0.32	−0.25
6	−0.52	−0.64	**−0.63**	**−0.47**
7	−0.48	−0.6	−0.47	−0.19
8	−0.29	−0.51	−0.32	−0.23
9	**−0.64**	−0.65	−0.55	−0.37
10	−0.55	−0.51	−0.42	−0.26
11	−0.61	−0.65	−0.5	−0.39
12	−0.2	−0.27	−0.08	−0.07
Annual	−0.75	−0.81	−0.52	−0.44

which can be stimulated by ice retreat and SAT increase and vice versa. Francis et al. (2005) found that the end of summer maximal retreat of the ice edge is influenced predominantly by downwelling longwave flux anomalies moreso than by advected heat flux, insolation and wind forcing anomalies.

To show the relationship between summer SAT and ice extent more clearly, we used the sum of positive daily SAT over each summer. Variations in the sum series (positive daily SAT) are greater than variations in mean summer SAT series and hence highlight the interannual variability of summer ice melting determined by the cumulative influence of heat fluxes on the ice. The sums of positive SAT were calculated from air temperatures at 38 stations in the marine Arctic, which includes the Arctic Ocean and adjacent sea areas covered with ice in winter (Fig. 2). Hence, the southern border of the marine Arctic is the mean winter sea ice edge. To estimate the influence of summer SAT on the accelerated decrease of SIE, we correlated the mean sum of summer SAT in the marine Arctic and SIE in the NH in September from satellite observations for 1979–2005 (Fig. 9). This figure shows a rapid shrinking of SIE during the last 5 years in September superimposed on a negative trend from the beginning of the series in 1979. The correlation (R = 0.85) reveals good agreement between variations of the SAT sum and SIE, including an accelerated change of both values. This acceleration likely reflects the influence of positive feedbacks evolving jointly with SAT and SIE.

Because ice thickness in the Arctic Ocean depends on air temperature and ice dynamics, it is believed that the warming effect may be seen more clearly in the thickness of land-fast ice (Polyakov et al. 2003). Figure 10 shows the time series of

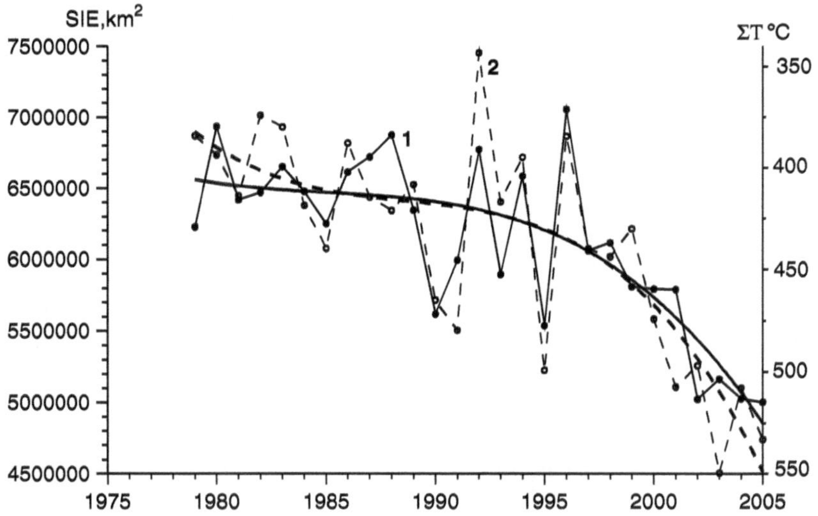

Fig. 9 Changes in September ice extent (1 – solid line) and mean sum of positive summer air temperature in the marine Arctic (2 – dashed line). Thick lines are polynomial of degree three trends. The scale of the SAT sum is inverted for comparison.

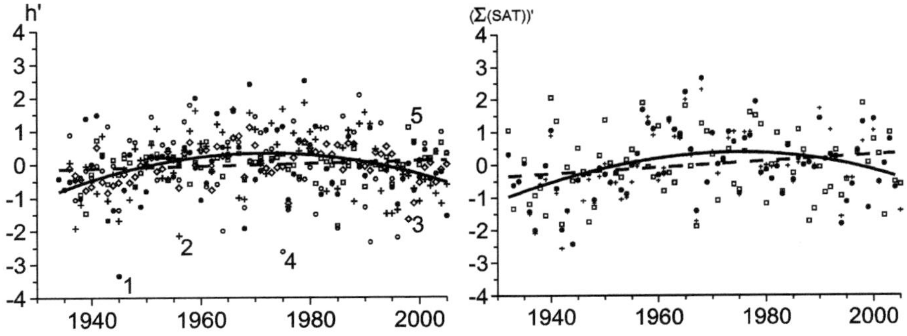

Fig. 10 Normalized maximal land-fast ice thickness (left panel) and the sum of winter SAT (right panel, the axis is inverted for comparison) from measurements at the Arctic stations along the shore of the Kara and Laptev seas (1 – Dickson, 2 – Sterligova, 3 – Cheluskin, 4 – Sannikova, 5 –Tiksi). The position of each station is shown as stars on the map from Fig. 2. The dashed line indicates a trend, and the solid line is a second-order polynomial fit.

maximal land-fast ice thickness up to 2005, from five offshore sites of the Kara and Laptev seas. Thickness values on the plot are normalized by subtracting the mean value and dividing by the mean square deviation. This reveals an insignificant positive linear trend and decreasing ice thickness at both the beginning and end of the time series. On the right panel normalized sums of negative winter temperature at three stations are shown; observations started in the 1930s. The insignificant linear trend points to an increase of the SAT sums in absolute magnitude. Correlation between the SAT sum and ice thickness lies in the range of –0.75 (Dickson) to –0.47 (Tiksi).

The absence of significant trends in the land-fast ice thickness in the Arctic seas that agrees with the SAT sums trend, points to an absence of a warming signal in the series of these parameters for 1934–2005. Such a signal however becomes apparent starting from the 1970s.

5 Conclusions

The time series of SIE in the northern hemisphere obtained from different data sets show significant disagreements between monthly mean SIE for certain months, and the best match for the mean annual SIE. They demonstrate however, better agreement after the 1950s and especially since 1979 with the onset of regular remote sensing observations from satellites. Mean values of monthly SIE in the northern hemisphere show minima in August–September. Mean square deviations reach maxima in July–August.

The distributions of the negative trend coefficients show a more significant decrease of summer SIE. Statistics of monthly SIE series for the Siberian Arctic seas show a similar distribution. SIE series for the Greenland and Barents seas show the minimum SIE in September. The maximum of MSD falls on March and May, respectively. This distinction from other Arctic seas results from variations of winter ice transport from the Arctic Basin (Kwok et al. 2005).

September SIE in the majority of the Siberian Arctic seas and in the Barents Sea reveal rapid shrinking during Arctic warmings in the 1920–1940s and 1990s. Negative trends in the SIE series for the entire period of observation is seen for all seas. But 95% significant trends are found for the East Siberian and Chukcha seas only.

Significant correlation between SAT and SIE series, especially between mean annual values, is due to opposite trends in both time series. After trend removal the maximum correlation occurs in June. This is related to SIE and SAT anomalies of the opposite sign, developing under the influence of June maximum solar irradiation, and amplified by heat advection in the atmosphere and SIE anomalies in the previous months. The importance of high insolation for the decrease of sea ice extent is found in experiments with a coupled climate model (Tuenter et al. 2005). The relationship between variations of winter air temperature and ice extent is weaker because winter ice extent anomalies depend on air temperature anomalies as well as on the area occupied by a freshened upper layer. For this reason winter ice cover is more stable when summer ice cover is shrinking (Comiso 2006).

Good agreement between variations of the sum of summer air temperature in the marine Arctic and sea ice extent is found. This results from the fact that the sum of air temperatures is an integral result of heat influxes decreasing ice extent. Influxes include incoming solar radiation, downwelling longwave radiation, and heat advected with southern winds. The impact of these heat fluxes on summer ice decrease depends on the Arctic region (Francis et al. 2005) and differs between models (Lindsay and Zhang 2005). The SAT influence on ice is supported by results of Rothrock and Zhang (2005) based on model simulation. They showed that the temperature-forced component of the Arctic ice volume has a significant downward trend of 3% per decade in comparison with a decrease of 4% per decade in total ice volume.

A strong influence of summer SAT on the accelerated decrease of SIE in the Arctic is confirmed by a high correlation (coefficient is 0.85) between the sum of summer SAT in the marine Arctic and SIE in September from satellite observations for 1979–2005. The acceleration of the decrease of SIE during the last 5 years likely reflects the influence of positive feedbacks jointly with SAT and SIE. It confirms that summer melting plays the most important role in the sea ice volume decrease. Several experiments with the global ice-ocean models (Goosse et al. 2004; Zhang et al. 2004) showed that the increase of Atlantic Water inflow in the 1990s could also affect thinning of the ice.

Regular measurements of sea ice vibrations carried out at the Russian "North Pole" drifting stations during the 1970s and 1990s reveal a decrease in ice thickness on the order of 8–10cm. The renewed observations in 2004–2005 revealed that the area-averaged perennial ice in the Arctic Basin decreased by 110cm relative to the 1990 value. The insignificant positive linear trend found for 1934–2005 in the thickness of land-fast ice agreed with the SAT sums trend and points to an absence of a warming signal in both parameters. Such a signal becomes apparent starting from the 1970s.

Acknowledgments We are grateful to anonymous referees for their reviews with many useful comments to improve the manuscript and to Tracy Ewen for significant corrections to the grammar in the text.

The studies have been supported by the Norwegian Ministry of Education and Research and Research Council of Norway through the Projects N179125/S30, by RFBR (project 06-05-64054) and INTAS (grant 03-51-4620).

References

Alekseev, G. V., V. F. Zakharov, N. E. Ivanov, and S. I. Kuzmina, 2005: Relation between changes of sea ice extent and air temperature at northern hemisphere. *Materiali glyaziologicheskich issledovanii*, **99**, 62–70 (in Russian).

Chapman, W. L. and J. E. Walsh, 1993: Recent variations of sea ice and air temperature in high latitudes. *Bull. Amer. Meteorol. Soc.*, **74**, 33–47.

Comiso, J. C., 2006: Abrupt decline in the Arctic winter sea ice cover. *Geophys. Res. Lett.*, **33**, L18504, doi:10.1029/2006GL027341.

Divine, D. V. and C. Dick, 2006: Historical variability of sea ice edge position in the Nordic Seas. *J. Geophys. Res.*, **111**, C01001, doi:10.1029/2004JC002851.

Francis, J. A., E. Hunter, J. R. Key, and X. Wang, 2005: Clues to variability in Arctic minimum sea ice extent. *Geophys. Res. Lett.*, 32:L21501, doi:10.1029.

Goosse, H., R. Gerdes, F. Kauker, and C. Ko Berle, 2004: Influence of the exchanges between the Atlantic and the Arctic on sea ice volume variations during the period 1955–1997. *J. Climate*, **17**, 1294–1303.

Johannessen, O., L. Bengtsson, M. Miles, S. Kuzmina, V. Semenov, G. Alekseev, A. Nagurnyi, V. Zakharov, L. Bobylev, L. Pettersson, K. Hasselmann, and H. Cattle, 2004: Arctic climate change: observed and modelled temperature and sea-ice variability. *Tellus*, **56A**, 328–341.

Johannessen, O. M., E. Shalina, and M. Miles, 2000: *NORSEX Sea Ice Concentration in the Arctic, 1978–1999 Including MY Ice Concentration During Winter Months*. NORSEX CD-ROM, Version 1.0.

Kwok, R., W. Maslowski, and S. W. Laxon, 2005: On large outflows of Arctic sea ice into the Barents Sea. *Geophys. Res. Lett.*, **32**, L22503, doi:10.1029/2005GL024485.

Lindsay, R. W. and J. Zhang, 2005: The thinning of Arctic Sea ice, 1988–2003: have we passed a tipping point? *J. Climate*, **18**:4879–4894.

Løyning, T. B., C. Dick, H. Goodwin, O. Pavlova, T. Vinje, G. Kjærnli, and T. Villinger, 2003: ACSYS Historical Ice Chart Archive (1553–2002). International ACSYS/CliC Project Office.

Melling, H., 2002: Sea ice of the northern Canadian Arctic archipelago. *J. Geophys. Res.*, **107**(C11), 3181, doi:10.1029/2001JC001102.

Melling, H. D., A. Riedel, and Z. Gedalof, 2005: Trends in the draft and extent of seasonal pack ice, Canadian Beaufort Sea. *Geophys. Res. Lett.*, **32**, L24501, doi:10.1029/2005GL024483.

Nagurny, A. P., V. G. Korostelev, and V. P. Abaza, 1994: A method for determination of effective sea ice thickness in the Arctic Basin for climate monitoring. *Bull. Russian Acad. of Sci.*, Physics/Suppl. Physics of vibration. Allerton Press, New York, 58, 168.

Nagurny, A. P., 2003: Sea ice thickness. In: *Arctic Environment Variability in Context of Global Change*. L. P. Bobylev, K. Ya. Kondratyev, and M. Johannessen Ola, eds., Published in association with Praxis Publishing, Springer, Chichester, UK, 160–171.

Nagurny, A. P., V. G. Korostelev, M. Johannessen Ola, P. Wadhams, and V. N. Smirnov, 2007: Multiyear variability of the area-averaged ice pack thickness in the Arctic Basin measured by elastic-gravity waves on the sea ice surface. *IEEE J. Oceanic Eng.* (in revision).

NSIDC News, 28 September 2005: Sea Ice Decline Intensifies. http://nsidc.org.

Polyakov, I., G. Alekseev, R. Bekryaev, U. Bhatt, M. Johnson, V. Karklin, D. Walsh, and A. Yulin, 2003: Variability and trends of air temperature and pressure in the maritime Arctic, 1875–2000. *J. Climate*, **16**, 2067–2077.

Rayner, N. A., D. E. Parker, E. B. Horton, C. K. Folland, L. V. Alexander, D. P. Rowell, E. C. Kent, and A. Kaplan, 2003: Global analyses of sea surface temperature, sea ice, and night marine air temperature since the late nineteenth century. *J. Geophys. Res.*, **108**(D14), 4407, doi:10.1029/2002JD002670.

Rothrock, D. A., Y. Yu, and G. A. Maykut, 1999: Thinning of the Arctic Sea ice cover. *Geophys. Res. Lett.*, **26**(23), 3469–3472.

Rothrock, D. A. and J. Zhang, 2005: Arctic Ocean sea ice volume: what explains its recent depletion? *J. Geophys. Res.*, **110**, C01002, doi:10.1029/2004JC002282.

Shapiro, I., R. Colony, and T. Vinje, 2003: April sea ice extent in the Barents Sea. *Polar Res.*, **22**, 5–10.

Tuenter, E., S. L. Weber, F. J. Hilgen, and L. J. Lourens, 2005: Sea-ice feedbacks on the climatic response to precession and obliquity forcing. *Geophys. Res. Lett.*, **32**, L24704, doi:10.1029/2005GL024122.

Vinje, T., 2001: Anomalies and trends of sea ice extent and atmospheric circulation in the Nordic Seas during the period 1864–1998. *J. Climate*, **14**(3), 255–267.

Vize V. Yu, 1944: The basics of long-term ice forecasts for the Arctic Seas. *Glavsevmorput*, 274 pp (in Russian).

Wadhams, P., 1997: Ice thickness in the Arctic Ocean: the statistical reliability of experimental data. *J. Geophys. Res.*, **102**(C13), 27951–27959.

Wadhams, P. and N. R. Davis, 2000: Further evidence of ice thinning in the Arctic Ocean. *Geophys. Res. Lett.*, **27**(24), 3973–3976.

Zakharov, V. F., 2003: Changes of the Arctic sea ice extent in XX century. *Meteorologiya i Gidrologiya*, **5**, 75–86 (in Russian).

Zhang, J., M. Steele, D. A. Rothrock, and R. W. Lindsay, 2004: Increasing exchanges at Greenland-Scotland ridge and their links with the North Atlantic oscillation and Arctic Sea ice. *Geophys. Res. Lett.*, **31**, L09307, doi:10.1029/2003GL019304.

The Evolving SST Record from ICOADS

S. D. Woodruff[1], H. F. Diaz[1], E. C. Kent[2], R. W. Reynolds[3], and S. J. Worley[4]

Abstract Sea surface temperature (SST) is a key oceanic variable – widely used for research, including global climate change assessments and atmospheric reanalyses. This paper reviews the evolution of the SST data and products available from the International Comprehensive Ocean-Atmosphere Data Set (ICOADS), since that project's inception in 1981. Climate-scale SST products based on ICOADS (or related in situ data) are also reviewed. Measurements of SST have been made since around the early 1800s from ships, augmented in recent decades by in situ measurements from buoys and other automated Ocean Data Acquisition Systems (ODAS). SST, unlike some other ICOADS variables such as surface air temperature or humidity, is observed from space with reasonable accuracy. However, without reference to in situ measurements most satellite-based SST products will contain large-scale biases due to varying atmospheric composition and imperfect instrumental calibration. ICOADS is vital to the removal of such biases, which are especially large following volcanic eruptions. We describe products combining in situ and satellite SSTs that exploit the strengths of each type of measurement, to yield both high resolution and high accuracy. Finally, we discuss future developments anticipated for ICOADS and SST products, such as further blending of metadata and enhanced product uncertainty assessments.

1 Introduction

The oceans – covering more than 70% of the Earth's surface – form a major component of the Earth's climate system. By storing energy and exchanging heat and momentum with the atmosphere, the oceans are key to understanding climate variability and change.

[1] NOAA Earth System Research Laboratory
Scott.D.Woodruff@noaa.gov, henry.f.diaz@noaa.gov

[2] National Oceanography Centre, Southampton, UK
elizabeth.c.kent@noc.soton.ac.uk

[3] NOAA National Climatic Data Center
Richard.W.Reynolds@noaa.gov

[4] National Center for Atmospheric Research, US
worley@ucar.edu

S. Brönnimann et al. (eds.), *Climate Variability and Extremes during the Past 100 Years.*

In situ meteorological and oceanographic observations can measure characteristics of this exchange, including ocean and air temperatures, winds, barometric pressure, humidity, waves, and clouds. Among these variables, sea surface temperature (SST) data and products are presently best developed and characterized. SST – the water temperature near the ocean's surface – is used for a wide range of research and monitoring activities, including global climate change assessments (e.g., Folland et al. 2001a) and assimilation into global atmospheric reanalysis products (e.g., Kistler et al. 2001).

This paper discusses the evolution of the SST data and products available from the International Comprehensive Ocean-Atmosphere Data Set (ICOADS) (Worley et al. 2005). This evolution is examined in three parts. Section 2 describes changes in the in situ observing system of ships, buoys, and other automated Ocean Data Acquisition Systems (ODAS), which contribute SST (plus many other) data to ICOADS. Section 3 discusses ICOADS, and improvements made to its SST products, since the inception of that project in 1981 (Woodruff et al. 1987). Section 4 reviews the development of higher-level climate-scale SST products based on ICOADS, including those blending remotely sensed with in situ observations. Finally, in Section 5 we discuss future developments in ICOADS and SST products, including further blending of data and metadata, enhanced product assessments of data error characteristics, and joint analyses of SST with other variables.

2 In situ SST Data

2.1 Ship Data

Seafarers and sailors – with their intimate connections with the ocean and its weather – have recorded textual and numerical weather descriptions in logbooks for centuries. Instrumental logbook data started becoming prevalent in the early 19th century, leading to the landmark 1853 Maritime Conference, convened in Brussels (Maury 1854). This, the first international meteorological conference, paved the way toward modern operational oceanography and marine meteorology, and marks the beginning of systematic observations from merchant, naval, and research vessels – now called Voluntary Observing Ships (VOS).

The Brussels conference prescribed an "Abstract Log" for reporting observations (Woodruff et al. 2005), and recommended the use of a wooden bucket to sample seawater for SST measurement. The extent to which maritime nations followed this recommendation may never be certain, because ships rarely reported the SST observing method, until the mid-20th century. However, Folland and Parker (1995) found an evolving bucket mixture including un-insulated canvas buckets described in national observing instructions from the 19th and 20th centuries. Eventually, as the effects of heat exchange with the atmosphere on the accuracy of early bucket measurements were recognized, ships transitioned to insulated (e.g., rubber) buckets (Fig. 1).

The transition from sail to steam propulsion on ships, around 1900, represents another important change in the SST record. On steam and later motorized propulsion ships, SST sensors could be placed conveniently on the engine cooling water

Fig. 1 (Top): Example of wooden bucket, 1891. (Courtesy of Scottish Maritime Museum and David Parker.) (Bottom, left to right): South African "Crawford" bucket, German metal and leather bucket, and UK Meteorological Office canvas bucket (courtesy of David Parker).

intake system, introducing measurements sampled at a variety of depths to the SST record. Some countries adopted this practice widely (e.g., USA) while others (e.g., UK) continued to make bucket measurements, believing them to be of greater accuracy. Since 1955, the World Meteorological Organization (WMO) has published SST measurement methods for VOS. Methods used include hull contact or through-hull sensors, radiation or bait tanks thermometers, and outboard trailing thermistors (Kent and Taylor 2006; Kent et al. 2007).

Ocean Weather Stations (OWS) were examples of specialized ships introduced around World War II that provided high-quality data at fixed locations in support of Numerical Weather Prediction, but were phased out after the 1970s. OWS Mike in the Norwegian Sea is the last remaining station. Research vessels (RV) can also contribute SST data either from underway sampling or from more specialized vertical temperature profiling instruments.

2.2 *ODAS Data*

Starting in the early 1970s, the USA and eventually some other nations deployed arrays of moored data buoys, generally around coastlines. Similarly to OWS, the development of these platforms was principally motivated by operational requirements such as weather prediction and navigation safety. However, for climate purposes they also provide time series of SST and other meteorological measurements, often hourly, but usually limited to near-coastal areas.

Drifting buoys, measuring SST plus sometimes either pressure, wind speed or air temperature, were introduced for the First Global GARP (Global Atmospheric Research Program) Experiment (FGGE) in 1978–1979. Their global coverage has increased through sustained international investments, and today it provides a large fraction of the overall SST data. Hull contact sensors are usually used to measure SST from drifting buoys, but as with moored buoys, the buoy and instrument designs can vary (Fig. 2).

Starting in 1984, the first elements of the Tropical Atmosphere Ocean (TAO) moored buoy array were deployed across the Pacific by the USA (McPhaden et al. 1998), with the westernmost mooring locations later expanded by Japan into the Triangle Trans Ocean Buoy Network (TRITON). The operational measurements of the combined TAO/TRITON array, presently consisting of some 70 moorings, include SST and air and subsurface temperatures. The Pilot Research Moored

Fig. 2 Examples of a moored (left) and drifting buoy (right) (images courtesy of NOAA National Data Buoy Center and the Data Buoy Cooperation Panel).

Array in the Tropical Atlantic (PIRATA) was initiated in 1997, and in the future a similar array may be added to the Indian Ocean.

Well before buoys were introduced, oceanographers had sampled the ocean depths using a variety of instruments deployed from RV and VOS. For example, the 1853 Abstract Log included space for daily readings of subsurface temperature. Advances in oceanography since then included the transition from mechanical to expendable bathythermographs, and most recently the widespread deployment of Argo autonomous subsurface profiling floats (see www.argo.net). Ocean profile data are increasingly used for operational ocean analysis and ocean forecasting, and profile temperatures sufficiently close to the surface can be used for SST.

2.3 *Observing System*

The evolution of the VOS/ODAS platform mixture, as illustrated in Figs. 3 and 4 is complicated. Changing ship sizes and propulsion, coupled with historic events such as the openings of the Suez and Panama Canals, in 1869 and 1914 respectively, have

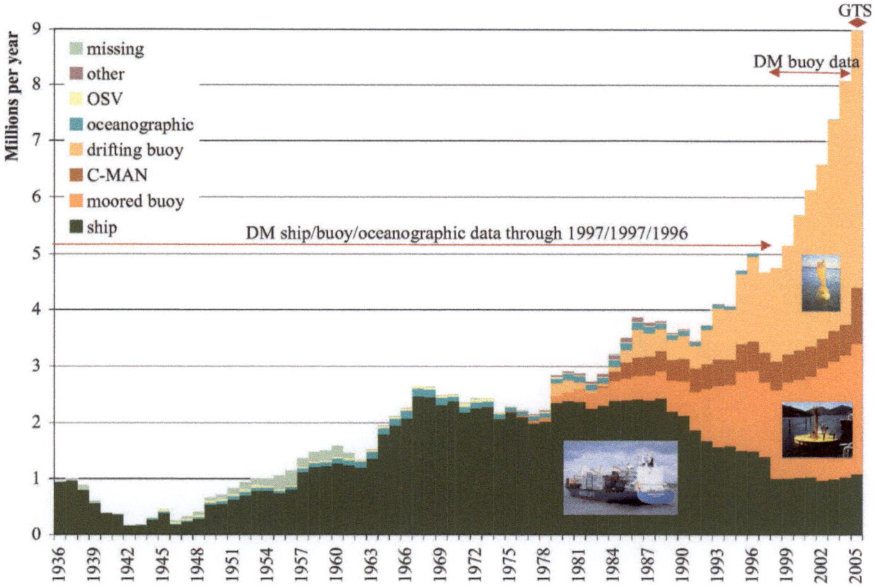

Fig. 3 Annual numbers of marine reports in ICOADS, stratified by platform type for 1936–2005 ("C-MAN" refers to ODAS reports from the US land-ocean boundary zone). The indicated delayed-mode (DM) sources have been blended through 1997 (or 1996 in the case of oceanographic data) and for 1998–2004, whereas 2005 is composed exclusively of real-time data from the Global Telecommunication System (GTS). The total in 2005 is approximately 14 million (not shown), owing to many GTS drifting buoy reports, which are consolidated by the providers of the DM data (i.e., by removal of duplicates plus some compositing of fragmentary receipts) (ship photo courtesy of www.ShipPhotos.co.uk).

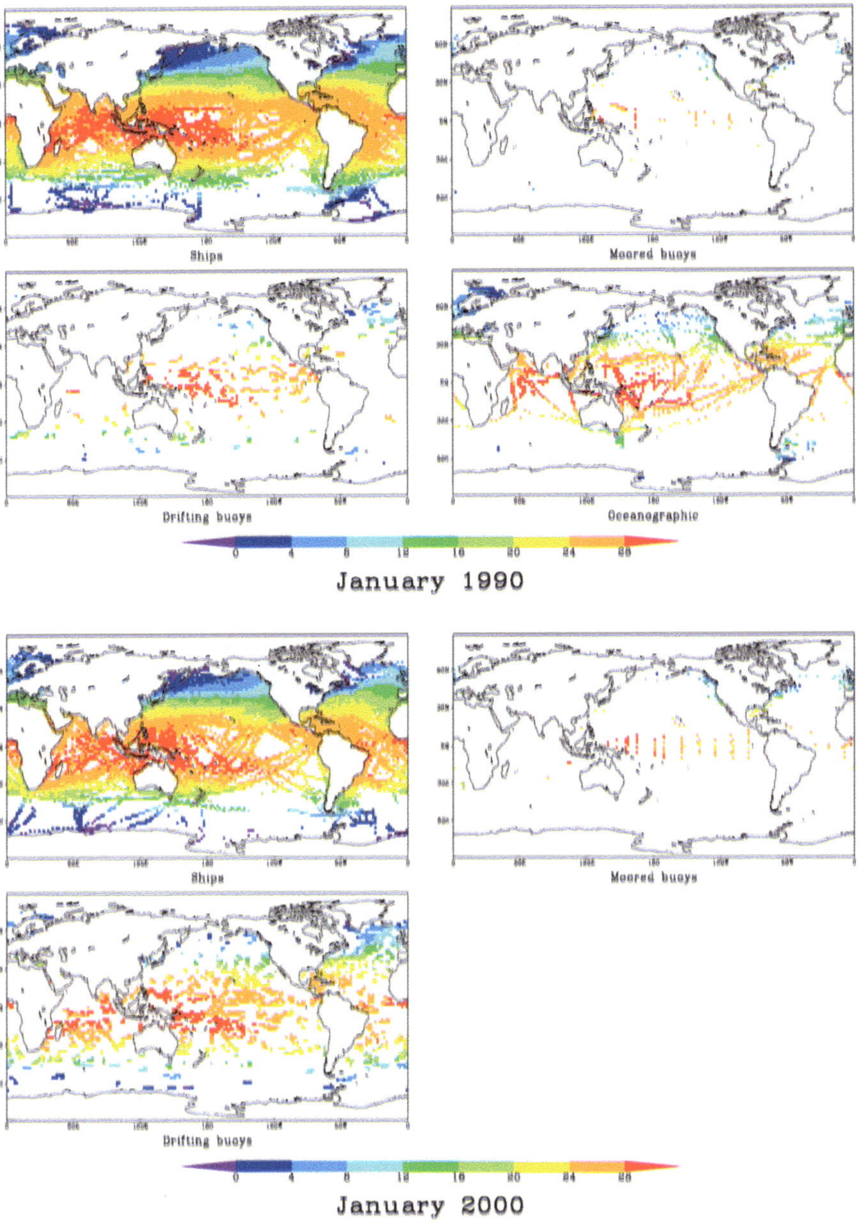

Fig. 4 A 2° monthly mean SST (°C) for January 1990 (upper) and January 2000 (lower) plotted separately by platform type: ships, moored buoys, drifting buoys, and oceanographic data. Note: As indicated in Fig. 3, oceanographic data are presently available in ICOADS only through 1996, hence the missing panel.

affected global patterns of spatial coverage (Woodruff et al. 1987). Increasing global commerce since the 18th century is reflected in a general growth trend in VOS traffic, punctuated by sharp data deficits around the two World Wars, and recent declines owing partly to security and cost considerations and partly to the increased use of satellite data for numerical weather prediction (Fig. 3) (Kent et al. 2006).

Diurnal sampling times have varied as reporting practices have changed. Early ship observations often made at fixed local times were gradually replaced by synoptic VOS observations (generally at 0, 6, 12, 18 UTC) around the mid-20th century (Bottomley et al. 1990), and later mixed with increasing numbers of buoy reports at all hours (Worley et al. 2005). These sampling changes were partly driven by the introduction of telecommunicated data into the digital record starting in 1966, transmitted at the synoptic hours to support weather forecasting. Earlier transmitted data were noisy but observations are now circulated more reliably on the Global Telecommunication System (GTS). All these variations pose complex challenges for creating spatially and temporally consistent products and data analyses.

3 ICOADS and SST

In 1981, cooperative work began between the US National Oceanic and Atmospheric Administration (NOAA) and the National Center for Atmospheric Research (NCAR) on the Comprehensive Ocean-Atmosphere Data Set (COADS). The COADS Release 1 observational data and summary products (1854–1979) were

Table 1 Summary of the temporal period and update scope, issue date, and references and update highlights, for selected ICOADS updates. Release 2.0 was designated in 2002, but the data were available under different release names by 2001. The concept of separate delayed-mode (DM) and real-time (RT) archives was introduced in conjunction with Release 2.1. The RT archive is strictly based on GTS data, whereas DM sources may include keyed ship logbook data, and ODAS archives that have received additional quality control (QC) and include more complete data receipts

Release name	Resultant period (update scope)	Issue date	References (update highlights)
Release 1	1854–1979	1985	Slutz et al. 1985; Woodruff et al. 1987
Release 1a	1854–1992 (1980–1992)	1993	Woodruff et al. 1993 1° and 2° standard and enhanced summaries
Release 2.0	1784–1997	2001–2002	Woodruff et al. 1998, 2003 Full replacement of Release 1 data/products
Release 2.1	1784–2002 (1998–2002)	2003	Worley et al. 2005; Woodruff et al. 2005 DM (–1997) and RT (1998–2002) archives
Release 2.2	1784–2004 (1998–2004)	2005	DM update for 1998–2004
Release 2.3	1784–2005 (2005)	2006	RT update for 2005

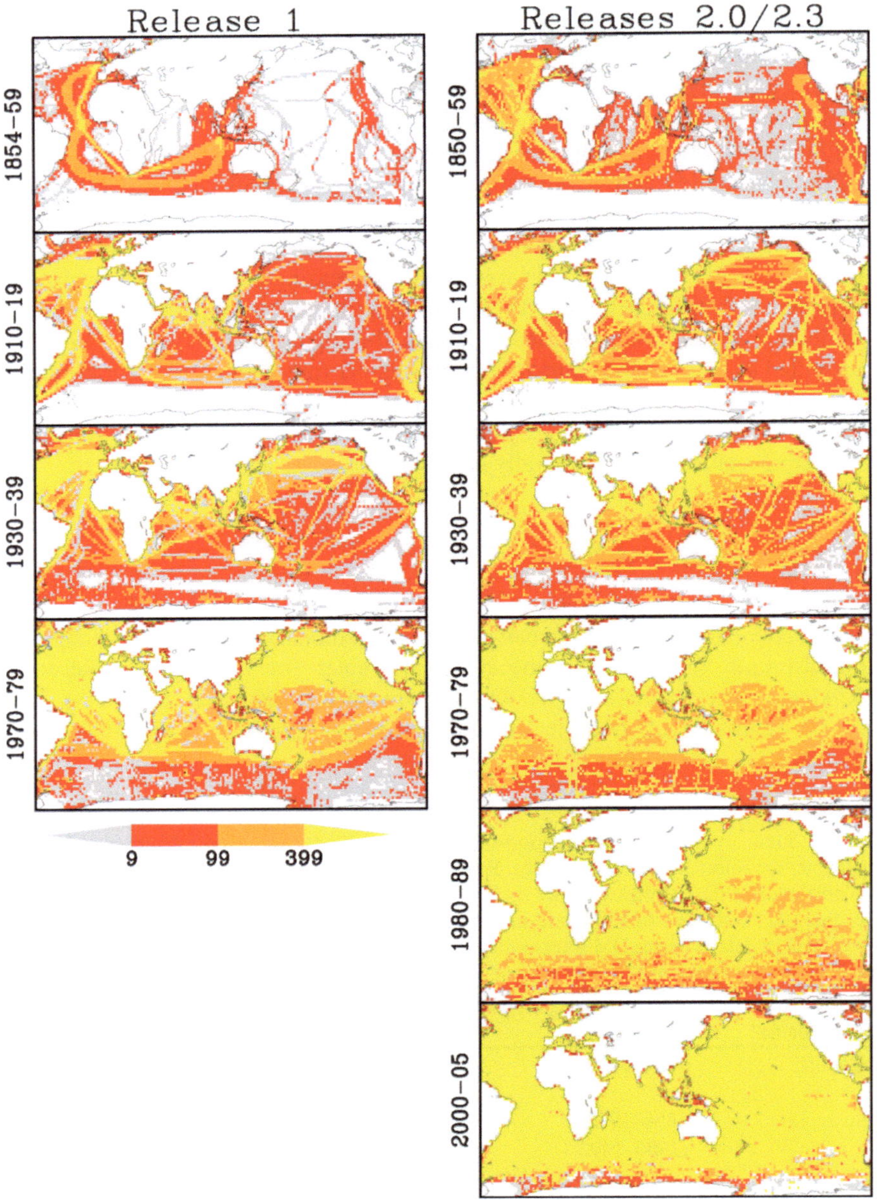

Fig. 5 Enhancements in ICOADS spatial SST coverage for selected decades, comparing Release 1 with Releases 2.0 (19th- and 20th-century decades, completed by 2002) and 2.3 (2000–2005, completed in 2006). Colors show the total number of SST observations per decade: 1–9 (gray), 10–99 (red), 100–399 (orange), more than 399 (yellow). Counts for the two partial decades (1854–59 and 2000–05) were adjusted (/6 × 10) to be comparable to the full decades.

completed in 1985 (Woodruff et al. 1987). Since Release 1, periodic updates (Table 1) have led to substantial improvements in coverage (Fig. 5), and the observations, products, and data access systems have been redesigned to better meet user requirements (Sects. 3.1 and 3.2).

A new name, International COADS (ICOADS), was adopted (Diaz et al. 2002; Parker et al. 2004) to recognize the multinational data input and the benefits gained from extensive international collaboration. ICOADS maintains continuity of identity with COADS, which has been extensively referenced in the scientific literature. ICOADS data, metadata, and products are freely offered to the research community and the general public via the project's web portal (icoads.noaa.gov).

3.1 Observational Improvements

The individual marine "reports" (i.e., the observed variables and metadata for a given VOS/ODAS, time, and position) for Release 1 comprised a small selection of fields, including SST and, when available, an SST measurement method indicator (SI), stored in a very concise packed-binary format. Updates since Release 1 transitioned through a more inclusive binary format, to the current International Maritime Meteorological Archive (IMMA) format, which preserves (in simple ASCII characters) a very complete set of uniformly formatted fields and also appends the data in their original format.

SST and SI have been augmented with additional IMMA metadata fields that can be important for interpretation of the SST data, including the platform type and source "deck" (originally referring to punched cards). For VOS, the ship and instrumental metadata published by WMO starting in 1955 (Kent et al. 2007), including depth of SST sensor beginning in 1995, are also being blended into the IMMA format. The current overall period of record is 1784–2005, at Release 2.3 (Table 1). While scattered SST observations exist in the early 1800s, significant increases in data density follow the Brussels conference that established recording standards in 1853.

3.2 Product Improvements

The year-month summary products for Release 1 comprised a selection of statistics (e.g., the mean and median) computed for observed and derived variables – including SST – using 2° × 2° latitude–longitude boxes. Prior to product computation, observations outside 3.5σ (standard deviations) relative to three distinct climatological periods (1854–1909, 1910–1949, 1950–1979) were "trimmed" and not used (Slutz et al. 1985). Among early update improvements (see Table 1) were higher-resolution summary products (1° × 1°) for the period 1960 onward when the sampling density supports this higher resolution.

The Release 1 trimming (or quality control; QC) limits were found to be too restrictive (Wolter et al. 1989; Wolter 1997) for some extreme climate events (e.g., 1982–1983 El Niño). In response, "enhanced" summaries were created, with the trimming limits expanded to 4.5σ (but still fixed around the Release 1 climatological medians) to partially account for larger environmental variability. Another characteristic of the enhanced summaries is that they include more platform types (ships plus most ODAS). Table 2 summarizes the SST statistical quantities and product categories currently available from ICOADS.

4 Higher-level SST Products

ICOADS offers a detailed set of statistics and products (Table 2) to summarize the SST data falling within each 2° or 1° box and year–month. However, the coverage of the products is limited to boxes containing SST data – with no interpolation to fill in temporal or spatial data gaps, and no analysis to smooth the statistics (Fig. 6). Moreover, the available summaries incorporate no adjustments for variations in observing practices. These deficiencies have been addressed with two main types of higher-level SST analyses, derived from the ICOADS (or closely related) in situ data.

Table 2 ICOADS statistics computed for each SST grid box containing at least one observation[a] (left). Monthly summary products, and their characteristics, available for SST, plus for 21 other observed and derived variables (see Worley et al. 2005 for additional details) (right columns)

Abbrev.	Statistic	2° × 2° (1800 forward)	1° × 1° (1960 forward)
s_1	1/6 sextile (estimate of $m - 1\sigma$)	Enhanced: Trimming: 4.5σ Platform mixture: VOS plus most ODAS	Enhanced: (same as 2° × 2°)
s_3	3/6 sextile (estimate of m)		
s_5	5/6 sextile (estimate of $m + 1\sigma$)		
m	Mean		
n	Number of observations		
σ	Standard deviation[b]	Standard: Trimming: 3.5σ Platform mixture: VOS only	Standard: (same as 2° × 2°)
d	Mean day-of-month of observations		
h_t	Fraction of observations in daylight		
x	Mean longitude of observations		
y	Mean latitude of observations		

[a]All the statistics are computed even with as few as one observation. For example, in that case the sextiles and the mean all are set to the single observed value, and the standard deviation to zero (Slutz et al. 1985).

[b]A separate estimate e of σ can be computed as $e = (s_5 - s_1)/2$.

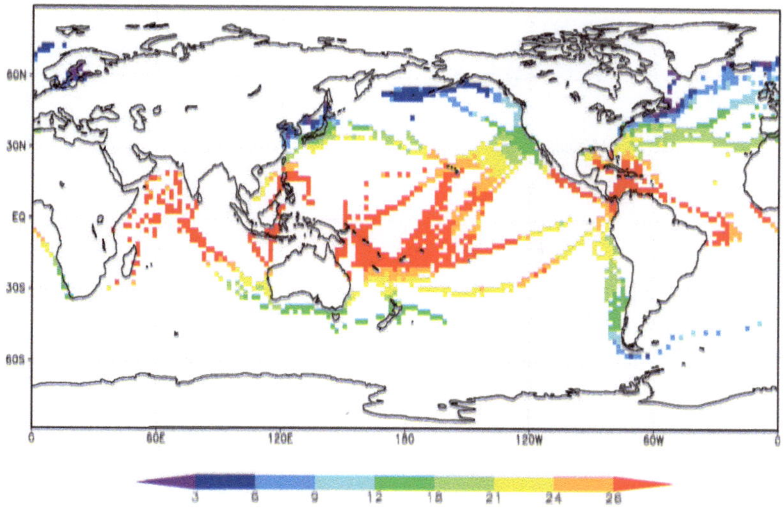

Fig. 6 ICOADS-enhanced monthly mean SST (°C) for December 1942.

4.1 Historical Reconstructions

The length and environmental significance of the in situ SST record makes it of great interest for climate change studies, even though the coverage is sparse in much of the 19th century and other periods such as the world wars. However, the full-length raw SST time series is not an accurate climate record. Around the time of Release 1 it was recognized that data heterogeneities introduced by observing system changes needed to be accounted for in climate studies (Barnett 1984; Wright 1986) – for example, the transition from bucket to intake SST measurement (Folland et al. 1984). Adjustments for such inhomogeneities have so far focused on the VOS data (e.g., Folland and Parker 1995; Hanawa et al. 2000; Smith and Reynolds 2002; Folland 2005; Kent and Kaplan 2006). Reconciling measurements from different observing systems (e.g., ODAS and VOS) is yet to be achieved (Fig. 7).

Coupled with the adjustment efforts, statistical techniques such as optimal interpolation (OI), averaging (OA), or smoothing (OS) have been used to estimate past global (or near global) coverage from limited amounts of observed data, with the key assumption that contemporary modes of variability apply historically (Hurrell and Trenberth 1999). High-latitude SST analyses depend critically on accurate estimates of time varying sea ice coverage, and have benefited from gradual improvements in sea ice concentration products. Finally, several analyses are starting to incorporate estimates of uncertainties and other statistical properties. These added fields are used to supplement the interpretation of the basic interpolated SST values.

One line of products developed by the UK began with the un-interpolated (5° resolution) Met Office Historical SST Dataset (MOHSST; Bottomley et al. 1990; Parker et al. 1995), followed by Global Sea Ice and SST (GISST), and now Hadley Centre Sea Ice and SST (HadISST1) and SST (HadSST2) datasets (Rayner et al.

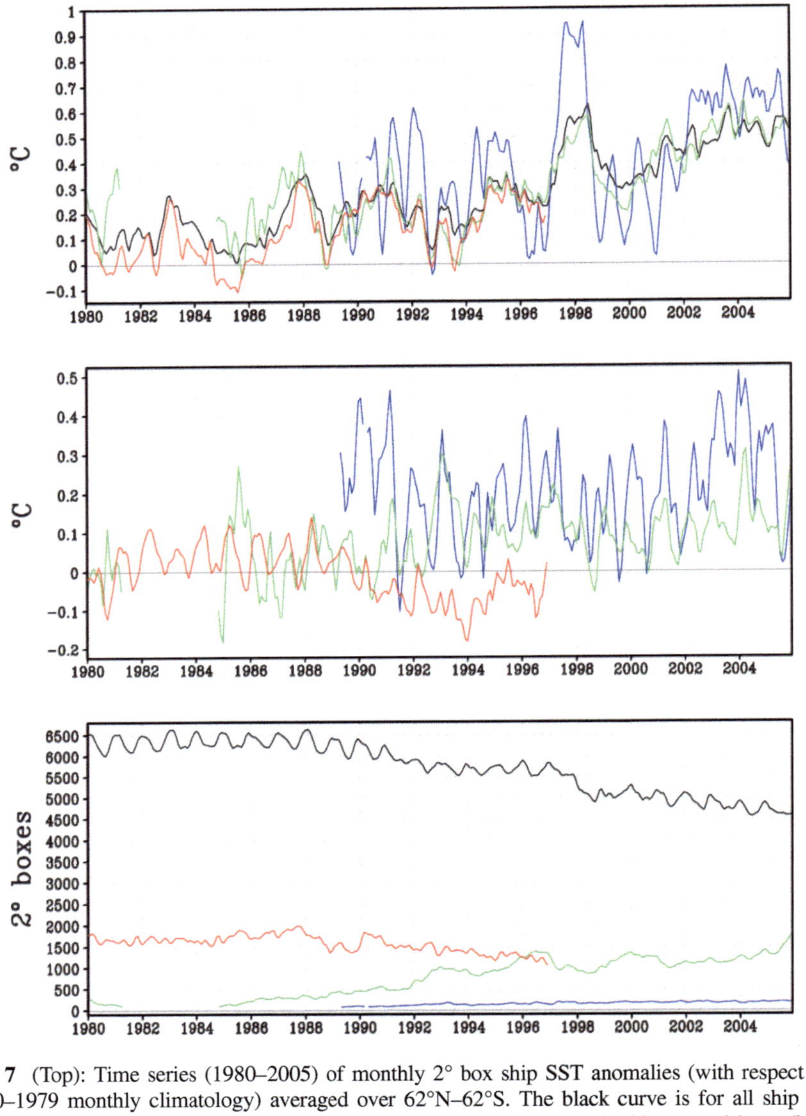

Fig. 7 (Top): Time series (1980–2005) of monthly 2° box ship SST anomalies (with respect to a 1950–1979 monthly climatology) averaged over 62°N–62°S. The black curve is for all ship SST boxes, with colored curves for three subsets: ship SST boxes colocated with moored buoy (blue), drifting buoy (green), or oceanographic (red) SST boxes. (Middle): Differences between the ship SST anomalies in the top panel, minus monthly average anomalies of the colocated moored buoy, drifting buoy, and oceanographic SST data (same colors). (Bottom): Monthly numbers of all ship (black) and collocated (same colors) 2° SST boxes. Note: Curves were smoothed, and required at least 80 colocated boxes, thus eliminating months with sparse data. As indicated in Fig. 3, oceanographic data are presently available in ICOADS only through 1996, hence the early ending points of those curves.

2003, 2006). HadISST1 employed all available (including some satellite) data, and estimated monthly 1° resolution patterns using OI. HadSST2 utilized the latest ICOADS data, and includes uncertainty estimates and other novel features such as flexible gridding (www.hadobs.org).

Another major product line, developed by NOAA, began with OA of seasonal averages (Smith et al. 1994), transitioned to a global monthly reconstruction (Smith et al. 1996), which in turn lead to Extended Reconstructed SST (ERSST) datasets (Smith and Reynolds 2003, 2004). The ERSST datasets, which are available at a monthly 2° resolution (www.ncdc.noaa.gov), include sea ice information similar to HadISST1, and were optimally reconstructed using variability characteristics from an advanced in situ and satellite analysis (Reynolds et al. 2002) (Fig. 8). The difference between these fields (Fig. 9) is an indication of the uncertainty that arises through different analysis techniques.

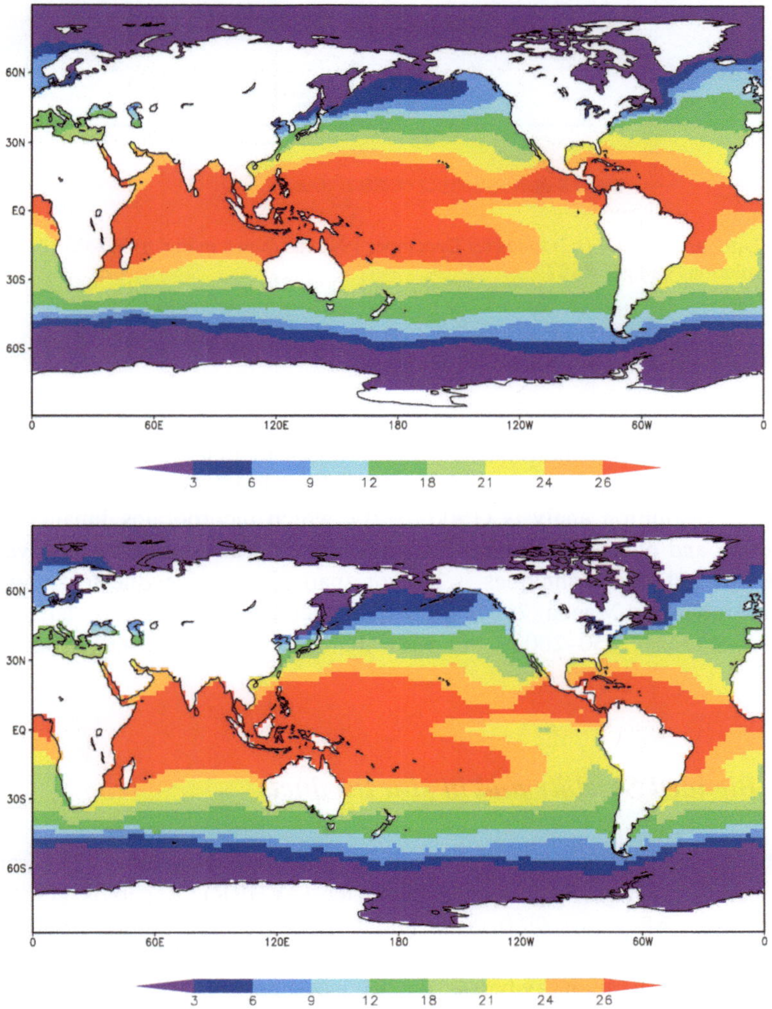

Fig. 8 HadISST (upper; 1° resolution) versus ERSST (lower; 2° resolution) SST (°C) fields for December 1942.

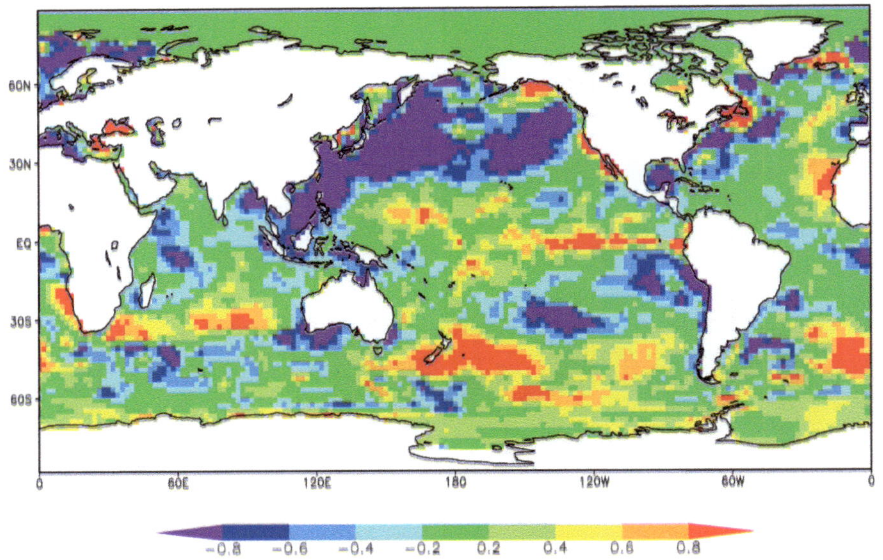

Fig. 9 Difference in SST (°C) fields for December 1942 (as shown in Fig. 8): HadISST (averaged to 2° resolution) minus ERSST.

Other noteworthy interpolated SST products include Kaplan et al. (1997, 1998) 5° resolution analyses (1856–1991; using MOHSST), and the Ishii et al. (2005) 1° resolution analysis (1900–2001), which incorporates Japanese Kobe Collection and other data not yet fully available in ICOADS. One disadvantage of the interpolation techniques is loss of spatial resolution due to smoothing, which Minobe and Maeda (2005) addressed by computing un-interpolated means at 1° resolution (1850–2002), but with QC beyond that currently available in ICOADS.

4.2 Blended Satellite and in situ Products

Reliable infrared (IR) satellite SST observations begin in November 1981 from the Advanced Very High Resolution Radiometer (AVHRR). Reynolds (1988) and Reynolds and Marsico (1993) made initial efforts to blend satellite and in situ observations. The in situ record has the advantages of a lengthy period and many independent observations, whereas the IR measurements provide dense spatial coverage at high resolution (~4 km) – but with the caveat that any bias (e.g., Reynolds et al. 1989; Reynolds 1993) will influence all data from that satellite instrument. In situ observations are vital to identify and remove large-scale biases in the AVHRR SST measurements, which are caused by aerosols such as Saharan dust or volcanic particles, as well as by imperfect calibration.

Reynolds and Smith (1994) was the first analysis (OI.v1) to optimize the advantages of both types of data. The latest analysis version (OI.v2) (Reynolds et al. 2002) is updated weekly at a 1° resolution with improved satellite bias corrections and sea ice conversions. More recently, other blended real-time products based on the AVHRR or other IR data have become available operationally (e.g., Thiébaux et al. 2003; O'Carroll et al. 2006).

Starting in late 1997, satellite data became available from the Tropical Rainfall Measuring Mission (TRMM) Microwave Imager (TMI), and since 2002 from the Advanced Microwave Scanning Radiometer aboard the Earth Observing System (AMSR-E) satellite. These and other planned microwave instruments have lower resolution (50 km) than the IR data, but are able to measure SST through many clouds and aerosols (Chelton and Wentz 2005). The microwave data should form an important ingredient for future climate-scale SST analyses (Stammer et al. 2003; Reynolds et al. 2004), partly also because they have different error characteristics than the IR data, and the independent sources may allow for corrections in regions lacking in situ data.

5 Future Directions

This paper has reviewed the status and evolution of the in situ SST data available from ICOADS, plus some higher-level climate-scale products. We conclude by suggesting a variety of desirable further improvements and directions for future work to help improve the SST record:

- Numerous additional data are available, or becoming available for blending into ICOADS (Woodruff et al. 2004, 2005; Worley et al. 2005). Attempts have been made to estimate the in situ data density needed to support blended satellite analyses (e.g., Reynolds et al. 2005). The prioritization of available sources of historical data for digitization and incorporation into ICOADS is essential. One obvious need is for more observations in data sparse periods, such as during the world wars.
- Continued blending of published VOS metadata for 1955 forward into ICOADS (Kent et al. 2007) remains a crucial related task to help improve our understanding of SST measurement methods. Also, it may be possible to tap new sources of historical VOS metadata, in the course of emerging digitization projects (e.g., of recently imaged UK Royal Navy 1938–1947 Ship's Logs), and international efforts should be supported to assemble and organize historical ODAS metadata.
- We recommend the continuing creation of SST products by independent methods, using the best available data and metadata, coupled with comparisons of the resulting products to better quantify the random and systematic uncertainties (e.g., Trenberth et al. 1992; Folland et al. 1993; Hurrell and Trenberth 1999;

Folland et al. 2001b; Kent and Taylor 2006). So far, for example, apparent SST biases between VOS and ODAS data have not been well quantified.

- With a few exceptions (e.g., basic ICOADS summaries, and Ishii et al. 2005) most products to date have used SST in isolation from other variables (e.g., air temperature, barometric pressure, and winds), which are reported by VOS and to a more limited extent by ODAS. We anticipate that cross-validations between the different variables, and VOS and ODAS platform types, will lead toward the creation of improved climate-quality products for SST and other variables.
- A related data quality issue is to better capitalize in ICOADS on the QC feedback information obtained both from operational weather prediction, and reanalyses, for example, one planned using only surface pressure (Compo et al. 2006). However, models can also be imperfect, especially when observations are sparse, and the resulting information would need to be carefully compartmentalized and documented within ICOADS.

Acknowledgments We are grateful to S. Lubker for preparation of the figures, to J.R. Keeley and D.E. Harrison for advice, and to D. Parker and an anonymous reviewer for many helpful comments and suggestions on the paper. The original COADS project, and the continuing US contribution toward the new international database, ICOADS, is the result of a cooperative project between NOAA – specifically its Earth System Research Laboratory (ESRL), its National Climatic Data Center (NCDC), and the Cooperative Institute for Research in Environmental Sciences (CIRES, conducted jointly with the University of Colorado) – and the National Science Foundation's National Center for Atmospheric Research (NCAR).

References

Barnett, T. P., 1984: Long-term trends in surface temperature over the oceans. *Mon. Wea. Rev.*, **112**, 303–312.

Bottomley, M., C. K. Folland, J. Hsiung, R. E. Newell, and D. E. Parker, 1990: *Global Ocean Surface Temperature Atlas 'GOSTA.'* Meteorological Office/MIT, Boston, MA, 20 + iv pp, 313 plates.

Chelton, D. B. and F. J. Wentz, 2005: Global microwave satellite observations of sea surface temperature for numerical weather prediction and climate research. *Bull. Am. Meteorol. Soc.*, **86**, 1097–1115.

Compo, G. P., J. S. Whitaker, and P. D. Sardeshmukh, 2006: Feasibility of a 100-year reanalysis using only surface pressure data. *Bull. Am. Meteorol. Soc.*, **87**, 175–190.

Diaz, H., C. Folland, T. Manabe, D. Parker, R. Reynolds, and S. Woodruff, 2002: Workshop on advances in the use of historical marine climate data. *WMO Bull.*, **51**(4), 377–380.

Folland, C. K., 2005: Assessing bias corrections in historical sea surface temperature using a climate model. *Int. J. Climatol.*, **25**, 895–911.

Folland, C. K. and D. E. Parker, 1995: Correction of instrumental biases in historical sea surface temperature data. *Q. J. Roy. Meteorol. Soc.*, **121**, 319–367.

Folland, C. K., D. E. Parker, and F. E. Kates, 1984: Worldwide marine temperature fluctuations, 1856–1981. *Nature*, **310**, 670–673.

Folland, C. K., R. W. Reynolds, M. Gordon, and D. E. Parker, 1993: A study of six operational sea surface temperature analyses. *J. Climate*, **6**, 96–113.

Folland, C. K. et al., 2001a: Observed climate variability and change. In: *Climate Change 2001: The Scientific Basis*, J. T. Houghton et al., eds., Contribution of Working Group I to the Third

Assessment Report of the Intergovernmental Panel on Climate Change, Cambridge University Press, Cambridge, 99–181.

Folland, C. K., N. A. Rayner, S. J. Brown, T. M. Smith, S. S. P. Shen, D. E. Parker, I. Macadam, P. D. Jones, R. N. Jones, N. Nicholls, and D. M. H. Sexton, 2001b: Global temperature change and its uncertainties since 1861. *Geophys. Res. Lett.*, **28**, 2621–2624.

Hanawa, K., S. Yasunaka, T. Manabe, and N. Iwasaka, 2000: Examination of correction to historical SST data using long-term coastal SST data taken around Japan. *J. Meteorol. Soc., Japan*, **78**, 187–195.

Hurrell, J. W. and K. E. Trenberth, 1999: Global sea surface temperature analyses: multiple problems and their implications for climate analysis, modeling, and reanalysis. *Bull. Am. Meteorol. Soc.*, **80**, 2661–2678.

Ishii M., A. Shouji, S. Sugimoto, and T. Matsumoto, 2005: Objective analyses of sea-surface temperature and marine meteorological variables for the 20th century using ICOADS and the Kobe Collection. *Int. J. Climatol.*, **25**, 865–879.

Kaplan, A., Y. Kushnir, M. A. Cane, and M. B. Blumenthal 1997: Reduced space optimal analysis for historical datasets: 136 years of Atlantic sea surface temperatures. *J. Geophys. Res.*, **102**, 27835–27860.

Kaplan, A., M. A. Cane, Y. Kushnir, A. Clement, and M. Blumenthal, B. Rajagopalan, 1998: Analyses of global sea surface temperature 1856–1991. *J. Geophys. Res.*, **103**, 18567–18589.

Kent, E. C. and A. Kaplan, 2006: Toward estimating climatic trends in SST. Part III: systematic biases. *J. Atmos. Oceanic Technol.*, **23**, 487–500, doi:10.1175/JTECH1845.1.

Kent, E. C. and P. K. Taylor, 2006: Toward estimating climatic trends in SST. Part I: methods of measurement. *J. Atmos. Oceanic Technol.*, **23**, 464–475, doi:10.1175/JTECH1843.1.

Kent, E. C., D. I. Berry, S. D. Woodruff, and P. K. Taylor, 2006: Voluntary observing ships: a vital observing system in decline. *CLIVAR Exchanges*, **11**(3), 15, 20–21.

Kent, E. C., S. D. Woodruff, and D. I. Berry, 2007: WMO Publication No. 47 metadata and an assessment of observation heights in ICOADS. *J. Atmos. Oceanic Technol.*, **24**, 214–234.

Kistler, R. et al., 2001: The NCEP–NCAR 50-year reanalysis. *Bull. Am. Meteorol. Soc.*, **82**, 247–267.

Maury, M. F., 1854: Maritime Conference held at Brussels for devising a uniform system of meteorological observations at sea, August and September, 1853. In: *Explanations and Sailing Directions to Accompany the Wind and Current Charts* (6th edn), E. C. Kent and J. Biddle, Philadelphia, PA, 54–96.

Minobe, S. and A. Maeda, 2005: A 1° monthly gridded sea-surface temperature dataset compiled from ICOADS from 1850 to 2002 and northern hemisphere frontal variability. *Int. J. Climatol.*, **25**, 881–894.

McPhaden, M. J., A. J. Busalacchi, R. Cheney, J. R. Donguy, K. S. Gage, D. Halpern, M. Ji, P. Julian, G. Meyers, G. T. Mitchum , P. P. Niiler, J. Picaut, R. W. Reynolds, N. Smith, and K. Takeuchi, 1998: The tropical ocean-global atmosphere (TOGA) observing system: a decade of progress. *J. Geophys. Res.*, **103**, 14169–14240.

O'Carroll, A. G., J. G. Watts, L. A. Horrocks, R. W. Saunders, and N. A. Rayner, 2006: Validation of the AATSR meteo product sea surface temperature. *J. Atmos. Oceanic Technol.*, **23**, 711–726.

Parker, D. E., C. K. Folland, and M. Jackson, 1995: Marine surface temperature: observed variations and data requirements. *Clim. Change*, **31**, 559–600.

Parker, D., E. Kent, S. Woodruff, D. Dehenauw, D. E. Harrison, T. Manabe, M. Mietus, V. Swail, and S. Worley, 2004: Second JCOMM workshop on advances in marine climatology (CLIMAR-II). *WMO Bull.*, **53**(2), 157–159.

Rayner, N. A., D. E. Parker, E. B. Horton, C. K. Folland, L. V. Alexander, D. P. Rowell, E. C. Kent, and A. Kaplan, 2003: Global analyses of sea surface temperature, sea ice, and night marine air temperature since the late nineteenth century. *J. Geophys. Res.*, **108**(D14), 4407, doi:10.1029/2002JD002670.

Rayner, N. A., P. Brohan, D. E. Parker, C. K. Folland, J. J. Kennedy, M. Vanicek, T. J. Ansell, and S. F. B. Tett, 2006: Improved analyses of changes and uncertainties in sea surface temperature measured in situ since the mid-nineteenth century: the HadSST2 dataset. *J. Climate*, **19**, 446–469.

Reynolds, R. W., 1988: A real-time global sea surface temperature analysis. *J. Climate*, **1**, 75–86.
Reynolds, R. W., 1993: Impact of Mount Pinatubo aerosols on satellite-derived sea surface temperatures. *J. Climate*, **6**, 768–774.
Reynolds, R. W. and D. C. Marsico, 1993: An improved real-time global sea surface temperature analysis. *J. Climate*, **6**, 114–119.
Reynolds, R. W. and T. M. Smith, 1994: Improved global sea surface temperature analyses using optimum interpolation. *J. Climate*, **7**, 929–948.
Reynolds, R. W., C. K. Folland, and D. E. Parker, 1989: Biases in satellite derived sea-surface temperature data. *Nature*, **341**, 728–731.
Reynolds, R. W., N. A. Rayner, T. M. Smith, D. C. Stokes, and W. Wang, 2002: An improved in situ and satellite SST analysis for climate. *J. Climate*, **15**, 1609–1625.
Reynolds, R. W., C. L. Gentemann, and F. Wentz, 2004: Impact of TRMM SSTs on a climate-scale analysis. *J. Climate*, **17**, 2938–2952.
Reynolds, R. W., H.-M. Zhang,, T. M. Smith, C. L. Gentemann, and F. Wentz, 2005: Impacts of in situ and additional satellite data on the accuracy of a sea-surface temperature analysis for climate. *Int. J. Climatol.*, **25**, :857–864.
Slutz, R. J., S. J. Lubker, J. D. Hiscox, S. D. Woodruff, R. L. Jenne , D. H. Joseph, P. M. Steurer, and J. D. Elms, 1985: *Comprehensive Ocean-atmosphere Data Set*; Release 1. NOAA Environmental Research Laboratories, Climate Research Program, Boulder, CO, 268 pp.
Smith, T. M. and R. W. Reynolds, 2002: Bias corrections for historical sea surface temperatures based on marine air temperatures. *J. Climate*, **15**, 73–87.
Smith, T. M. and R. W. Reynolds, 2003: Extended reconstruction of global sea surface temperatures based on COADS data (1854–1997). *J. Climate*, **16**, 1495–1510.
Smith, T. M. and R. W. Reynolds, 2004: Improved extended reconstruction of SST (1854–1997). *J. Climate*, **17**, 2466–2477.
Smith, T. M., R. W. Reynolds, and C. F. Ropelewsky 1994: Optimal averaging of seasonal sea surface temperatures and associated confidence intervals (1860–1989). *J. Climate*, **7**, 949–964.
Smith, T. M., R. W. Reynolds, R. E. Livezey, and D. C.Stokes, 1996: Reconstruction of historical sea surface temperatures using empirical orthogonal functions. *J. Climate*, **9**, 1403–1420.
Stammer, D., F. Wentz, and C. Gentemann, 2003: Validation of microwave sea surface temperature measurements for climate purposes. *J. Climate*, **16**, 73–87.
Thiébaux, J., E. Rogers, W. Wang, and B. Katz, 2003: A new high-resolution blended real-time global sea surface temperature analysis. *Bull. Am. Meteorol. Soc.*, **84**, 645–656.
Trenberth, K. R., J. R. Christy, and J. W. Hurrell, 1992: Monitoring global monthly mean surface temperatures. *J. Climate*, **5**, 1405–1423.
Wolter, K., 1997: Trimming problems and remedies in COADS. *J. Climate*, **10**, 1980–1997.
Wolter, K., S. J. Lubker, and S. D. Woodruff, 1989: Trimming – a potential error source in COADS. *Tropical Ocean-atmosphere Newsletter*, **51**, 4–7.
Woodruff, S. D., R. J. Slutz, R. L. Jenne, and P. M. Steurer, 1987: A comprehensive ocean-atmosphere data set. *Bull. Am. Meteorol. Soc.*, **68**, 1239–1250.
Woodruff, S. D., S. J. Lubker, K. Wolter, S. J. Worley, and J. D. Elms, 1993: Comprehensive ocean-atmosphere data set (COADS); Release 1a: 1980–1992. *Earth System Monitor*, **4**(1), 1–8.
Woodruff, S. D., H. F. Diaz, J. D. Elms, and S. J. Worley, 1998: COADS Release 2 data and metadata enhancements for improvements of marine surface flux fields. *Phys. Chem. Earth*, **23**, 517–526.
Woodruff, S. D., S. J. Worley, J. A. Arnott, H. F. Diaz, J. D. Elms, M. Jackson, S. J. Lubker, and D. E. Parker, 2003: COADS updates and the blend with the UK Met Office Marine Data Bank. In: *Advances in the Applications of Marine Climatology – The Dynamic Part of the WMO Guide to the Applications of Marine Meteorology*. WMO/TD-No. 1081 (JCOMM Technical Report No. 13), pp. 3–10 (on CD-ROM).
Woodruff, S., J. Elms, R. Reynolds, R. Garcia, F. Guo, S. Worley, and T. Yoshida, 2004: *Rescuing Marine Data*. World Climate News, WMO, Geneva, No 25: 10.
Woodruff, S. D., H. F. Diaz, S. J. Worley, R. W. Reynolds, S. J. Lubker, 2005: Early ship observational data and ICOADS. *Clim. Change*, **73**, 169–194.

Worley, S. J., S. D. Woodruff, R. W. Reynolds, S. J. Lubker, and N. Lott, 2005: ICOADS Release 2.1 data and products. *Int. J. Climatol.*, **25**, 823–842.

Wright, P. B., 1986: Problems in the use of ship observations for the study of interdecadal climate changes. *Mon. Wea. Rev.*, **114**, 1028–1034.

Upper-air Temperature Trends: Current Problems and Some Recent Results

A. M. Sterin[1], V. M. Khan[2], and K. G. Rubinshtein[2]

Abstract The free atmosphere is an essential part of the climate system, and the processes in the free atmosphere are strongly interrelated with the processes in other climate system components. Temperature in the free atmosphere is better studied than other parameters defining the upper-air climate, but nevertheless there exist numerous problems in the upper-air temperature trends. This paper gives the overview of the current status of radiosonde based empirical studies of the upper-air temperature climate changes. Problems such as data availability, data quality and completeness, inhomogeneity detection in the upper-air temperature time series, temperature trend estimates, overall patterns of upper-air trends are discussed in the paper. Special attention is given to the problem of whether the reanalysis outputs can reproduce trends over long periods in the upper-air temperature series.

1 Introduction

Since 1930, when the first radiosonde in the world was launched by Professor Pavel Molchanov in St. Petersburg (Leningrad), former USSR, much effort has been made to make radiosonde observations covering the globe, as well as to improve their temporal regularity. The worldwide radiosonde observations, first of all, are provided with the main goal of operational weather prediction. However, they could also be useful for climate studies, in addition to other sources of upper-air data. Other sources such as aircrafts (providing observations irregularly in time and space), or satellites (which provide observations for a shorter period, with better horizontal, but worse vertical, resolution than the radiosondes), are valid for climate research community, but in no way can substitute radiosonde data in upper-air temperature climate research.

[1] Russian Research Institute for Hydrometeorological Information – World Data Center (RIHMI-WDC), sterin@meteo.ru

[2] Hydrometeorological Center of Russian Federation (HMC RF) khan@mecom.ru, rubin@mecom.ru

S. Brönnimann et al. (eds.), *Climate Variability and Extremes during the Past 100 Years.*

Upper-air temperature trends are strongly connected to the problem of global warming (IPCC 1996, 2001), but their patterns are somewhat different from those on the land surface (Hurrell and Trenberth 1998) and carry a large uncertainty. Therefore the patterns of long-term trends in upper-air temperature series have become the focus of numerous discussions.

This paper is written with two main goals in mind. The first goal is to summarize the patterns of upper-air temperature trends and express a vision of the current status of upper-air temperature trend related problems. The second goal, much more modest than the first one, is to concentrate on several relatively new results by the authors using radiosonde data for upper-air temperature trend studies, including comparisons between temperature trends in radiosonde data and trends in temperature from reanalysis outputs.

In no way is this paper able to provide a comprehensive review of the upper-air temperature-related problems, rather it partly reviews several strengths and weaknesses in the current status of the empirical study of upper-air temperature climate patterns, which are supplemented by some authors' recent findings.

2 Current Status of Empirical Upper-air Temperature Trend Studies

2.1 Radiosonde and Satellite Data Relevant to Upper-air Temperature Trend Studies

In this section, we briefly summarize the information about radiosonde and satellite data that can be used to study upper-air temperature trends, especially their strengths and weaknesses, and provide references to existing radiosonde and satellite data collections and upper-air temperature series data sets relevant to trend studies.

Currently there are about 1,000 stations worldwide which provide moderately long periods of observations. Not all observational sites produce continuous series, for example, some stations had interruptions in soundings, so the continuity of the series is not ideal. The stricter the requirements applied to the length and continuity of series, the fewer the number of stations that meet these criteria. As a way to resolve this, a reasonable compromise is needed between the length of the observing period, data completeness, continuity, as well as spatial coverage.

Two main problems of spatial coverage of the atmosphere by radiosonde data can be outlined. The first problem is that geographical coverage of the world by radiosonde observations acceptable for the climate studies is essentially limited to the land areas of the northern hemisphere, while the vast ocean areas, as well as the southern hemisphere, have much weaker coverage. The second problem of radiosonde data coverage is the vertical coverage. Poorer sampling in the stratosphere compared to the troposphere increases the uncertainty. This is most critical in the early part of the record, and, as a result, may bias stratospheric temperature trend estimates.

At least two different global radiosonde data sets are known:

- The Integrated Global Radiosonde Archive (IGRA) (Durre et al. 2006); a project by the National Climatic Data Center (NCDC) and NOAA, available from NCDC. The IGRA data set is updated by current data via the global telecommunications system (GTS) on a regular basis. Together with daily data, monthly mean values are calculated and are made available from the same source.
- Based on the data collected from the GTS, the Russian Research Institute for Hydrometeorological Information – World Data Center (RIHMI-WDC), is compiling a global archive of daily radiosonde data, named AEROSTAB. This data set is also updated on a regular basis and is archived after the Complex Quality Check (CQC).

Along with daily data, monthly station statistics for the global radiosonde network are produced by RIHMI-WDC, which constitute MONADS (MONthly Aerological Data Set).

In addition to the above data sets, there are special radiosonde temperature data products:

- The LKS data set of upper-air temperature for a selected set (about 80) of radiosonde stations (Lanzante et al. 2003a, b). The LKS data set is part of RATPAC ("Radiosonde Atmospheric Temperature Products for Assessing Climate") (Free and Seidel 2005).
- The HadAT grid-point data set for upper-air temperature, produced by the Hadley Centre of the UK (Thorne et al. 2005).

Both products passed through a bias detection and removal process which is described below. Other upper-air temperature products are based on the microwave sounding instruments (MSU and AMSU) on board the NOAA polar orbiting satellites. These series reach back to 1979 and are updated on regular basis (see Mears, this volume). The physical background of obtaining series of tropospheric and stratospheric temperatures based on MSU observations is given by Spencer and Christy (1990).

At least two different temperature data sets are available based on these satellite observations:

- The data set provided by University of Alabama in Huntsville (UAH; Christy et al. 2000).
- The data set provided by Remote Sensing Systems, Inc (RSS; Mears et al. 2003).

Though the primary satellite measurements used for obtaining both of these series are the same, their processing is different, and, as a result, the values of decadal trends calculated from these series differ from each other as discussed below. Unlike radiosonde observations, the satellite observations provide even coverage of the globe, including ocean areas and areas in Tropics and in southern hemisphere, but do not provide the detailed vertical resolution, as it is provided by radiosondes.

Both radiosonde and satellite upper-air temperature time series are known to suffer from artificial steplike changes, which can originate from changes of instruments, data transfer or processing algorithms, site locations (for radiosondes), orbit decays and instrumentation differences (for satellites), etc. Since these changes, referred to as inhomogeneities, affect the trend values, the problem of detection, estimation of magnitude, and removal of artificial changes in the climate time series are among the most crucial problems of the present-day climatology.

Each meteorological parameter requires a specific approach to detect and estimate inhomogeneities in the corresponding time series. Such approaches are well known for surface teperature and surface precipitation time series. However, the paper of Gaffen et al. (2000) contains a wide and, probably, comprehensive, list of reasons why the approaches applicable to land-surface meteorological parameters are less effectively used for the upper-air radiosonde-based meteorological parameters, the upper-air temperature in particular.

As concluded by Gaffen et al. (2000), neither metadata alone nor statistical techniques alone is sufficient for adequately detecting inhomogeneities in upper-air data. Techniques specifically aimed to detect, estimate, and eliminate inhomogeneities in the upper-air temperature time series include, among others:

- Use of physical models for various types of sondes and temperature sensors (for radiosondes)
- Use of adjustments which take into consideration the satellite orbital decays, and changes in MSU instruments (for satellite data)
- Use of intercomparisons between the radiosonde-based and satellite-based upper-air temperature series (keeping in mind that both types of data have independent and different sources of inhomogeneities)
- Use of metadata to determine the candidate dates when inhomogeneities could possibly be introduced
- Use of statistical techniques for detection steplike changes in the series
- Efforts of Lanzante et al. (2003a, b) and of Thorne et al. (2005), to use a combined technique based on metadata and statistical inference, which produced the LKS data set and HadAT, respectively
- Effort of Sterin (2004a) to use a combined metadata and statistical instrument approach to detect inhomogeneities in upper-air temperature station time series and to reestimate trends after eliminating these candidate inhomogeneities (thus providing "what-if" analysis of homogeneity effects on trends)
- Effort of Haimberger (2005, 2007) to use the ECMWF ERA-40 reanalysis outputs and operational analysis outputs for homogenization of radiosonde temperature time series (this method enables automated corrections to be applied to radiosonde data)

The problem of inhomogeneity detection and bias removal in the upper-air temperature time series, however, remains open for further investigations. A paper by Free et al. (2002) summarizes the results of special effort to compare existing techniques of inhomogeneity detection in the upper-air radiosonde temperature series (this effort was assessed in 2000 at special workshop at NCDC, Asheville, USA), and contains a conclusion that no single method (and no single output of so called "bias-free" upper-air temperature series) could be recommended for the future use, in particular for the use in upper-air temperature climate trend analysis. The difficulties in efforts in this field are, among other factors, related to the fact that such detection procedures nontrivial and time-consuming. All the more, several of them (but not all) require much manual effort, presume a lot of "art," and, thus, are difficult to formalize and unlikely to be reproduced by research teams other than the original authors of the methods. Essential progress in constructing upper-air temperature inhomogeneity detection and removal procedures, as well as in preparing so called bias-free radiosonde products, have been achieved in recent years by UKMO (Thorne et al. 2005), by Haimberger (2005, 2007), by Free and Seidel (2005). Recognizing this, we need to note, however, that the process of bias removal, for the upper-air time series of temperature, must be a careful and well-weighted effort, because of strong natural variability of the processes in the atmosphere (e.g., Henderson-Sellers et al. 1998). Otherwise, signals of these processes could be easily suspected as biases and erroneously removed. In all such efforts, a step back to previous versions of time series must be provided.

2.2 *Overview of Upper-air Temperature Trend Patterns*

In this section, we describe main features of upper-air temperature trends, which are shown to be sensitive to many factors, so that it is recommended to use an aggregate measure of trend estimates, especially for the troposphere where the trend signal is much weaker and less certain than the signal of stratospheric cooling.

Since the early publications based on radiosonde data processing, two main features of trends in the upper-air temperature series were reported:

- A strong cooling in the lower stratosphere
- A warming in the troposphere, which is, however, not as evident as the stratospheric cooling

The numerical values of these trends, as well as standard errors of these values, are shown to be very sensitive to several factors. The list of these factors, extensively discussed by Gaffen et al. (2000), and later by Sterin (2004b), includes:

- Choice of radiosonde observational network (for radiosonde data)
- Method of primary data processing
- Method of inhomogeneity detection, estimation, and elimination (if any) that was applied before trend estimation (for radiosonde data and for satellite data)
- Exact period of time series taken for trend calculation
- Statistical techniques used for trend calculation (using traditional least squares regression method or robust regression estimates – see Huber (1981), Hoaglin et al. (1983), Lanzante (1996), Yohai (1987))

The time series beginning date usually taken is the late 1950s – early 1960s (beginning of most regular global observations), or 1979 (beginning of satellite series). Trends are usually calculated in aggregation for the globe, for hemispheres, as well as for large latitude belts.

In the middle of the 1990s the controversy between surface temperature and tropospheric temperature trends was noted. The tropospheric temperature trends estimated from UAH satellite data were shown to be negative (about −0.03 K/decade for the globe), while the warming on the surface was shown to be much larger (about + 0.21°C/decade for the globe for the period since 1979). Note that radiosonde troposphere trends for series beginning in the early 1960s demonstrated warming which, however, was mainly attributed to steplike warming in middle of 1970s rather than to gradual temperature increase, therefore the controversy considered just the trends for satellite era (beginning in 1979). This phenomenon triggered a vast discussion on possible reasons for this difference (see, e.g., Hurrell and Trenberth 1998, for a detailed description and possible explanations for this discrepancy, including both physical differences between the thermal processes on the surface and in the troposphere, different effects of ocean versus land on satellite and in situ surface temperature, and some details of MSU data processing, which could,as artifacts, affect the trend values in the series). Seidel et al. (2004) contains a vast intercomparison of trend values in several well-known radiosonde and satellite MSU/AMSU upper-air temperature series.

Figure 1, taken from Seidel et al. (2004), demonstrate the main patterns of upper-air temperature trends, their dependency on source of data, period of trend estimate, as well as the degree of uncertainty in their values due to different sources

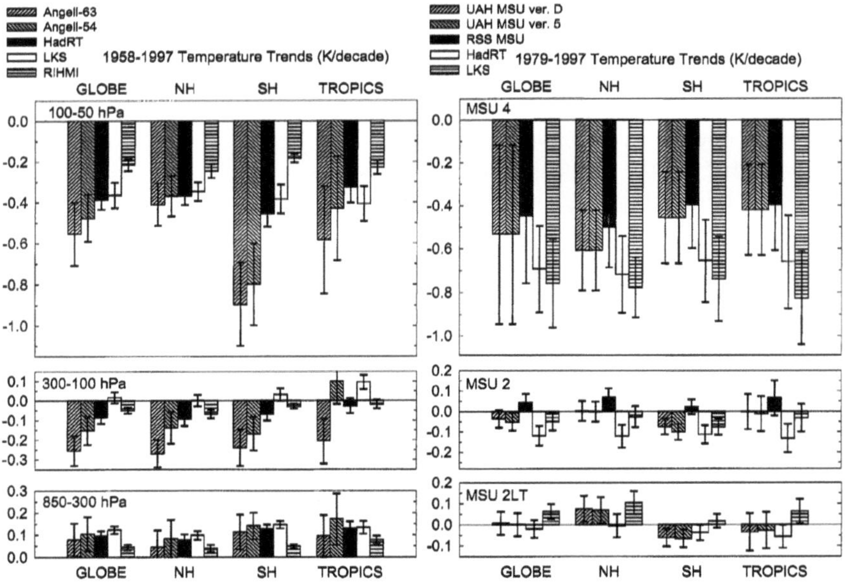

Fig. 1 Temperature trends (K/decade) in the series of upper-air temperature anomalies for the globe, both hemispheres, and the tropical zone (20°S–20°N), as calculated from various sources of radiosonde and satellite data (Seidel et al. 2004). Left: trends for the 1958–1997 period for the vertical layers 100–50 hPa, 300–100 hPa and 850–300 hPa, right: trends for the 1979–1997 period for MSU4, MSU2 and MSU2LT satellite data as well as for radiosonde data recalculated to these layers. The confidence intervals shown are the ±1 standard error uncertainty estimates. (Copyright 2004 American Meteorological Society).

of data and processing procedures. Note that, among the sensitivity factors listed above, choice of period of trend calculation has a large impact on conclusions of trend values. Figure 1 contains trend values for series ending in 1997 (the latest year for which all compared series were available at that time). However, the addition of only one year (1998, year of strong ENSO) to the series beginning in 1979 makes a change in tropospheric trends, in particular, for UAH data, that is a switch from slight negative trend to slight positive trend (however, both are not significant, Sterin 2004a, 2006). In the early 2000s, other research groups (Mears et al. (2003), from RSS, and Vinnikov and Grody (2003)) independently processed the original MSU/AMSU data and derived satellite temperature series which, unlike the UAH series, yielded troposphere temperature trends that were in better agreement with the surface warming.

Considering essential differences of trends in the upper-air temperature, Seidel et al. (2004) recommend that, for trend estimation, an aggregate measure based on different data sets should be taken. In contrast to trends, some other large-scale climate signals, such as the Southern Oscillation, the Quasi-Biennial Oscillation in the stratosphere, post-volcanic stratospheric warming episodes, or the steplike tropospheric warming in 1976/77, which can be evaluated from the same upper-air temperature series, are shown to be manifested in a more similar way in all data sources.

Santer et al. (2005) studied scaling ratios of upper-air to surface temperature characteristics in the tropics for month to month and annual variability and for interdecadal variability. For this study, the outputs of 19 models used for IPCC Fourth Assessment Report, radiosonde and satellite data, surface data, as well as theoretical considerations, were jointly used. In studies of the expected amplification of warming from surface to troposphere, the main conclusion is that the source of discrepancies in trends lies in the radiosonde and UAH satellite data, which are consistent with models and with basic thermodynamic considerations on monthly and annual scales, but are inconsistent with them on an interdecadal scale due to residual errors. Unlike other data sources mentioned in this paper, the RSS satellite data are shown to be consistent with theoretical considerations and with model outputs on interdecadal scales as well.

An additional comparison of global upper-air temperature trends calculated from different sources is contained in Table 1 along with additional trend estimates for surface temperature (see Sterin 2006). This table additionally demonstrates two details that could be of interest:

1. The negative trend in the stratosphere is considerably smaller for the 1979–2005 period compared to the period 1979–1998. Possible reasons for this could be the cessation of ozone depletion in early 2000s and the effect of sensitivity of stratospheric temperature to water vapor concentrations (de Forster and Shine 1999).
2. The UAH series trends for the troposphere for 1979–2005 depend on the statistical methods used. While for these series a trend of +0.05 K/decade is found on the basis of standard least-squares technique, four different robust regression trend techniques, independently applied, demonstrate trend estimates ranged from +0.12 to +0.14 K/decade, significantly different from zero. This is closer to the global troposphere trend estimates based on satellite series of other

Table 1 Comparison of root-mean-square values of temperature anomalies (σ, in K), lag 1 (1 month) autocorrelation values ($\tau(1)$), and trend values (K/decade), for the global series of temperature anomalies in the troposphere (subscript TR) and in the lower stratosphere (ST) and for the surface temperature (SF).

			Trend, K/decade				
Series	σ	τ (1)	OLS	M	MM	S	LTS
RIH_{TR} (1979–2005)	0.13	0.69	0.04 (0.03)	0.03	0.03	0.03	0.03
JON_{SF} (1979–2005)	0.18	0.66	0.17 (0.03)	0.17	0.17	0.17	0.17
UAH_{TR} (1979–2005)	0.20	0.69	0.05 (0.04)	0.12	0.11	0.11	0.14
RSS_{TR} (1979–2005)	0.23	0.77	0.19 (0.05)	0.19	0.19	0.19	0.22
RIH_{ST} (1979–1998)	0.33	0.88	−0.43 (0.10)	−0.44	−0.43	−0.43	−0.41
UAH_{ST} (1979–1998)	0.50	0.81	−0.54 (0.17)	−0.57	−0.56	−0.55	−0.52
RSS_{ST} (1979–1998)	0.45	0.96	−0.50 (0.19)	−0.44	−0.43	−0.43	−0.40
RIH_{ST} (1979–2005)	0.39	0.86	−0.40 (0.06)	−0.40	−0.38	−0.37	−0.36
UAH_{ST} (1979–2005)	0.49	0.81	−0.42 (0.10)	−0.40	−0.39	−0.38	−0.34
RSS_{ST} (1979–2005)	0.43	0.95	−0.36 (0.11)	−0.29	−0.27	−0.26	−0.22
RIH_{TR} (1958–2005)	0.14	0.73	0.05 (0.01)	0.05	0.05	0.05	0.06
JON_{SF} (1958–2005)	0.22	0.76	0.12 (0.02)	0.12	0.12	0.12	0.14
RIH_{ST} (1958–2005)	0.41	0.86	−0.26 (0.03)	−0.26	−0.26	−0.26	−0.26

Trends are estimated by the following regression methods: OLS: Ordinary least squares, with standard error of trend estimates considering the autocorrelations in series given in parenthesis; M: Robust M method (Huber 1981); MM: Robust MM method (Yohai 1987); S: Robust S method (Rousseeuw 1984); LTS: Robust Least Trimmed Squares method (Rousseeuw and Yohai 1984).
The source of series are RIH: RIHMI-WDC, radiosonde; UAH: University of Alabama, Huntsville, satellite; RSS: Remote Sensing Systems, Inc., satellite; JON: surface temperature global anomalies (Jones et al. 2005). The years in brackets indicate the beginning and ending of the series.

authors (e.g., RSS estimates of +0.19 K/decade for this period and Vinnikov estimates), as well as to the estimates of surface temperature trends for this period (+0.17 K/decade). However, this effect of UAH trend sensitivity to statistical techniques does not remove known general problems related to UAH tropospheric temperature series.

3 Comparisons Between Air Temperature Trends from Radiosonde Data and Reanalyses Outputs

This section aims to provide a comparison between long-period trends in the upper-air temperature obtained from radiosonde data and from reanalysis outputs. A broad description of reanalysis projects and their value for climate study is given by Uppala et al. (this volume), so this general material will not be contained in the paper.

Numerous publications demonstrate that reanalyses reproduce relatively well the inter-annual and high-frequency variations in the atmosphere. However, it is not fully clear if they can reproduce the inter-decadal changes, such as trends. Several publications assess how the reanalyses reproduce upper-air temperature statistics, including long-period trends (Basist and Chelliah 1997, Rubinshtein and Sterin 2002, Khan et al. 2003, Santer et al. 2004).

In one of the early publications on this, Basist and Chelliah (1997) provided comparison between tropospheric temperature series derived from NCAR/NCEP reanalysis (NNR), NCEP operational analysis fields and UAH MSU series, and demonstrated that NNR represents a substantial improvement over the operational analysis for monitoring climate variability. On the other hand, although some recognized problems and possible limitations in the reanalysis temperatures were detected, the comparison between the MSU and reanalysis spatial anomalies showed good agreement.

Santer et al. (2004) provided a vast study of trends in upper-air temperatures obtained from reanalyses, from MSU and UAH satellite data, as well as from the model outputs. The results of this comprehensive intercomparison demonstrate several main points, among them the following are of particular interest for this review:

- The possibility of detecting tropopause height increase from ERA-40 second-generation reanalysis
- The possibility of attributing long-period tropopause height changes to tropospheric and stratospheric temperature changes, which are manifested in a similar way in ERA-40, satellite data, and model output
- An improved consistency of the upper-air temperature statistics, including patterns of trends in upper-air temperature, between second generation ERA-40 reanalysis and MSU data, compared to earlier NCEP-50 reanalysis. The difference between temperature trend spatial patterns obtained from NCEP reanalysis and from other sources studied in the paper is shown to be essential.

Note that the published comparisons between tropospheric and stratospheric temperature trends based on reanalysis outputs and upper-air temperature data involve MSU satellite data more than radiosonde data.

The choice of outputs from reanalysis projects is not limited by ECMWF and NCAR/NCEP projects. For the current moment, several global meteorological centers provide reanalysis output products: ERA-40 (Uppala et al. 2005), NNR (Kalnay et al. 1996), NCEP/DOE (Kanamitsu et al. 2002), GEOS1 (Schubert et al. 1993), JRA-25 (http://www.jreap.org/others/others-e-top.html). However, NNR and ERA-40 outputs are the most popular and most comprehensive, each encompassing a period of 40 years or more. These two reanalyses were chosen in our own study for temperature trend comparison with observed radiosonde air temperature data and with each other.

We keep in mind that both reanalyses include assimilated satellite temperature retrievals from the Microwave Sounding Unit Channel 4 (MSU4) dataset and radiosonde data and hence do not constitute an independent source. However, in regions where observations are sparse, various reanalysis outputs can disagree with observations and with each other, indicating properties of physical consistency of numerical weather prediction model that is used. Note that radiosonde temperature series in no way could be considered as "truth" in any trend comparisons, due to known problems with the radiosonde data homogeneity.

The comparison of observed data and reanalysis outputs requires consideration of many possible "dimensions" of such a comparison (vertical and horizontal spatial dimensions, temporal dimensions such as period of trend calculation, etc.). Each researcher who encounters these problems needs to introduce and to use aggregate estimates of similarity in reproducing the trends. Such aggregate estimates of similarity between the trends in the upper-air temperature series, obtained from the radiosonde station data time series, and trends in the series constructed from the reanalysis outputs, are suggested and studied below in this section.

To evaluate quantitatively the generalized measure of consistency between the temperature trends for points of numerous radiosonde stations and numerous levels at each station, calculated from station observational radiosonde data, and trends, calculated from reanalysis (or model) outputs, the Relative Operating Characteristics (ROC) technique (Mason and Graham 2002) has been applied in this study. This score is recommended by the WMO as an obligatory part in standardized verification system of weather forecasts.

The approach of using ROC characteristics is as follows: First, the values of temperature from reanalysis outputs were recalculated to station points by bilinear interpolation from the four surrounding grid points. From these reanalysis colocated values, trends (*b*) were calculated. The values of *b* were grouped into three categories termed:

NEGATIVE (for $b < -1$ SE)
NEUTRAL (for -1 SE $\le b < +1$ SE),
POSITIVE ($b \ge +1$ SE).

SE here denotes the standard error of the trend at given level and a given station, calculated from reanalysis outputs. Categories were determined as above for values of trends and their standard deviations calculated from station data.

Second, a 3 × 3 contingency table was constructed by comparing the categories for the observational data with the reanalysis data, and Hit Rates and the False Alarm Rates (HR and FAR) were determined. Third, HR and FAR were calculated for each relative sample frequency P_n, which is defined as the ratio of sample size to the total number of observations in the row. The graph of HR_n versus FAR_n forms the ROC curve. The normalized area under the ROC curve represents a summary statistics of a verified data set. A perfectly verified system has an area of 1.0, and, in contrast, a no-skill set is indicated by an area less than or equal to 0.5. For each category, the percentages of stations with ROC score above 0.5 and less than or equal to 0.5 were determined.

Special attention has been paid to definition of periods for calculation of trends. We have taken into consideration two different periods: from 1964 to 1998 and from 1979 to 1998. The first period begins from 1964, because the radiosonde observations started to be more reliable since this date. The second period represents the satellite era and coincides with the Atmospheric Models Intercomparison Project (AMIP-2) period.

For intercomparing NNR and ERA-40, we used the 1958–2000 period (see Fig. 3). Such a comparison enables a better understanding of the difference between the two most popular assimilations systems in reproducing long-term tendencies of air temperature. Vertical profiles of zonally averaged trends of monthly mean temperature for both periods 1979–1998 and 1964–1998 from NNR output and from radiosonde station temperature data set were analyzed in Khan et al. (2003).

Table 2 contains the aggregated ROC scores for air temperature trends, first using ERA-40 and second with the NNR data. Considering the extended period from 1964 to 1998, it is shown that, for categories of negative or positive trend values, the trend estimates from both sources have significantly higher ROC scores than those related with neutral trend category. The NNR scores are slightly lower than the ERA-40 scores. For period 1979–1998, the ROC values for category of negative trends are significantly larger then those for positive and neutral trend categories over the tropics and the southern extratropics. Most of the negative trends are associated with a low stratosphere cooling. In total, for both periods, values of ROC scores are similar, while being smaller for the second period for the southern extratropics.

The spatial distribution of ROC values between air temperature trends from ERA-40 reanalysis and station observed data, calculated separately at each station of the global network, is shown in Fig. 2. ERA-40 reproduces the upper-air temperature trends well in the high latitudes of both hemispheres, especially over Western Europe and the USA, where the density of radiosonde stations is maximal. Over the territory of Russia, a mixed picture is observed: the point stations with high values of

Table 2 Aggregated ROC scores between upper-air temperature trends from outputs of ECMWF reanalysis, outputs of NCEP/NCAR reanalysis, and station monthly statistics based on observation data

Station data vs	Negative	Neutral	Positive	All categories
Northern extratropics (20°N–90°N), 1964–1998				
ECMWF	0.79	0.61	0.79	0.78
NCEP/NCAR	0.75	0.57	0.76	0.74
Tropical area (20°S–20°N), 1964–1998				
ECMWF	0.81	0.58	0.79	0.64
NCEP/NCAR	0.73	0.59	0.71	0.60
Southern extratropics (90°S–20°S), 1964–1998				
ECMWF	0.84	0.62	0.75	0.81
NCEP/NCAR	0.80	0.60	0.68	0.77
Northern extratropics (20°N–90°N), 1979–1998				
ECMWF	0.77	0.67	0.78	0.74
NCEP/NCAR	0.77	0.60	0.68	0.75
Tropical area (20°S–20°N), 1979–1998				
ECMWF	0.75	0.58	0.60	0.67
NCEP/NCAR	0.67	0.60	0.59	0.62
Southern extratropics (90°S–20°S), 1979–1998				
ECMWF	0.82	0.63	0.54	0.74
NCEP/NCAR	0.79	0.61	0.55	0.65

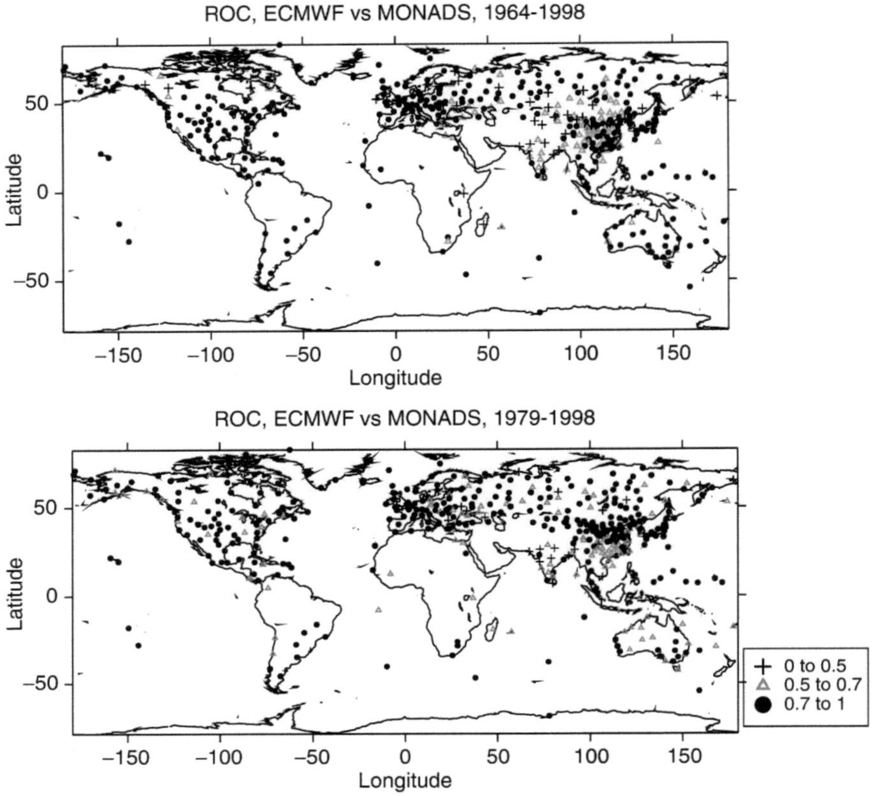

Fig. 2 Geographical distribution of grouped ROC scores between air temperature trends calculated from ECMWF reanalysis and station radiosonde monthly (MONADS) data.

ROC scores are mixed in with locations where the ROC scores are rather poorly satisfactory. The patterns of lowest values of ROC scores are observed in the regions of Southeast Asia, Tibet, and in the Himalayas. This may be due to complex topography of the region and to the low quality and insufficient quantity of the radiosonde data in the region.

In the southern hemisphere, the number of radiosonde stations is quite limited, except Australia. In Australia, the satisfactory agreement between trends in comparison is confirmed for extended period of observations from 1964 to 1998. For the shorter period of 1979–1998, an improvement in simulation of trends by ERA-40 is noted in Southeast Asia. But in the same period, a decreasing of number of stations with satisfactory agreement of trends can be observed in Australia. This fact is confirmed by previous results of authors (Khan et al. 2003; Rubinshtein and Sterin 2002), where the discrepancies between air temperature in the free atmosphere were analyzed.

Figure 3 shows latitude-height sections of trends calculated from 10° zonally averaged air temperature data for period from 1958 to 2000 for two reanalyses.

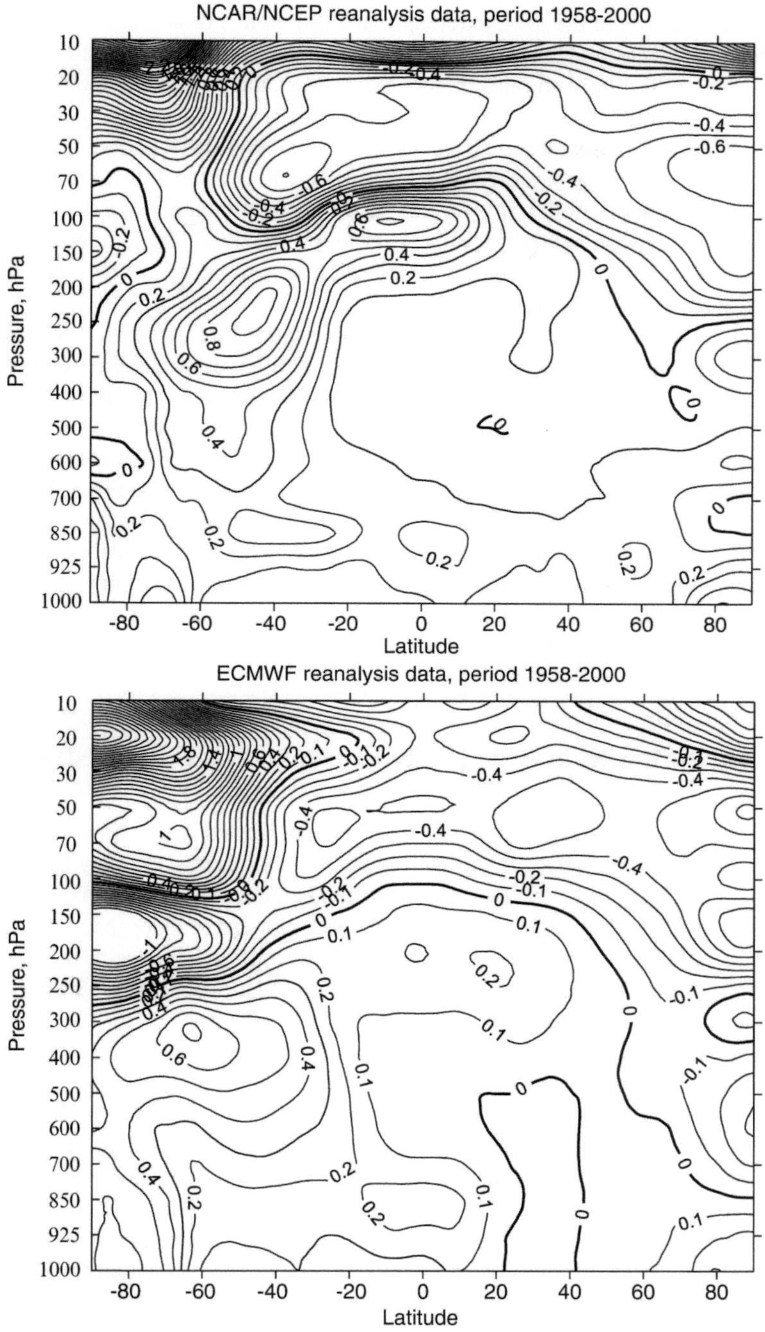

Fig. 3 Latitude-height sections of air temperature trends (K/decade, period of 1958–2000) with 10° zonal averaging for NCAR/NCEP reanalysis (top) and ECMWF reanalysis (bottom).

Widespread cooling of the lower stratosphere throughout a broad region is confirmed in both reanalysis outputs. The cooling minima from ERA-40 is located in the south polar region at 200 hPa with rates of about −1.0 K/decade. Enhanced cooling in NNR is found at the southern tropics near ~30°S at 70 hPa.

The results of Khan et al. (2003) show that, for the 1964–1998 period, the minimum in stratospheric cooling in the NNR output is −0.8 K/decade. From the radiosonde data, the stratospheric cooling rate for this period is slightly higher. For the 1979–1998 period, the lower stratospheric cooling is more pronounced: the temperature trends reach −1.3 K and −1.4 K/decade along the latitude section, from the NNR outputs and radiosonde data, respectively.

In Fig. 3 the transition from warming to cooling (highlighted by the zero isoline) is at a lower altitude in ERA-40 than in NNR. The lower-stratospheric cooling in the southern hemisphere is more pronounced in ERA-40. The trend estimates from both reanalysis sources demonstrate positive trends of monthly mean temperature in the southern middle troposphere with warming rates up to 0.8 K/decade. However, we need to take into account the fact that in this region the observational radiosonde network density is very low. The fewest discrepancies in the temperature trends from the two types of data are found in the free atmosphere in the northern hemisphere, where the radiosonde network density is high. Both reanalyses reveal low troposphere warming up to 0.2 K/decade at very high latitudes of the northern hemisphere. This warming is more pronounced from NNR outputs than from ERA-40 outputs.

4 Concluding Remarks: Some Problems for Future Studies

In this paper we concentrated on the temperature trends in the free atmosphere. High sensitivity of the values of climate signals, such as trends, to many factors, such as period and exact length of the series, statistical techniques of analysis, etc., makes it necessary to provide more detailed analyses. Wider groups of independent researchers should be involved in this analysis, and explanation of differences in the results of their analyses may become a key to better understanding of the upper-air temperature processes related to climate.

Several directions of future studies of upper-air temperature trends can be outlined. The first is to provide a more detailed study of trends in upper-air temperature, involving more complete databases, longer series, radiosonde and satellite series for comparison, as well as various statistical techniques for trend calculations.

The second is to study the long-period tendencies in intra-annual and intra-seasonal variability in the atmosphere (i.e., trends in measures of variability in addition to trends in "averaged" status of temperature). This is an important issue in context of extreme phenomena statistics for climate studies. Only a few papers consider this important issue for the upper-air temperature variability (e.g., Iskenderian and Rosen 2000).

The third is to extend the intercomparison, that was currently performed only between the observed trends in upper-air temperature and the trends obtained from few reanalysis outputs, with some involvement of model outputs. More wide intercomparison should involve additional reanalysis outputs (such as JMA reanalysis), as well as wide set of Global Circulation Models (GCM) outputs. Such a triple intercomparison (observational data, reanalysis, and GCM outputs) could give a more sound answer to the question of whether reanalysis and climate models can reproduce trends in the atmosphere. For the GCMs, the results of such triple intercomparison will constitute a new important General Circulation Model (GCM) output diagnostic characteristic, which can be used in the atmospheric model intercomparison experiments. As a preliminary effort, the authors used this diagnostic characteristic for two different Russian GCMs (Dianski and Volodin 2002; Rubinshtein and Egorova 2000). The main results of this effort, as being more oriented to GCM specialists, will be published separately.

The fourth direction is to consider the other meteorological parameters in the free atmosphere measured by radiosonde observations, which surely also deserve attention in the context of analyses suggested above for the upper-air temperature.

Acknowledgments Partial support by the NATO ESP CLG grant 981842 and RFBR grants by Russian Basic Research Foundation, Projects 04-05-64681, 06-05-64104, 04-05-64151, 05-05-08025, 07-05-00740, 07-05-00264 is highly appreciated. Authors appreciate the decisive recommendations of Dr. Stefan Bronnimann and of an anonymous reviewer that improved the paper, and efforts of Andrea Grant in editing the English text.

References

Basist, A. N. and M. Chelliah, 1997: Comparison of tropospheric temperatures derived from the NCEP/NCAR Reanalysis, NCEP operational analysis, and the Microwave Sounding Unit. *Bull. Am. Meteorol Soc.*, **78**, 1431–1447.

Christy, J. R., R. W. Spencer, and W. D. Braswell, 2000: MSU tropospheric temperatures: dataset construction and radiosonde comparisons. *J. Atmos. Ocean Tech.*, **17**, 1153–1170.

de Forster, P. M. and P. K. Shine, 1999: Stratospheric water vapour changes as a possible contributor to observed stratospheric cooling. *Geophys. Res. Lett.*, **26**, 3309–3312.

Dianski, N. A. and Volodin YeM, 2002: The reproducibilityof modern climate by model of atmosphere-ocean general circulation model (in Russian). *Izvestiya RAN, Fizika Atmospheri I Okeana*, **38**, 824–840.

Durre, I., R. S. Vose, and D. B. Wuertz, 2006: Overview of the integrated global radiosonde archive. *J. Climate*, **19**, 53–68.

Free, M. et al., 2002: Creating climate reference datasets. CARDS workshop on adjusting radiosonde temperature data for climate monitoring. *Bull. Am. Meteorol Soc.*, **83**, 891–899.

Free, M. and D. J. Seidel, 2005: Causes of differing temperature trends in radiosonde upper air data sets. *J. Geophys. Res.*, **110**, D07101.

Gaffen, D. I., M. A. Sargent, R. E. Habermann, and J. R. Lanzante, 2000: Sensitivity of tropospheric and stratospheric temperature trends to radiosonde data quality. *J. Climate*, **13**, 1776–1796.

Haimberger, L., 2005: Homogenization of radiosonde temperature time series using ERA-40 analysis feedback information. ERA-40 Project Report Series, No. 22, 73 pp.

Haimberger, L., 2007: Homogenization of radiosonde temperature time series using innovation statistics. *J. Climate*, **20**, 1377–1403.

Henderson-Sellers, A. et al., 1998: Tropical cyclones and global climate change: a post-IPCC assessment. *Bull. Am. Meteorol Soc.*, **79**, 19–38.

Hoaglin, D., F. Mosteller, and J. Tukey, 1983: *Understanding Robust and Exploratory Data Analysis*. Wiley, New York, 447 pp.

Huber, P. J., 1981: *Robust Statistics*, Wiley, New York.

Hurrell, J. W. and K. E. Trenberth, 1998: Difficulties in obtaining reliable temperature trends: reconciling the surface and satellite microwave sounding unit records. *J. Climate*, **11**, 945–967.

IPCC, 1996: *Climate Change 1995: The Science of Climate Change*. Contribution of Working Group I to the Second Assessment Report of the Intergovernmental Panel on Climate Change, J. T. Houghton, L. G. Meira Filho, B. A. Callander, N. Harris, A. Kattenberg, and K. Maskell, eds., Cambridge University Press, Cambridge, UK/New York, 572 pp.

IPCC, 2001: *Climate Change 2001: The Scientific Basis*. Contribution of Working Group 1 to the Third Assessment Report of the Intergovernmental Panel on Climate Change. Cambridge University Press, Cambridge, UK/New York, 881 pp.

Iskenderian, H. and R. D. Rosen, 2000: Low-frequency signals in midtropospheric submonthly temperature variance. *J. Climate*, **13**, 2323–2333.

Kalnay, E. et al., 1996: The NCEP/NCAR 40-year reanalysis project. *Bull. Am. Meteorol Soc.*, **77**, 437–471.

Kanamitsu, M., W. Ebisuzaki, J. Woolen, J. Potter, and M. Fiorino, 2002: NCEP–DOE AMIP-II reanalysis (R-2). *Bull. Am. Meteorol Soc.*, **83**, 1631–1643.

Khan, V. M., A. M. Sterin, and K. G. Rubinshtein, 2003: Estimates of upper-air temperature trends from the reanalysis and radiosonde data (in Russian). *Meteorologia and Gidrologia*, **12**, 5–18.

Lanzante, J. R., 1996: Resistant, robust and non-parametric techniques for the analysis of climatic data. Theory and examples, including the applications to historical radiosonde station data. *Int. J. Climateatol*, **16**, 1197–1226.

Lanzante, J., S. Klein, and D. Seidel, 2003a: Temporal homogenization of monthly radiosonde temperature data. Part I: methodology. *J. Climate*, **15**, 224–240.

Lanzante, J., S. Klein, and D. Seidel, 2003b: Temporal homogenization of monthly radiosonde temperature data. Part II: trends, sensitivities and MSU comparison. *J. Climate*, **15**, 241–262.

Mason, S. J. and N. E. Graham, 2002: Areas beneath the relative operating characteristics (ROC), and levels (ROL) curves: statistical significance and interpretation. *Q. J. Roy. Meteorol Soc.*, **128**, 2145–2166.

Mears, C. A., M. C. Schabel, and F. J. Wentz, 2003: A reanalysis of the MSU channel 2 tropospheric temperature record. *J. Climate*, **16**, 3650–3664.

Rousseeuw, P. J., 1984: Least median of squares regression. *J. Am. Stat. Assoc.*, **79**, 871–880.

Rousseeuw, P. J. and V. Yohai, 1984: Robust regression by means of S estimators. In: *Robust and Nonlinear Time Series Analysis, Lecture Notes in Statistics*, **26**, J. Franke, W. Härdle, and R. D. Martin, eds., Springer, Berlin Heidelberg, NY, 256–274.

Rubinshtein, K. G. and A. M. Sterin, 2002: Comparison of the reanalysis results with aerological data (in Russian). Izvestiya RAN, Fizika Atmospheri I Okeana, **38**, 301–315.

Rubinshtein, K. G. and Egorova Ye. N, 2000: The estimates of annual cycle reproducibility for the land surface and for the atmosphere byGCM (in Russian). *Trudy Gidrometcentra Rossii*, **333**, 41–98.

Santer, B. D. et al., 2004: Identification of anthropogenic climate change using a second-generation reanalysis. *J. Geophys. Res.*, **109**, D21104.

Santer, B. D. et al., 2005: Amplification of surface temperature trends and variability in the tropical atmosphere. *Science*, **309**, 1551–1556.

Seidel, D. J. et al., 2004: Uncertainty in signals of large-scale climate variations in radiosonde and satellite upper-air temperature datasets. *J. Climate*, **17**, 2225–2240.

Schubert, S. D., R. B. Rood, and J. Pfaendtner, 1993: An assimilated dataset for Earth science applications. *Bull. Am. Meteorol Soc.*, **74**, 2331–2342.

Spencer, R. W. and J. R. Christy, 1990: Precise monitoring of global temperature trends from satellites. *Science*, **247**, 1558–1562.

Sterin, A. M., 2004a: Sensitivity of the estimates of temperature trends in the troposphere and lower stratosphere from radiosonde data: 2. Detection of inhomogeneities in monthly-resolution series (in Russian). *Meteorologia and Gidrologia*, **6**, 5–22.

Sterin, A. M., 2004b: Sensitivity of the estimates of temperature trends in the troposphere and lower stratosphere from radiosonde data: 1. Selection of data set, time series length and analysis method (in Russian). *Meteorologia and Gidrologia*, **5**, 21–36.

Sterin, A. M., 2006: Climatic changes of the upper-air temperature: problems of empirical studies (in Russian). Vichislitelnye Technologii 11, Special issue, part 3, 6–15.

Thorne, P. W., D. Parker, S. F. B. Tett, P. D. Jones, M. McCarthy, H. Coleman, P. Brohan, and J. R. Knight, 2005: Revisiting radiosonde upper-air temperatures from 1958 to 2002. *J. Geophys. Res.*, **110**, D18105, doi:10.1029/2004JD005753.

Uppala, S. M. et al., 2005: The ERA-40 re-analysis. *Q. J. Roy. Meteor Soc.*, **131**, 2961–3012.

Uppala, S., A. Simmons, D. Dee, P. Kållberg, and J.-N. Thépaut, 2007: Atmospheric reanalyses and climate variations (this volume).

Vinnikov, K. and N. Grody, 2003: Global warming trend of mean tropospheric temperature observed by satellites. *Science*, **302**, 269–272.

Yohai, V. J., 1987: High breakdown point and high efficiency robust estimates for regression. *Annals of Statistics*, **15**, 642–656.

Atmospheric Reanalyses and Climate Variations

S. Uppala[1], A. Simmons[1], D. Dee[1], P. Kållberg[2], and J.-N. Thépaut[1]

Abstract The Earth's climate has traditionally been studied by statistical analysis of observations of particular weather elements such as temperature, wind and rainfall. Climatological information, usually expressed as long-term averages and variability, is then presented over a geographical area or at a single location and time series of these quantities or of the observations themselves are examined for evidence of warming, more-frequent severe storms, and so on.

A powerful new approach to climate analysis has emerged in recent years. It applies the tools and techniques of modern everyday weather forecasting in a process called reanalysis. The products, reanalyses, have applicability far beyond that of traditional climate information. Reanalyses have become established as an important and widely utilized resource for the study of atmospheric and oceanic processes and predictability also over the data sparse polar regions. They are used in a range of applications that require a comprehensive record of the state of either the atmosphere or its underlying land and ocean surfaces. The reanalysis products, unlike their operational counterparts, do not suffer from inhomogeneities introduced by changes in the data assimilation system. Thus they are in principle better suited for use in studies of low frequency variability and climate trends that complement studies of climate change based on individual instrumental records and climate-model simulations.

Climate quality requirements can be met by reanalyses for the decades with good upper-air data coverage by satellites or at least radiosonde data. The possibility of extending reanalyses to cover earlier periods when only surface observations are available in reasonable numbers (e.g., from the 1850s to the 1930s) is nevertheless of interest, and has been explored in pilot studies comparing analyses with good coverage of satellite and other upper-air data with analyses using only surface-pressure observations.

[1] European Centre for Medium Range Weather Forecasts
Sakari.Uppala@ecmwf.int
Adrian.Simmons@ecmwf.int
Dick.Dee@ecmwf.int
Jean-Noel.Thepaut@ecmwf.int

[2] The Swedish Meteorological and Hydrological Institute
Per.Kallberg@smhi.se

S. Brönnimann et al. (eds.), *Climate Variability and Extremes during the Past 100 Years.*

1 Dynamical Climate Analysis

In operational real-time forecasting systems the latest ground- and satellite-based observations are combined with a short forecast based on earlier observations to create the initial state for a new forecast. The initial state describes the atmosphere and the geophysical properties of land and ocean surfaces. In reanalysis, the observations collected in past decades are supplied to a modern forecasting system, which is configured to make optimal use of the historical data. Atmospheric and surface conditions are reconstructed for each day of the period. Reanalysis differs from the traditional climate approach in that it processes a wide variety of observations simultaneously, using the physical laws embodied in the forecast model and knowledge of the typical errors of forecasts and observations to interpret conflicting or indirect observations and fill gaps in observational coverage. The choice of the data assimilation system is usually limited to the operational scheme at the reanalysis centre, but as will be discussed later there is a growing need to design an analysis scheme suitable for climate purposes. The first two reanalyses were carried out in the USA; at NCEP (National Centers for Environmental Prediction) for the period starting in 1948 and continuing in near-real time, Kalnay et al. (1996), and at NASA for the period 1980–1996, Schubert et al. (1995). These were followed by two European reanalyses, both performed at ECMWF: ERA-15 (1979–1993), Gibson et al. (1997) and ERA-40 (1957–2002), Uppala et al. (2005). Japan also has recently completed a reanalysis of the period 1979–2004, Onogi et al. (2006).

2 Historical Global Observations for Reanalyses

The evolution of the global observing system is shown schematically in Fig. 1 for the last 150 years with a view to the future. Before 1979 there is a marked asymmetry of the "mass" of observations between the two hemispheres, both near the surface and in the free atmosphere. Since 1850, the surface parameters have been observed more regularly so that climatic as well as regional synoptic analyses have been possible. A comprehensive reconstruction of daily mean sea level pressure has been done for the European–North Atlantic region for 1850–2003 by the EMULATE project, Ansell et al. (2006). The exposure of the historical products to the limitations and uncertainties in observed values and the time variations in coverage should not be underestimated, as detailed in the IPCC Third Assessment Report (2001).

Since 1972, after the launch of the first polar orbiting satellite carrying an instrument (VTPR) that sounded through the depth of the troposphere and stratosphere, the atmosphere has been more uniformly and symmetrically measured. In 1979, the year of the global weather experiment (FGGE, WMO 2004), an observing system comprising two polar orbiting satellites with microwave and infrared sounding instruments and five geostationary weather satellites was introduced.

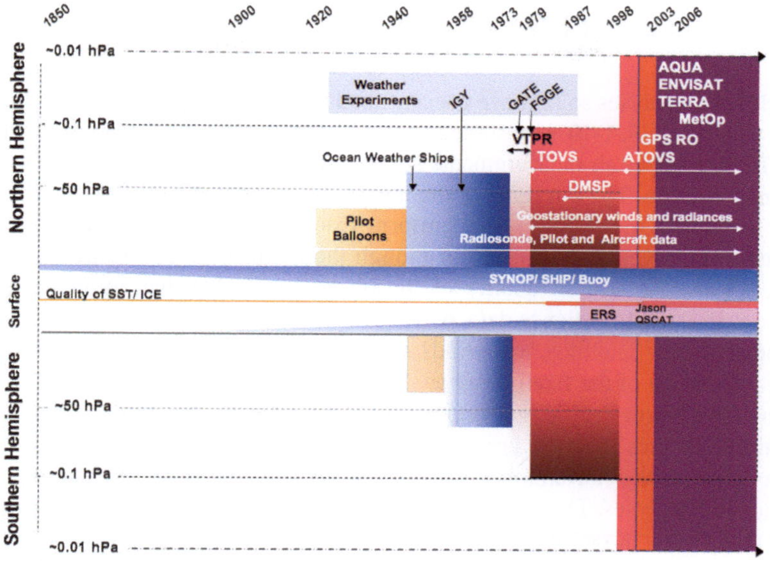

Fig. 1 Observing system evolution presented schematically. Until 1920 only surface observations, which originate from land stations and ships, are available. Pilot balloons (brown) start around 1920 and radiosondes (blue) around 1940 in the northern hemisphere and respectively around 1946 and 1943 in the southern hemisphere. Aircraft data starts in the 1960s, in the northern hemisphere first. The IGY year 1958 marks the beginning of the synoptic conventional global observing system. The era of satellite instruments start in 1972 with VTPR. In 1979 a system of two polar orbiting and five geostationary satellites became a part of the World Weather Watch. Generations of new instruments have been introduced from then on with a significant increase of sounder capabilities from 1998 onwards.

This system proved successful and became a long-term component of the global observing system with further significant upgrades in 1987, 1991 and 1998. In recent years there has been a large increase of advanced scanning and limb radiometers, which not only measure the dynamic and hydrodynamic parameters, but also the greenhouse gases and atmospheric constituents high up to the mesosphere for monitoring and data-assimilation purposes. The evolving observing system, the availability of observations and the analysis methods used form the basis for the quality of the assimilated products, both for traditional climate analysis and for dynamical climate reanalysis. Externally produced sea-surface temperature and sea ice data sets are important components in reanalyses, since they interact through the model physics with the data-assimilation. They have better quality and higher time resolution during the period of satellite observations from 1981 onwards. Accurate sea-surface analyses have been shown to be especially important when limited atmospheric observations are available, Kanamitsu and Hwang (2006). For future reanalyses during the data sparse historical periods this will be a challenge.

3 The Quality of ERA-40 Analyses

The global reanalyses so far have concentrated on the periods with enough digitized upper air data. ERA-40 was started in 1958, the International Geophysical Year (IGY), when upper air measurements at the main synoptic hours (00, 12 UTC) first became available. The coverage of the northern hemisphere by radiosondes is relatively good and to a large extent uniform from 1958 onwards, and their quality gradually improves over time. Even though the number of radiosondes decreased in the 1990s, the overall quality of the observing system has improved due to increases in satellite radiance data, winds deduced from successive images recorded by geostationary satellites, and winds and temperatures from commercial aircraft. The contribution of the fixed weather ships to the observing system over the North Atlantic and North Pacific was important before 1979. In the southern hemisphere and in the Tropics over the large ocean areas, and over the Antarctic, Africa and South America, the conventional observing system alone is not sufficient to produce high-quality analyses. Therefore, the year 1979 brings a more dramatic improvement in the quality of analyses in the southern hemisphere than in the northern hemisphere.

ERA-40 used the First Guess at Appropriate Time (FGAT) version of the variational 3D-Var analysis method with a six hour analysis window and a T159L60 model. It obtains the analyses at 00, 06, 12 and 18 UTC by minimizing iteratively (until convergence) the distance of the current model state (all model variables) to the observations and to the background (or "first guess"). The background + 9 h forecast starts from the previous analysis and observations ± 3 h around the analysis time are used. The distance to the background is calculated at the time of observation (FGAT), which is especially important for satellite data but also for other asynoptic observations such as aircraft data and the drifting sea-surface instruments.

The quality of both analyses and observations can be inferred from the comprehensive statistics of observation-minus-background and observation-minus-analysis differences (the background and analysis fits to data) that have been archived for each analysis time through the ERA-40 period. Figure 2 shows time series of background and analysis fits to surface-pressure measurements from land stations and ships (included in the 12UTC SYNOP and SHIP reports) for the extratropical northern and southern hemispheres. The background fit in the northern hemisphere is closer, more confined and more stable than in the southern hemisphere. A general improvement in the fit occurs over the period of ERA-40. Changes in data coverage can affect this, as increased coverage in the subtropics, where variance is lower, tends to reduce values in plots such as these in which no area weighting is applied. This probably explains the shift for the southern hemisphere at the beginning of 1967, for example. However, the improvement in the background fit at the end of 1978 is a consequence of the major improvement of the overall observing system mentioned before. It is particularly marked in the southern hemisphere, where a distinct improvement can be seen also at the beginning of 1973 when satellite sounder data (VTPR) were first used. The data fits improve gradually from 1979 onwards, more so for the southern than for the northern hemisphere, and become

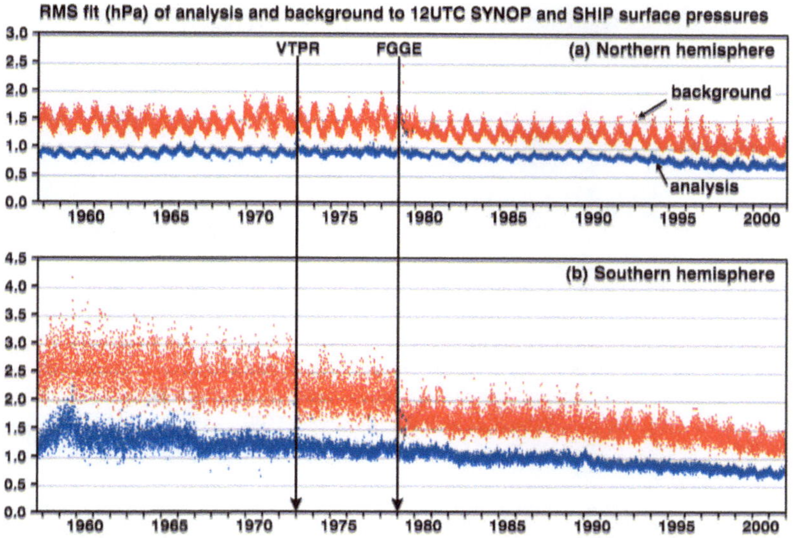

Fig. 2 Daily values of the root-mean-square background (red) and analysis (blue) fits to 12UTC SYNOP and SHIP measurements of surface pressure (hPa) over the extratropical northern (upper) and southern (lower) hemispheres.

similar for the two hemispheres by the end of the period. These and other results presented in Uppala et al. (2005) indicate a general improvement over time of the post-FGGE observing system, especially for the southern hemisphere.

Examination of global mass statistics provides further evidence of improvement over time in the quality of the surface pressure analyses, and indicates a measure of consistency over time of the analysed water content of the atmosphere for the latter half of the reanalysis period (Trenberth and Smith 2005). Prior to 1973 the diagnosed pressure of dry air, which is the difference of mean global surface pressure and mean global water vapour pressure, and should be almost constant in time, shows relatively large fluctuations in ERA-40, with a standard deviation of about 0.3 hPa, and a mean value that is about 0.3 hPa higher than for subsequent years. Fluctuations are much smaller from 1973 onwards, and smaller still from 1979 onwards. Not only is there a degree of consistency over the final two decades of ERA-40 between the directly analysed annual cycle in net atmospheric water-vapour content and the analysed annual cycle in global-mean surface pressure, there is also a degree of consistency in the analyses of longer-term variations. Trenberth and Smith (2005) include comparisons with the NCEP/NCAR reanalyses. The results from ERA-40 are clearly superior from 1973 onwards.

The skill of daily 10-day 500 hPa forecasts for the period 1958–2001 is presented in Fig. 3 for the northern hemisphere and for the Australia/New Zealand region, since over the latter area the verifying analysis has sufficient quality over the whole period in the southern hemisphere. The skill reflects the improving

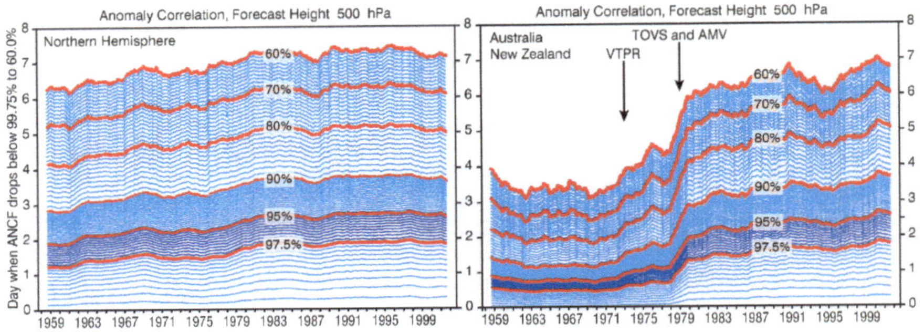

Fig. 3 The skill of daily 500 hPa forecasts in the northern hemisphere (left) and in Australia/New Zealand (right) for 1958–2001. Each curve represents a 2-year moving average of the time in the 10-day forecast, when the anomaly correlation drops below a given percentage value; lines 100–90% with 0.25% interval, from 90% down with 1% interval.

performance of the ERA-40 data assimilation seen in Fig. 2. Over the northern hemisphere the combination of the Ocean Weather Ships (before 1980s), satellites (after 1979) and the land based observations have made it possible to produce good quality forecasts throughout the period. We can see the skill improving slowly until 1997, but during the following years we can see the skill drop slightly. The most likely explanation for the drop is not the deterioration of the observing system, but variation in predictive skill, since similar drop was observed in ECMWF operations in 1999, Simmons and Hollingsworth (2002).

In the southern hemisphere the improvement due to the use of satellite data is clear in 1973 and becomes dramatic in 1979, due to improved satellite observations combined with increasing numbers of aircraft and buoy data.

4 Low-frequency Variability

Prior to the advent of reanalyses, operational analyses produced for weather prediction were used in general circulation studies. However, due to frequent updates of the forecasting system it was difficult to study interannual variability and it was impossible to study climate change. There has also been an increasing need to understand trends in the global atmosphere. In the late 1980s TOVS data, in particular MSU channels, have been used to study trends in the troposphere and the lower stratosphere. These studies have also revealed discrepancies between observations and the operational analyses (see Hurrell and Trenberth 1992 and Oort and Liu 1993). In this sense high expectations were expressed concerning the proposed reanalyses projects in the late 1980s.

In climate research the long-term variations in the quality and coverage of observations need to be taken into account. The same applies to reanalysis products, which not only depend on a variety of input observations but also on the assimilation system including the assimilating atmospheric model. The use of reanalyses for helping to document and understand climatic trends and low-frequency variations is therefore complicated and a matter of some debate. Problems arise partly because the atmospheric models that carry the assimilated observational information forward in time have biases. If observations are abundant and unbiased, they can correct the bias of a background model when assimilated. In reality, however, observational coverage varies over time, observations are themselves prone to bias, and these biases change over time as well. These factors introduce trends and low-frequency variations in analyses that are mixed with the true climatic signals. ERA-40 is no exception, as shown for example by Simmons et al. (2004), who present a variety of evidence pointing to a cold tropospheric bias in the analyses for the extratropical southern hemisphere in the pre-satellite years, contributing to an excessive global-mean warming trend in ERA-40 noted by Bengtsson et al. (2004).

The ERA-40 analyses nevertheless clearly have a role to play in studies of trends and low-frequency variations, especially for the period from 1979 onwards. A feature of the data assimilation is that it is multivariate and that it makes use of all observations and the model-generated background simultaneously in the analysis and in the quality control. The power of data assimilation is demonstrated in an application by Haimberger (2005), who identified and corrected inhomogeneities in historical radiosonde temperature records using the fit of the ERA-40 background forecasts to these data for individual stations. The resulting bias corrections will be used as additional information in future reanalyses and also for direct analysis of trends based on the corrected radiosonde database.

Simmons et al. (2004) compare monthly-mean anomalies in surface air temperature from ERA-40 with values from processed station climate data (see Fig. 4) and discuss some aspects of the vertical consistency of the analyses. The ERA-40 temperature trend agrees very well with the CRU data (Jones and Moberg 2003) from the late

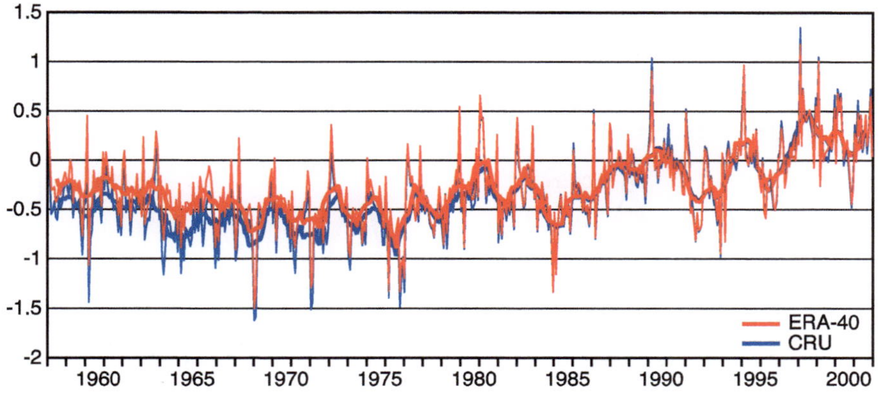

Fig. 4 Trends in 2 m temperature time series represented by temperature anomalies from the 1987–2001 mean for the ERA-40 reanalysis and for the CRU analysis. Monthly means and 12-months running means are shown (Simmons et al. 2004).

1970s onwards, indicating that the quality of the reanalysis temperatures is good. Earlier, however, the ERA-40 temperatures are systematically too warm, the main reason being that not all observations used in the CRU analysis were available for ERA-40.

Santer et al. (2004) show reasonable agreement between estimates of layer-mean tropospheric and lower stratospheric temperature changes from ERA-40 and independent processings of the microwave radiance record (see Fig. 5; Bengtsson et al. 2004). Before 1979 the lack of stratospheric satellite data causes the lower stratospheric analysed temperatures towards the model state and they cannot therefore be used for estimation of trends. Santer et al. (2004) also show consistent evolution of tropopause height between ERA-40 and climate-model simulations that include anthropogenically influenced change. These studies have also shown clear improvement of ERA-40 over the earlier NCEP/NCAR reanalysis. Simmons et al. (2004) also discuss how both reanalyses have helped identify erroneous station data in the near-surface temperature climate record.

The time consistency is important in climate applications. The background forecast used for data assimilation provides a baseline for detecting and correcting some inhomogeneities in observations. In ERA-40 adjustments or corrections were applied to radiosonde temperatures and to all satellite radiances.

Radiosonde temperatures were corrected from 1980 onwards for estimated daytime biases related to radiative heating. As sufficiently comprehensive metadata were not generally available, stations were separated into groups designed to represent different countries or areas assumed to use similar types of sonde at any given time. Mean differences between background forecasts and observations were accumulated for each group of stations over at least 12 months for different classes of solar elevation. Since the background forecast can have a systematic error which does not depend on the solar elevation angle, the mean error for all classes was subtracted from the bias computed for each class to provide a correction for radiation effects. For some station groups the complete bias was removed.

The ERA-40 made use of calibrated Level-1c satellite radiances. During the data assimilation the model equivalent was calculated for each scan position using a

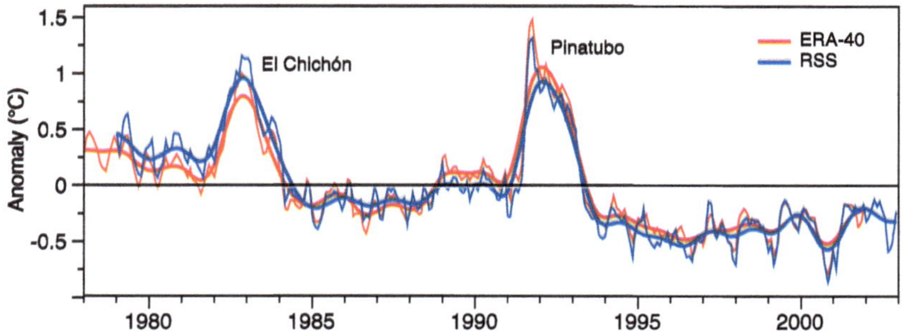

Fig. 5 Time series of global mean monthly anomalies in lower stratospheric temperatures from ERA-40 and from Remote Sensing Systems/ MSU data (Santer et al. 2004).

radiative transfer model. Due to errors in the radiative transfer model, the atmospheric model and the observations, adjustments are needed to remove the systematic scan-angle- and air-mass-dependent errors and satellite-to-satellite inconsistencies. The applied adjustments were stable and usually lasted the lifetime of the satellite. In direct use of satellite radiances the orbital drift has to be taken into account. In data assimilation, the background forecast is defined globally during the analysis window, so that orbital drift is not an issue. Similarly the problem of stratospheric contamination discussed in Fu et al. (2004) is not an issue in the data assimilation, since the background forecast is defined for the whole weighting function and therefore also the tail of weighting function contributes to the analysis.

5 Improving the Reanalyses

Instead of being viewed as a one-off effort, reanalysis has come to be seen as an iterative process, where developments in modelling, data-assimilation techniques, data-rescue efforts and computing power, and data and experience from earlier reanalyses, are utilized to produce successive reanalyses of increasing quality.

ECMWF has recently begun the ERA-Interim reanalysis, Simmons et al. (2007), which will run from 1989 onwards using 4-dimensional variational data assimilation (4D-Var) with a 12 h analysis window and a T255L60 model, which includes significant upgrades, for example, to the physical parametrization and humidity analysis since ERA-40. The ERA-40 observations are used as input and the radiosonde records are homogenized based on Haimberger (2005). A new set of wave height altimeter data is used throughout the period. Variational bias adjustment is applied to satellite radiances. In 4D-Var, the model state at the beginning of the assimilation window is optimized in order to minimize the distance to the background and to the observations together with their error statistics. The minimization is done iteratively by using the forecast model, its adjoint and the sensitivity to the initial state of the mismatch of the forecast state to observations within the analysis window. The analysis in 4D-Var is a snapshot of the final model integration from the optimized model state in the beginning of the analysis window.

Once it reaches the present day, ERA-Interim will continue in close to real time as ECMWF Climate Data Assimilation System (ECDAS). It will serve as an intermediate reanalysis between ERA-40 and the next extended ECMWF reanalysis. To illustrate the evolving data assimilation and modelling capability at ECMWF, the verification of tropical wind forecasts for 1989–1990 from the ECMWF operational forecasts at the time, ERA-15, ERA-40 and a preliminary version of the ERA-Interim system are shown in Fig. 6. We can see, for example, that the data assimilation system today provides a 4.5-day wind forecast with the quality of a 1-day forecast from the operations in 1989.

Clear improvements in the analyses/ forecasts of the tropical winds can be seen from ERA-15 to ERA-40 and from ERA-40 to ERA-Interim. The improved quality of the tropical analysis can also be seen in the global precipitation for year 2000 in Fig. 7.

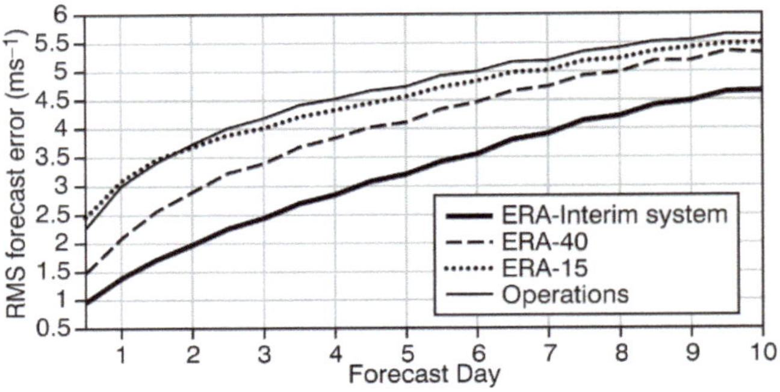

Fig. 6 Tropical wind root-mean-square forecast errors (m/s) at 850 hPa, averaged over forecasts from 12UTC on the 1st and 16th of each month of 1989 and 1990, for the preliminary ERA-Interim system, ERA-40, ERA-15 and the corresponding error from the old ECMWF operational forecasts for 1989–1990. Forecasts from each system are verified against analyses from the same system.

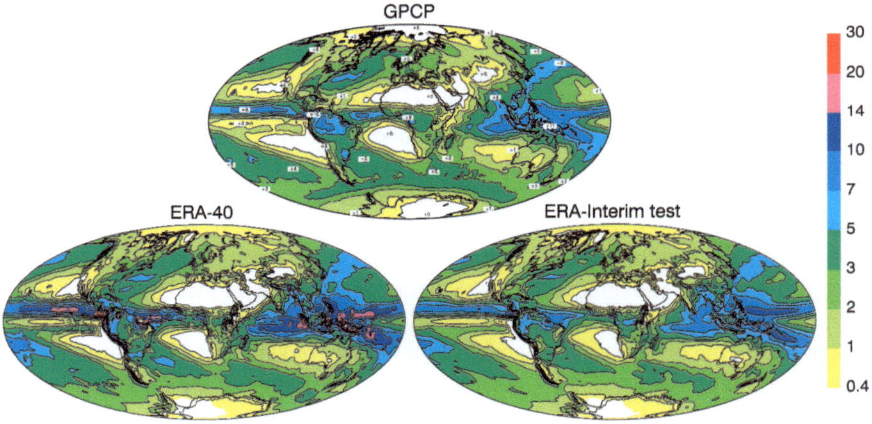

Fig. 7 The mean daily total precipitation (mm/day) for year 2000. GPCP (top), ERA-40 (bottom left) and ERA-Interim test assimilation (bottom right).

The precipitation in ERA-40 over the tropical oceans has been reduced in the ERA-Interim test assimilation and is now closer to the GPCP data. Over the Amazon basin and tropical Africa we can also see an improvement in ERA-Interim. At high latitudes, e.g., Europe, the ERA-Interim precipitation has increased and also here it agrees better with the GPCP estimates. The main improvement to the precipitation comes from the new formulation of humidity analysis and the upgraded model physics. The global hydrological balance, precipitation-minus-evaporation, is much closer to zero in ERA-Interim mainly due to the reduced precipitation over the tropical areas.

6 Data-Assimilation in Historical Periods

The potential of reanalysis is illustrated well by revisiting one of the most severe storms of the 20th century over Europe, which occurred on 31 January/1 February 1953 causing the greatest surge on record for the North Sea as a whole (see Jung et al. 2004, 2005). In the early 1950s the observing system had good coverage from upper air soundings including the ocean weather ships, which were located in areas of critical importance. Moving to the 1940s and further back to the 1930s, the observing system is increasingly confined to the surface and before the1930s it contains surface observations only. For the characterization of circulation patterns for the period 1850–2003, Ansell et al. (2006) have created daily European–North Atlantic mean sea-level pressure fields using a Reduced Optimal Interpolation technique. They used several observational data-sources blended with gridded northern hemispheric analysis fields available from 1881, and they have shown that the earlier periods can indeed be analysed using appropriate statistical methods.

Compo et al. (2006) suggest that there is scope to reanalyse past periods when only surface observations were available, provided the assimilation system is "dynamical" enough (4D-Var or Ensemble Kalman Filter) and adequately tuned towards past observational coverage and quality.

Using the ECMWF data assimilation system the relative merits of a 3D-Var and 4D-Var methods in the presence of few observations have been demonstrated by Thépaut (2006) for the period 4 December 2004–25 February 2005. The surface pressure observations were reduced to those from the stations of the GCOS (Global Climate Observing System) Surface Network (GSN), with buoy data excluded. The CONTROL 3D-Var and 4D-Var assimilations, which used all the conventional and satellite observations that were abundantly available during the period, both performed quite similarly. To account for the degradation of the observing system, the background error covariances used in the assimilation were tuned to reflect the poorer accuracy of analysis and short range forecasts. The sea-surface temperature analyses were not degraded to the quality of the past periods in the "surface pressure only" assimilation.

One of the main findings is the clear superiority of 4D-Var over 3D-Var for the degraded observing system. The 3D-Var analysis can underestimate or even miss low-pressure systems over oceans. Conversely, 4D-Var, for the same cases, captures all the weather patterns present in the CONTROL, even in data-void areas such as the Pacific. The different behaviour between 3D-Var and 4D-Var is even more pronounced at 500 hPa (see Fig. 8). The realism of the 4D-Var analysis as compared with the CONTROL is quite remarkable, and demonstrates the ability of 4D-Var to transfer information aloft in a dynamically consistent way. By exploiting the time dimension of the surface pressure observations, 4D-Var is able to extract surface pressure tendency information and to propagate this information using the model dynamics.

While the 24 h 500 hPa RMS forecast error for the CONTROL 4D-Var assimilation is about 10 m, the RMS error is about 40 m for the "surface pressure only" assimilation

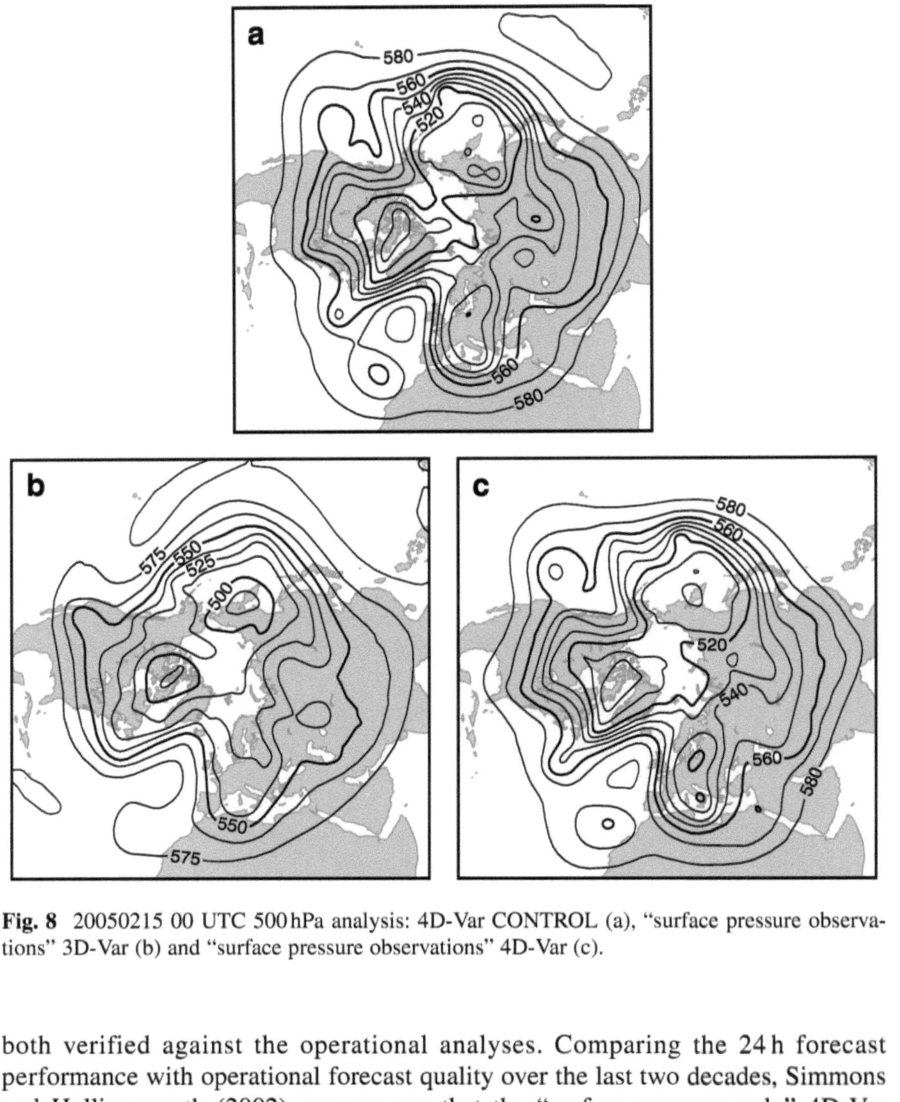

Fig. 8 20050215 00 UTC 500 hPa analysis: 4D-Var CONTROL (a), "surface pressure observations" 3D-Var (b) and "surface pressure observations" 4D-Var (c).

both verified against the operational analyses. Comparing the 24 h forecast performance with operational forecast quality over the last two decades, Simmons and Hollingsworth (2002), we can see that the "surface pressure only" 4D-Var forecasts over the northern hemisphere have about the same skill as the operational 24 h forecasts had during 1981–1983 in the southern hemisphere.

4D-Var, when properly tuned, is able to extract and transfer information from data-dense to data-void areas in a dynamically consistent way, providing reasonable mid- to upper-tropospheric analyses and short-range forecasts. 4D-Var if applied to use regional observations such as described in Ansell et al. (2006) may also be able to create realistic analyses in surrounding areas up- and downstream from the observations. However, to be fully convincing, the experiments presented above should be more directly tailored towards historical periods, in particular to evaluate the breaking point beyond which a sensible dynamical analysis becomes impossible and also the used sea surface temperatures should be more realistic for the period.

7 Summary

Reanalyses have proved very useful in climate applications. Their quality varies in time and depends on all aspects of the data assimilation system, its optimization and the quality of observing system. Applications of reanalyses have to take this into account. New generations of extended "climate reanalyses" are likely to exploit possibilities in data assimilation that are not available in daily operational weather prediction. For example, a reanalysis system can be designed to make use of observations taken over a period after the analysis time as well as before. Utilization of the available observational information may be improved by lengthening the time window in a weak-constraint 4D-Var system. New approaches for handling observational and model biases also show promise for use in reanalysis.

In order to better detect the signal of climate variability, it has been proposed to create a space-and-time homogeneous common observation data set and use this as input for reanalysis. In reality the quality of data such as radiosondes has changed significantly over the years, and even if a common input data could be achieved, the quality would be deteriorated to the quality of reanalyses in the early years if not worse. Instead studies to measure the impact of each major observing system are needed to compare the reanalysis quality in different periods and to better understand the effect of observing system changes on the estimated trends.

In contrast with climate model integrations, reanalyses allow phenomenological studies and quantitative climate variability assessments over the past years. Climate model simulations give time consistent signals from the past to the future under different scenarios. Unlike the reanalysis data-assimilation, these are not affected by artificial discontinuities in the time-varying observing system. Therefore climate simulations and reanalyses are complementary and each serves to validate the other.

Parallel to the direct use of observations in climate studies, reanalyses have a potential during the historical periods to exploit the sparse surface observing system and to extend it to describe the multivariate atmospheric circulation. Here also the two approaches help to validate each other.

Data rescue efforts, such as carried out by NCDC (National Climatic Data Center, USA), NCAR (National Atmospheric Research Center), Brönnimann et al. (2005), are important and especially reanalyses could benefit not only from historical pressure data, but also wind data when available. Even if those data represent different parameters and are scattered in space and time, they can help to anchor the reanalysis with an advanced multivariate data assimilation method such as 4D-Var. The homogenization or reprocessing of the conventional and satellite data records will further improve the quality of the reanalysis products. Parallel to the data recovery it is very important for the reanalyses to obtain high quality sea surface temperature and ice analyses, which are large efforts on their own.

We have seen that by iterating reanalyses and by using more up-to-date methods the quality of reanalyses can measurably be seen to improve. It can be foreseen in the light of the current worldwide use of reanalyses that their importance will grow in the coming climate change assessments.

References

Ansell, T. J. et al., 2006: Daily mean sea level pressure reconstructions for the European North Atlantic region for the period 1850–2003. *J. Climate*, **19**, 2717–2742, doi:10.1175/JCLI3775.1.

Bengtsson, L., S. Hagemann, and K. I. Hodges, 2004: Can climate trends be calculated from reanalysis data? *J. Geophys. Res.*, **109**, D11111, doi:1029/2004JD004536.

Brönnimann, S., G. P. Compo, P. D. Sardeshmukh, R. Jenne, and A. Sterin, 2005: New approaches for extending the twentieth century climate record. *Eos*, **86**, 2–7.

Compo, G., J. S. Whitaker, and P. D. Sardeshmukh, 2006: Feasibility of a 100-year reanalysis using only surface pressure data. *Bull. Am. Meteorol. Soc.*, **87**, 175–190.

Fu, Q., C. M. Johanson, S. G. Warren, and D. J. Seidel, 2004: Contribution of stratospheric cooling to satellite-inferred tropospheric temperature trends. *Nature*, **429**, 55–58.

Gibson, J. K., P. Kållberg, S. Uppala, A. Nomura, A. Hernandez, and E. Serrano, 1997: ERA Description. ECMWF ERA-15 Project Report Series, 1. www.ecmwf.int/ publications.

Haimberger, L., 2005: Homogenization of radiosonde temperature time series using ERA-40 analysis feedback information. ECMWF ERA-40 Project Report Series, 23. www.ecmwf.int/ publications.

Hurrell, J. W. and K. W. Trenberth, 1992: An evaluation of monthly mean MSU and ECMWF global atmospheric temperatures for monitoring climate. *J. Climate*, **5**, 1424–1440.

IPCC (2001) *Climate Change 2001: The Scientific Basis*. J. T. Houghton et al., eds., Cambridge University Press, Cambridge.

Jones, P. D. and A. Moberg, 2003: Hemispheric and large-scale surface air temperature variations: an extensive revision and update to 2001. *J. Climate*, **16**, 206–223.

Jung, T., E. Klinker, and S. Uppala, 2004: Reanalysis and reforecast of three major European storms of the 20th century using the ECMWF forecasting system. I: analyses and deterministic forecasts. *Meteorol. Appl.*, **11**, 343–361.

Jung, T., E. Klinker, and S. Uppala, 2005: Reanalysis and reforecast of three major European storms of the 20th century using the ECMWF forecasting system. II: ensemble forecasts. *Meteorol. Appl.*, **12**, 111–122.

Kalnay, E., M. Kanamitsu, R. Kistler, W. Collins, D. Deaven, L. Gandin, M. Iredell, S. Saha, G. White, J. Woollen, Y. Zhu, M. Chelliah, W. Ebisuzaki, W. Higgins, J. Janowiak, K. C. Mo, C. Ropelewski, J. Wang, A. Leetmaa, R. Reynolds, R. Jenne, and D. Joseph, 1996: The NCEP/ NCAR 40-year reanalysis project. *Bull. Am. Meteorol. Soc.*, **77**, 437–471.

Kanamitsu, M. and S. O. Hwang, 2006: The role of sea surface temperature in reanalysis. *Mon. Wea. Rev.*, **134**, 532–552.

Onogi, K., J. Tsutsui, H. Koide, M. Sakamoto, S. Kobayashi, H. Hatsushika, T. Matsumoto, N. Yamazaki, H. Kamahori, K. Takahashi, S. Kadokura, K. Wada, K. Kato, R. Oyama, T. Ose, N. Mannoji, and R. Taira, 2006: The JRA-25 Reanalysis. *J. Meteorol. Soc., Japan* (submitted).

Oort, A. H. and H. Liu, 1993: Upper-air temperature trends over the globe, 1958–1989. *J. Climate*, **6**, 292–307.

Santer, B. D., T. M. L. Wigley, A. J. Simmons, P. W. Kållberg, G. A. Kelly, S. M. Uppala, C. Ammann, J. S. Boyle, W. Brüggemann, C. Doutriaux, M. Fiorino, C. Mears, G. A. Meehl, R. Sausen, K. E. Taylor, W. M. Washington, M. F. Wehner, and F. J. Wentz, 2004: Identification of anthropogenic climate change using a second-generation reanalysis. *J. Geophys. Res.*, **109**, D21104, doi:10.1029/2004JD005075.

Schubert, S., C.-K. Park, C.-Y. Wu, W. Higgins, Y. Kondratyeva, A. Molod, L. Takacs, M. Seablom, and R. Rood, 1995: A multi-year assimilation with the GEOS-1 system: overview and results. In: *NASA Technical Report Series on Global Modelling and Data Assimilation*, No. 6, M. J. Suarez, ed., NASA, Goddard Flight Center, Greenbelt, MD.
Simmons, A. J., P. D. Jones, V. da Costa Bechtold, A. C. M. Beljaars, P. W. Kållberg, S. Saarinen, S. M. Uppala, P. Viterbo, and N. Wedi, 2004: Comparison of trends and low-frequency variability in CRU, ERA-40 and NCEP/NCAR analyses of surface air temperature. *J. Geophys. Res.*, **109**, D24115, doi:10.1029/2004JD005306.
Simmons, A. J. and A. Hollingsworth, 2002: Some aspects of the improvement in skill of numerical weather prediction. *Q. J. Roy. Meteorol. Soc.*, **128**, 647–677.
Simmons, A., S. Uppala, D. Dee, and S. Kobayashi, 2007: ERA-Interim: new ECMWF reanalysis products from 1989 onwards. *ECMWF Newsletter*, **110**, 25–35. www.ecmwf.int/publications/newsletters.
Thépaut, J. N., 2006: Assimilating only surface pressure observations in 3D and 4DVAR. *Proceedings of the ECMWF/GEO Workshop on Atmospheric Reanalyses*, ECMWF. www.ecmwf.int/publications.
Trenberth, K. E. and L. Smith, 2005: The mass of the atmosphere: a constraint on global analyses. *J. Climate*, **18**, 864–875, doi:10.1175/JCLI-3299.1.
Uppala, S. M., P. W. Kållberg, A. J. Simmons, U. Andrae, V. da Costa Bechtold, M. Fiorino, J. K. Gibson, J. Haseler, A. Hernandez, G. A. Kelly, X. Li, K. Onogi, S. Saarinen, N. Sokka, R. P. Allan, E. Andersson, K. Arpe, M. A. Balmaseda, A. C. M. Beljaars, L. van de Berg, J. Bidlot, N. Bormann, S. Caires, F. Chevallier, A. Dethof, M. Dragosavac, M. Fisher, M. Fuentes, S. Hagemann, E. Hólm, B. J. Hoskins, L. Isaksen, P. A. E. M. Janssen, R. Jenne, A. P. McNally, J.-F. Mahfouf, J.-J. Morcrette, N. A. Rayner, R. W. Saunders, P. Simon, A. Sterl, K. E. Trenberth, A. Untch, D. Vasiljevic, P. Viterbo, and J. Woollen, 2005: The ERA-40 re-analysis. *Q. J. Roy. Meteorol. Soc.*, **131**, 2961–3012, doi:10.1256/qj.04.176.
WMO (2004) *Bull. 2004*, **53**(3).

Constructing Climate Quality Atmospheric Temperatures from Satellite Microwave Measurements

C. A. Mears

Abstract Satellite-borne microwave sounding instruments have been making measurements of the temperature of the Earth's atmosphere for several decades. In order to construct a single atmospheric temperature data set from these measurements, data from a number of satellites must be combined, since each satellite operated only during a small part of the longer time period. If the combined data set is to be of sufficient quality to evaluate changes on the decadal or longer time scales, a number of calibration issues and time-varying biases must be addressed, and their effects removed from the data to the extent possible. Other sources of atmospheric temperature data, such as in situ measurements made by radiosondes and the output of the various reanalysis efforts, have not been demonstrated to be of high-enough quality to validate the satellite data. Because of this, satellite data is typically intercalibrated using a detailed analysis of data from periods of simultaneous operation by two or more satellites. When this type of calibration is used, long periods of simultaneous observation are needed to reduce uncertainties in the calibration procedure.

1 Introduction

The temperature of the Earth's atmosphere can be measured on a global scale using a network of in situ instruments, such as radiosondes, or remotely sensed, using satellite-borne radiometers. In situ measurements have limited spatial coverage, particularly over large areas of the oceans, and are subject to a number of problems related to changing instrument types, configurations, and measurement practices (Lanzante et al. 2003; Thorne et al. 2005). Starting in late 1978, data are available from microwave measurements made by a series of polar-orbiting satellites operated by the United States National Oceanic and Atmospheric Administration (NOAA). Nine Microwave Sounding Unit (MSU) instruments have flown, followed by four

Remote Sensing Systems, Santa Rosa, CA, USA
mears@remss.com

S. Brönnimann et al. (eds.), *Climate Variability and Extremes during the Past 100 Years.*

Advanced Microwave Sounding Unit (AMSU) instruments. Both instruments measure the thermal emission from oxygen molecules, thus inferring the average temperature of a thick layer of the atmosphere. The height of the measured layer is determined by the opacity of the atmosphere for the microwaves being observed, and thus can be chosen by adjusting the microwave frequency (see Sect. 3 for a discussion of the vertical extent of the measurements made by an example channel). The instruments make measurements over a swath several hundred kilometers wide by scanning from side to side to measure the upwelling microwave radiation on either side of the satellite's orbital track.

Since each satellite in the series typically operates for only a few years, data from the various satellites must be merged together to assemble a temperature record for the entire period. The merging procedure is complicated by two issues. First, the instruments in the series suffer from a number of calibration problems and time varying biases that must be addressed during the merging process. Second, other sources of air temperature data are not accurate enough to serve as reference standards to assist in assessing these problems and biases. Instead, scientists inter-calibrate the different instruments by carefully analyzing data collected when two or more instruments were operating at the same time. The differences between these measurements can be studied to characterize the nature of the calibration differences and time varying bias. Once characterized, these effects can be removed from the data, and the data assembled into a single data set suitable for studying long-term changes in the atmospheric temperature.

2 An Example: Tropospheric Temperature

The remainder of this chapter will focus on the data merging process for MSU channel 2 (MSU2), which measures at a frequency of 53.74 GHz, and the corresponding AMSU channel, channel 5 (AMSU5). However, the issues discussed here, and the techniques used to characterize and adjust for their effects are similar for other channels on the MSU instrument, and, in a general sense, to many other types of satellite data. MSU2 primarily measures emission from the troposphere, so it can be loosely thought of as a measure of average tropospheric temperature. Between 80% and 85% of the total weight arises from the troposphere, with about 10% of the emission coming from the stratosphere, and the remaining 5–10% coming from the surface. Since the troposphere is expected to warm in response to anthropogenic greenhouse gases, measurements from this channel have been the focus of considerable attention.

Three groups have offered complete analyses of the entire data set and have arrived at somewhat different conclusions regarding the extent of tropospheric warming over the satellite era (Christy et al. 2003; Mears et al. 2003; Vinnikov et al. 2005). The three groups generally agree about the types of calibration and bias issues that need to be addressed, but differ in the detailed approaches used to characterize and remove these effects. The three most important issues are:

1. Overall offsets between different satellites. Typically the average values measured are different from each other by a few tenths of a degree C.
2. Time-dependent calibration errors that are correlated with the temperature of the reference target used for on-orbit calibration of the instruments. Each scan cycle, the instrument views the reference target, and deep space to perform a two-point calibration of the radiometer response and thus remove slow drifts in the radiometer gain or offset. If the response of the instrument is nonlinear, or the measured temperature of the target contains errors (Grody et al. 2004), this procedure will result in small errors that depend on the temperature of the reference target.
3. Time-dependent biases related to long-term drifts in the local time of day that the measurements are made. The NOAA satellites are in nominally "sun-synchronous" polar orbits that are configured to pass over each latitude at the same local time each day. Drifts in the orbital parameters lead to drifts in this local measurement time over the life of the satellite. If adjustments are not made for this drift, the diurnal (day-night) cycle in atmospheric temperature is aliased into the long-term record (Christy et al. 2000; Mears et al. 2002).

When AMSU data with MSU data are merged, a fourth issue becomes important:

4. Small differences in measurement frequencies and bandwidths lead to small differences in the vertical weighting functions, which leads to location and time-of-year dependent differences in the measured temperatures.

And finally, when we use the weighted difference between measurements made at different view angles to remove the stratospheric weight from the data product, a fifth issue become important.

5. The decay in orbital height over the lifetime of the satellite leads to long-term increases in the earth incidence angle of each measurement that is not made in a straight down direction. Larger incidence angles result in longer radiation paths through the atmosphere, which raise the vertical weighting function slightly in the atmosphere. For the troposphere, this leads to an artificial cooling of the data set that must be removed (Wentz and Schabel 1998).

In Fig. 1, I show the results of the calibration procedure of some of the adjustments listed above applied to data from the NOAA-11 and NOAA-12 satellites, the seventh and eighth satellites in the series. I choose to show this pair because the long time period of overlap makes the results of the adjustments easy to see. The long overlap period also makes this pair one of the pairs for which the adjustments are best characterized. Shorter overlaps make the analysis more challenging and prone to error, and no overlap period, of course, would make it impossible. The adjustments shown here are those found by our group – as noted above, different research groups have used different approaches to perform the needed adjustments, resulting in different final results. In Fig. 2, I show the anomaly time series found by the three groups, as well as the difference time series between these results on an expanded scale.

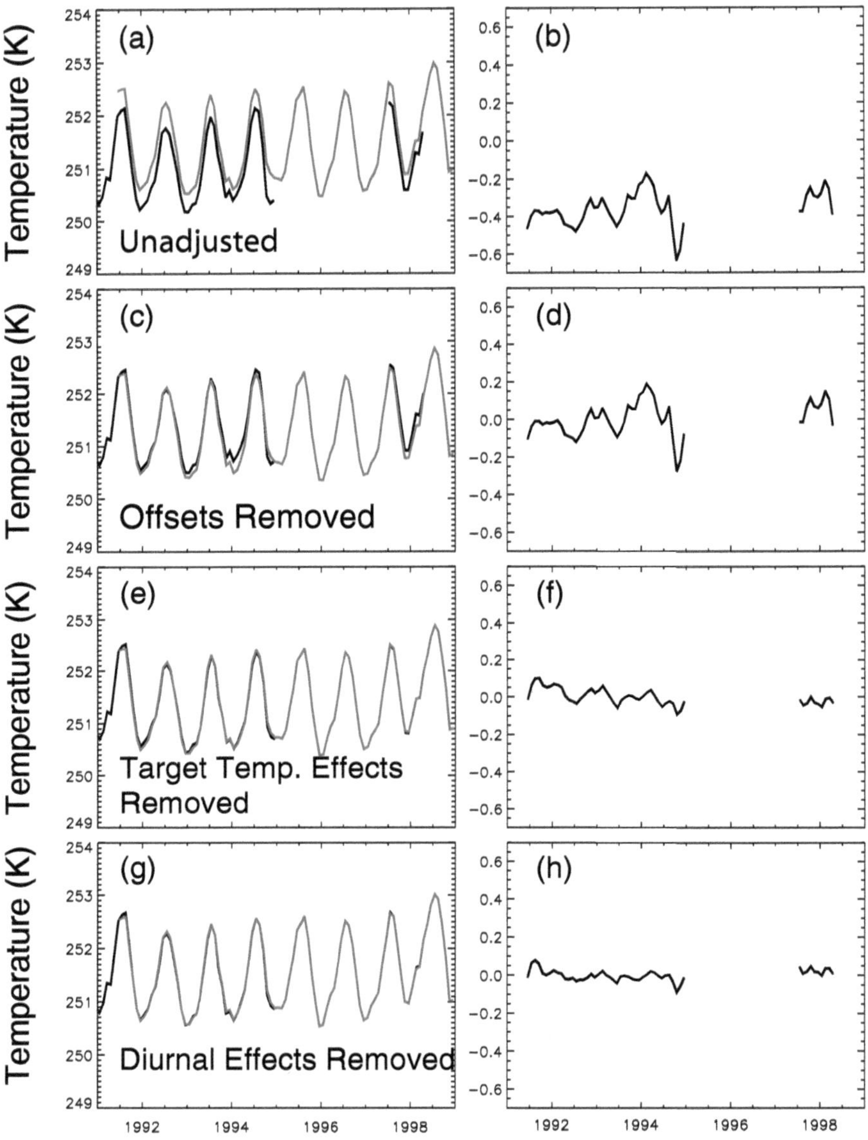

Fig. 1 An example of the effects of the MSU/AMSU calibration procedure. The left column shows the temperatures measured by MSU2 on the NOAA-11 (black) and the NOAA-12 (gray) satellites, and the right column shows the difference between the temperatures. (a) and (b) show the unadjusted data. In (c) and (d) the overall offset has been removed, but large fluctuations remain which are due to calibration errors caused by changes in the temperature of the calibration target, which are removed in (e) and (f). Note that a small slope remains in the difference time series, which is caused by the diurnal cycle aliasing into the time series because of drifting measurement time. The diurnal cycle is removed in (g) and (h).

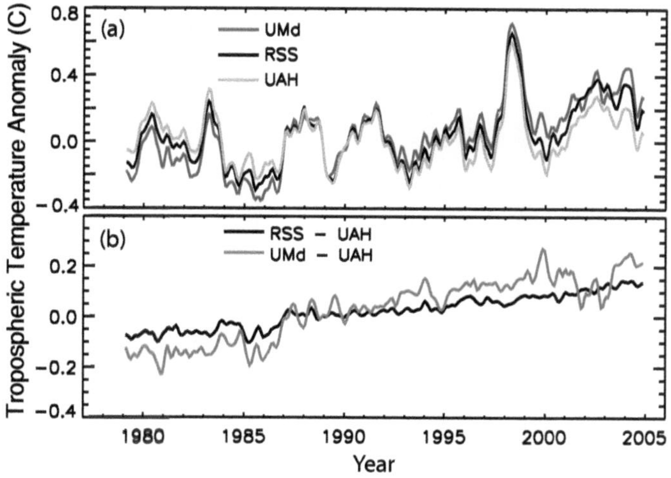

Fig. 2 Time series of global tropospheric temperature assembled by the three groups that have analyzed the entire MSU/AMSU data set. Since each group began with identical data from the satellite instruments, differences in the final data are due to the difference method used to remove the biases and calibration errors present in the raw data.

3 Accounting for Small Differences in Measurement Frequency

There are small differences in the measurement frequency between MSU channel 2 and the AMSU channel closest to MSU2, AMSU5. AMSU5 is at a slightly lower frequency, and thus is further away from the center of the oxygen line complex near 59 GHz. This results in a slightly lower opacity for the AMSU channel, and the AMSU-weighting function is centered slightly lower in the atmosphere at a given incidence angle. In Fig. 3, I plot the vertical weighting functions for MSU2 and AMSU5 for each of the viewing angles used by the satellites (remember, both instruments scan from side to side). The weighting functions that peak closest to the surface correspond to the nadir view. The AMSU-weighting functions peaks several hundred meters closer to the surface, with correspondingly higher contributions from surface emission. Because the average vertical rate of change of temperature depends both on location and time of year, the difference between the nadir temperature measurements for MSU2 and AMSU5 are location and season dependent, which makes it complicated to combine data from these two instruments. Current versions of the MSU/AMSU data from Remote Sensing Systems (RSS v2.1) and University of Alabama, Huntsville (UAH v5.1) use empirical methods to directly remove these differences, while the data set from University of Maryland (UMd), does not yet include AMSU data. More sophisticated methods to match the vertical

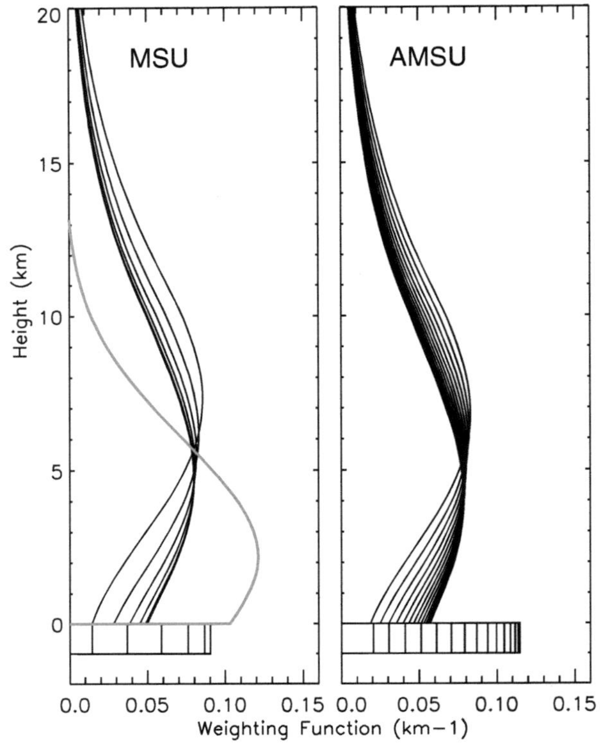

Fig. 3 Vertical weighting functions for each view angle for MSU channel (left), and AMSU channel 5 (right) for land scenes. The rectangular area at the surface represents the relative weight for surface emission. Also plotted is the weighting function for MSU TLT, a weighted difference of MSU views, shown in gray on the left plot.

response of AMSU to MSU, including using weighted combinations of incidence angles, or using data from other AMSU channels, are currently being investigated by both RSS and UAH.

4 Removing the Stratospheric Influence from MSU2

As noted above, MSU2 (and AMSU5) are influenced by stratospheric temperatures, which complicates simple assignment of these temperatures to the troposphere. The stratosphere is cooling rapidly, presumable due to a combination of increases in greenhouse gases and ozone depletion. This cooling, when added into the tropospheric data with the appropriate weight (~10%) obscures any tropospheric warming signal. For this reason, techniques have been developed to mathematically remove the stratospheric weight from the data. The first method, developed by the

Table 1 Global temperature trends for the 1979–2004 time period found by different groups. The spread in values between different groups reflects the effects of the different choices made while merging the data from different satellites together. The uncertainty estimates reflect purely statistical uncertainty, that is, how well the global temperature time series fit a straight line, but do not include estimates of uncertainty internal to the merging process. These values are from Lanzante et al. (2006)

	Global trend (K/decade)
UAH MSU2	0.05 ± 0.08
RSS MSU2	0.13 ± 0.08
UMd MSU2	0.20 ± 0.07
UAH TLT	0.12 ± 0.08
RSS TLT	0.19 ± 0.08
UAH T*	0.12 ± 0.09
RSS T*	0.19 ± 0.09

UAH group (Spencer and Christy 1992; Christy et al. 2000), uses a weighted difference of near-nadir and near-limb views to removed the weight associated with the stratosphere. This also have the effect of extrapolating the effective weighting function downward, and increasing the contribution from surface emission. The vertical weighting function, usually called "TLT" for "temperature lower troposphere," for this data product is plotted as the gray curve in Fig. 3. The main drawback of this method is that because of the differencing procedure, any noise in the data tends to be amplified significantly. A second method was developed by Fu and coworkers (Fu et al. 2004; Fu and Johanson 2005), who used data from the stratospheric measurements made by MSU channel 4 to remove the stratospheric influence from MSU2 – the results of this procedure are called T^*. Both methods produced data sets with larger long-term trends than the simple MSU2/AMSU5 data, as expected. In Table 1, we show the long-term trends for each of the data sets discussed so far.

5 Discussion of Uncertainty in Long-term Behavior

The construction of long-term data sets is complicated, with many steps to be performed and choices to be made during the process. This complicated nature makes it difficult to unambiguously determine the uncertainty in the final product. The uncertainty can be partitioned into two parts, the "internal uncertainty," which is the statistical uncertainty internal to the merging process *once the method has been chosen*, and the "structural uncertainty" which is due to *differences in methodology* used to construct different results. Structural uncertainty can often dominate other sources of uncertainty in climate data sets, and is often unappreciated until two or more research groups perform analysis of the same raw data (Thorne et al. 2005;

Mears et al. 2006). This is especially true for satellite data sets, where the huge number of observations reduces many sources of internal error when large-scale averages are performed. For example, for MSU channel 2, the internal uncertainty in the global trends calculated from the RSS product was estimated to be about 0.03 K/decade ($2 - \sigma$) (Mears et al. 2003), significantly less that the difference between the estimates in Table 1 from different groups. Note that this internal uncertainty is different from the uncertainty show in Table 1, which found by evaluating how well the final time series fits a straight line. Because the time series found by different groups are highly correlated, the "goodness of fit" uncertainties shown in Table 1 cannot be used to evaluate statistical significance of the difference between different time series, but only whether or not the trend in a specific time series is significantly different from zero.

6 Extending the Microwave-based Temperature Data Sets

In this section, I will discuss prospects for extending the data sets forward and backward in time, and upward in the atmosphere.

6.1 Backward in Time

It would clearly be beneficial to extend the microwave data back in time – in particular, it would be exciting to have microwave data that extends across the "climate regime shift" that may have occurred in the late 1970s. Though there are several microwave sounding instruments that flew on satellites before the first MSU instruments, there are no overlapping periods of operation with MSU, and the earlier instruments used different measurement frequencies. Since a long period of overlap is necessary to perform the detailed analysis needed to make an accurate merge of the data, I have concluded that extending the data set backward in time is nearly impossible.

6.2 Forward in Time

Despite the issues involved with merging the newer AMSU data with data from the previous MSU series of satellites, it should be possible to extend these data sets into the future. Continued research is needed to ensure that the MSU/AMSU merge is as accurate as possible. Currently, there are 5 AMSU instruments in operation, (3 on NOAA platforms, 1 on NASA's AQUA satellite, and one on the recently launched European Met-Op satellite), so that unexpected failure of 1 or more satellites should not fatally compromise the integrity of the data. Further into the future, the Advanced cross-Track Microwave Scanner expected to fly on the NPOESS preparatory project (NPP), as well as the National Polar Orbiting Environmental Satellite System (NPOESS) platforms preserve the functions of the AMSU instruments.

6.3 Upward in the Atmosphere

Currently, microwave derived, climate quality data sets exist up to the lower stratosphere, which was measured by MSU channel 4. The AMSU instruments contain 5 additional channels that measure higher in the stratosphere. The length of the AMSU data record is now approaching 10 years – soon it will be time to begin assembling the data from these higher channels into climate quality data sets. This will complement the earlier data from the Stratospheric Sounding Units (SSU), a series of infrared sounders.

References

Christy, J. R., R. W. Spencer, and W. D. Braswell, 2000: MSU tropospheric temperatures: dataset construction and radiosonde comparisons. *J. Atmos. Ocean Tech.*, **17**, 1153–1170.

Christy, J. R., R. W. Spencer, W. B. Norris, W. D. Braswell, and D. E. Parker, 2003: Error estimates of version 5.0 of MSU-AMSU bulk atmospheric temperatures. *J. Atmos. Ocean Tech.*, **20**, 613–629.

Fu, Q. and C. M. Johanson 2005: Satellite-derived vertical dependence of tropospheric temperature trends. *Geophys. Res. Lett.*, **32**, L10703.

Fu, Q., C. M. Johanson, S. G. Warren, and D. J. Seidel, 2004: Contribution of stratospheric cooling to satellite-inferred tropospheric temperature trends. *Nature*, **429**, 55–58.

Grody, N. C., K. Y. Vinnikov, M. D. Goldberg, J. T. Sullivan, and J. D. Tarpley, 2004: Calibration of multi-satellite observations for climatic studies: Microwave Sounding Unit (MSU). *J. Geophys. Res.*, **109**, D24104, doi:10.1029/2004JD005079.

Lanzante, J., S. Klein, and D. Seidel2003: Temporal homogenization of monthly radiosonde temperature data. Part I: methodology. *J. Climate*, **16**, 224–240.

Lanzante, J. R., T. C. Peterson, F. J. Wentz, and K. Y. Vinnikov, 2006: What do observations indicate about the change in temperatures in the atmosphere and at the surface since the advent of measuring temperatures vertically. In: *Temperature Trends in the Lower Atmosphere: Steps for Understanding and Reconciling Differences*, T. R. Karl, S. J. Hassol, C. D. Miller, and W. L. Murray, eds., A Report by the Climate Change Science Program and the Subcommittee on Global Change Research, Washington, DC, 47–70.

Mears, C. A., M. Schabel, F. J. Wentz, B. D. Santer, and B. Govindasamy, 2002: Correcting the MSU middle tropospheric temperature for diurnal drifts. *Proceedings of the International Geophysics and Remote Sensing Symposium III*, 1839–1841.

Mears, C. A., M. C. Schabel, and F. J. Wentz, 2003: A reanalysis of the MSU channel 2 tropospheric temperature record. *J. Climate*, **16**, 3650–3664.

Mears, C. A., C. E. Forest, R. W. Spencer, R. S. Vose, and R. W. Reynolds, 2006: What is our understanding of the contribution made by observational or methodological uncertainties to the previously reported vertical differences in temperature trends. In: *Temperature Trends in the Lower Atmosphere: Steps for Understanding and Reconciling Differences*. T. R. Karl,

S. J. Hassol, C. D. Miller, and W. L. Murray, eds., A Report by the Climate Change Science Program and the Subcommittee on Global Change Research, Washington, DC, 71–88.

Spencer, R. W. and J. R. Christy, 1992: Precision and radiosonde validation of satellite gridpoint temperature anomalies. Part II: a tropospheric retrieval and trends during 1979–1990. *J. Climate*, **5**, 858–866.

Thorne, P. W., D. E. Parker, J. R. Christy, and C. A. Mears, 2005: Causes of differences in observed climate trends. *Bull. Am. Meteorol. Soc.*, **86**, 1437–1442.

Thorne, P. W., D. E. Parker, S. F. B. Tett, P. D. Jones, M. McCarthy, H. Coleman, and P. Brohan, 2005: Revisiting radiosonde upper-air temperatures from 1958 to 2002. *J. Geophys. Res.*, **110**, D18105.

Vinnikov, K. Y., N. C. Grody, A. Robock, R. J. Stouffer, P. D. Jones, and M. D. Goldberg, 2005: Temperature trends at the surface and in the troposphere. *J. Geophys. Res.*, **111**, D03106.

Wentz, F. J. and M. Schabel, 1998: Effects of satellite orbital decay on MSU lower tropospheric temperature trends. *Nature*, **394**, 661–664.

Total Ozone Observations During the Past 80 Years

S. Brönnimann[1], C. Vogler[1], J. Staehelin[1], R. Stolarski[2], and G. Hansen[3]

Abstract Ozone plays a key role in the physics and chemistry of the atmosphere. Total ozone, that is, the amount of ozone in an air column, is therefore a variable of vital climatic and environmental importance. The operational measurement of total ozone reaches back to the pioneering work of G. M. B. Dobson in the 1920s. Here, we give a brief overview of total ozone observations during the past 80 years, including the development of ground-based monitoring networks as well as the more recent satellite sensors. We summarize the measurement techniques, the available data as well as issues related to quality and comparability.

1 Introduction

The amount of ozone in the Earth's atmosphere over a given location (total ozone), which mainly reflects stratospheric ozone, is an important variable in various contexts. For instance, ozone plays a key role in the physics and chemistry of the atmosphere, especially the radiation budget (Solomon 1999). The most prominent changes in total ozone during the past decades, that is, the anthropogenically induced ozone decline from 1970 to present and the appearance of a springtime ozone hole over Antarctica since the early 1980s, have been well documented using ground-based and satellite data. Another important aspect of total ozone is its close relation to the circulation of the stratosphere (Dobson and Harrison 1926; Staehelin et al. 2001), which is of particular interest as total ozone has been observed since

[1] Institute for Atmospheric and Climate Science ETH Zürich, Switzerland
stefan.bronnimann@env.ethz.ch
christian.vogler@alumni.ethz.ch
johannes.staehelin@env.ethz.ch

[2] NASA Goddard Space Flight Center, Greenbelt, MD, USA
Richard.S.Stolarski@nasa.gov

[3] Norwegian Institute for Air Research (NILU), Tromsø, Norway
ghh@nilu.no

the 1920s. Apart from providing a reference against which the anthropogenically induced ozone decline since the 1970s can be measured, historical total ozone measurements also carry information on the circulation variability of the stratosphere prior to the onset of anthropogenic perturbation.

In this paper we give an overview of the available total ozone measurements. We start with a brief discussion of various measurement techniques, including ground-based and space-borne observations. Then we give a chronological overview of the available data and issues related to data quality and re-evaluation. Finally, we end with an assessment of the relevance of these data for climate research and point to open problems and questions. For more detailed information on the history of ozone research and respective references, the reader is referred to Fabry (1950), Dobson (1968), Simmons (1990), Stolarski (2001), Staehelin et al. (2001), and Brönnimann et al. (2003).

2 Measurement Techniques

2.1 Ground-Based Observations

Most of the ground-based techniques that were used to measure ozone during the past 100 years are based on photometry in the UV wavelength range of approximately 300–340 nm, where atmospheric ozone is the most important absorber (see also Fabry 1950). The basic principle is to measure the radiance at two wavelengths, one of which is strongly absorbed by ozone and the other much less so. Total ozone is then calculated by comparing the measured intensity ratio with the intensity ratio outside the Earth's atmosphere, additionally considering the light path through the atmosphere (and the absorbing layer), the absorption coefficients for ozone, and the attenuation by other wavelength dependent processes such as scattering. Different light sources can be used such as direct sunlight, light scattered at the zenith (even under cloudy conditions) as well as light from the moon or stars.

Charles Fabry and Henri Buisson were the first to measure the atmospheric ozone amount using UV photometry around 1920 (Fabry 1950). Their instrument was a double spectrograph with a photographic plate as a detection device. In the mid-1920s, G. M. B. Dobson designed a much simpler instrument, which (instead of a double spectrograph) used a filter and a Féry prism (Dobson 1968). The registration used photographic plates, from which the intensities at given wavelengths were measured with a microdensitometer, a time-consuming procedure. A few years later, Dobson developed a new instrument with photoelectric detection (Dobson 1968). By alternatively passing light from one of two wavelength selecting slits to the detector and then equalizing the two detector currents by repositioning optical wedges, the intensity ratio could be measured directly. The new instrument allowed the measurement of total ozone within a few minutes for most weather conditions. This was necessary in order to make such measurement useful for operational weather forecasting, as was anticipated at that time (see Simmons 1990). Around 1947, Dobson improved the instrument by employing a photomultiplier,

Fig. 1 Two Dobson spectrophotometers in Arosa in the early 1950s (from: LKO, MeteoSwiss).

increasing the signal-to-noise ratio. This instrument still is the standard for total ozone observations. Figure 1 shows two Dobson instruments during an intercomparison at Arosa, Switzerland, in the 1950s.

Starting in the 1970s, another instrument type was developed (Brewer instruments, see Kerr et al. 1985) which is based on the same measurement principles but uses a holographic grating for selecting different wavelengths in the 290–372 nm region. In contrast to the old Dobson instruments, Brewer instruments are fully automated. Dobson and Brewer spectrophotometers are standard instruments of two ground-based networks operated since the 1970s under the auspices of the World Meteorological Organisation (WMO).

Another type of instruments uses filters to select the wavelengths. Such an instrument was used at Arosa already in the early 1920s (Staehelin et al. 1998). Filter instruments were extensively used to monitor ozone in the former Soviet Union since the 1950s (Bojkov et al. 1994). However, these measurements are considered to be less reliable compared with Dobson instruments and can have significant errors related to the large field of view, a large bandwidth, and aerosol effects.

Total ozone can also be derived from its weak absorption in the visible wavelength range. Early attempts in the 1930s and 1950s (e.g., Fabry 1950) could not catch on. The absorption signal is weaker, while at the same time there are strong interfering factors (water vapour, aerosols). With more advanced techniques, the SAOZ instrument (Pommereau and Goutail 1988) today is again making use of ozone absorption in the visible, using a holographic grating to select wavelengths in the 300–600 nm range. This instrument also allows total ozone observations at polar sites in winter, when Dobson and Brewer measurements are strongly restricted.

Solar and stellar spectra have not only been measured with the purpose of deriving total ozone, but also for other purposes (e.g., astrophysics). Several authors tried to derive total ozone retroactively from such data. Griffin (2006) showed that photographically recorded stellar spectra, of which there are hundreds of thousands worldwide (reaching back more than 100 years), can sometimes be used to derive total ozone by UV photometry. Angione and Roosen (1983) used data from the solar network of the Smithsonian Institution in the first half of the 20th century. They were obtained with spectrobolometers, which photographically record the solar spectrum between 0.35 and 2 μm. While there are very serious quality issues, reprocessing these recordings might allow a glimpse at ozone variability during the first part of the 20th century, when no or few other measurements were available.

2.2 *Satellite-based Observations*

Similar techniques as used for ground-based measurement can also be used from space, which can provide near-global information. Mostly, UV spectrophotometry is used, but light from the visible range can also be used, as well as emission lines in the thermal infrared. A nadir viewing or scanning mode is mostly used to measure total ozone. In this geometry the light source is solar light that is reflected at the Earth's surface or backscattered in the atmosphere, or thermal emission from the Earth's surface and the atmosphere. The basic principle is the same as for ground-based observations, but with additional difficulties such as due to Earth's albedo and clouds. In addition, satellite drift and instrument degradation pose serious problems, all the more as no direct calibration is possible.

The first observations of total ozone from a satellite were obtained in 1969 by the Infrared Interferometer Spectrometer (IRIS) on Nimbus-3 (Prabhakara et al. 1971), which measured thermal radiance from the Earth at 5–25 μm (including the 9.6 μm ozone band). However, the instrument failed only a few weeks after launch. A year later, Nimbus-4 was launched with the Backscatter Ultraviolet radiometer (BUV), which measured the UV spectrum at 12 wavelengths in the 250–340 nm region and a non-absorbing band at 380 nm that was used to retrieve total ozone (see Heath et al. 1973). However, due to battery power issues, the measurements were not continuous after the first 2.5 years.

The first comprehensive satellite observations were started in 1978 with the Nimbus-7 satellite, which carried a Total Ozone Mapping Spectrometer (TOMS) instrument and a Solar Backscatter Ultraviolet (SBUV) instrument. Other TOMS and SBUV instruments were flown on succeeding satellites, providing more than 25 years of continuous near-global records of total ozone. In 1995 the Global Ozone Monitoring Experiment (GOME) was launched. Its successor, GOME 2 has recently been launched on the MetOp satellite. In 2004, the successor instrument to TOMS, the Ozone Monitoring Instrument (OMI) was launched on the Aura satellite.

3 Data

With respect to the availability of total ozone data, the past 80 years can be subdivided into three periods that are separated by the International Geophysical Year (IGY) and the start of Nimbus-7 in November 1978.

For the first period, the amount of data is very limited, but some observations can be found (see Brönnimann et al. 2003). In the 1920s, total ozone was measured mainly by the two groups around Fabry and Dobson at several locations in Europe in order to study total ozone variability and its relation to atmospheric circulation (Dobson and Harrison 1926). In the late 1920s, Dobson relocated his instruments to more distant places in order to measure the global ozone distribution. However, most of these series cover only 1 or 2 years, which makes the re-evaluation difficult. It should be noted that the observations had not been made for long-term monitoring; the quality required for reliable trend analysis was never intended (Simmons 1990).

Additional ozone data can be obtained from other spectra. As an example, Fig. 2 (left) shows total ozone data as a function of season for Table Mountain, California, for the 1920s and 1930s. Data from different instrument types (Dobson spectrophotograph, spectrobolometer, stellar spectra from nearby Mount Wilson, see Brönnimann 2005; Griffin 2006) are shown. Note that there are systematic differences in absolute values as well as large differences in their variability. Still, characteristic features of the seasonal ozone cycle known from more recent data (Fig. 2 right) such as the sudden drop in June, the peak in October, or the seasonality of day-to-day variability appear both in the Dobson and in the sprectrobolometer series. Particularly the Dobson data seem to be in good shape so that one can have confidence at least in the short-term variability.

Continuous observations were only performed at very few sites, and there are even fewer high-quality data series. The most famous total ozone series is the one

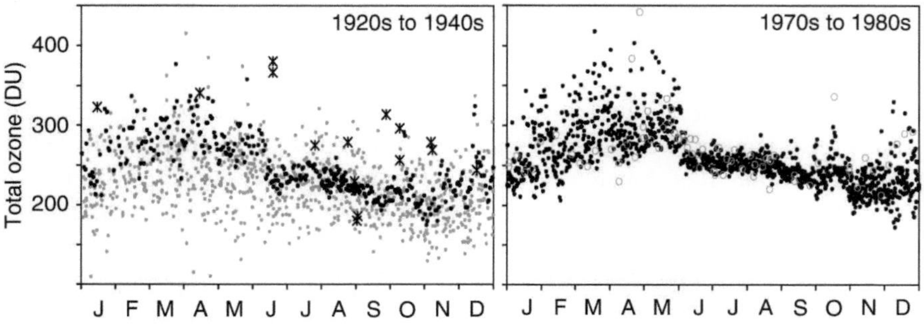

Fig. 2 Total ozone at Table Mountain, California, as a function of season from three different total ozone data sets from the 1920s to 1940s (left, black dots = Dobson Féry spectrograph, 1928–1929 (Brönnimann 2005), grey dots = spectrobolometer at 574 nm, 1925–1948 (Brönnimann 2005), stars = stellar spectra from nearby Mt. Wilson, 1935–1941 (Griffin 2006)) and from two satellite-based data sets from the 1970s and 1980s (right, black dots = TOMS Version 8 overpass, 1978–1981, grey circles = BUV, 1970–1972, within 1° lon/lat of Table Mountain).

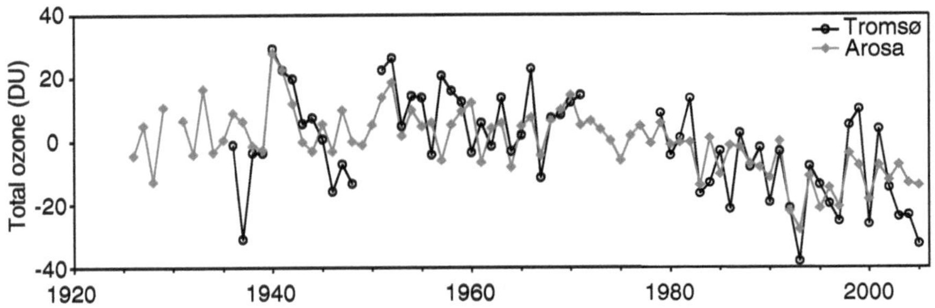

Fig. 3 Annual series of total ozone anomalies (with respect to 1961–1990) at Tromsø (Hansen and Svenøe 2005) and Arosa (Staehelin et al. 1998). The Tromsø data were supplemented with TOMS Version 8 data after 1978 in order to fill gaps (see Brönnimann et al. 2004 for details).

from Arosa, Switzerland, that begins in 1926 (Staehelin et al. 1998). In addition, the series from Tromsø in the Norwegian Arctic (since 1935, Hansen and Svenøe 2005) has recently been re-evaluated. Figure 3 shows time series of annual mean values of these two data series (Tromsø data were supplemented with TOMS in order to fill gaps). These long time series allow addressing interannual to decadal variability in ozone. The ozone decline from 1970 to around 1995 is clearly visible in these series and can be judged against the variability of the first part of the record. Note also the particularly pronounced peak in both series in 1940–1942, demonstrating that natural variability on interannual and even multiannual time scales can be very large. This peak was shown to be related to a strong El Niño event (Brönnimann et al. 2004).

Starting around 1950, a European Dobson total ozone network was finally established under the International Ozone Commission, but re-evaluation of these data is difficult because important information is missing (see Brönnimann et al. 2003). One of the series, a very interesting series from the high Arctic (Longyearbyen 1950–1962) could be re-evauated based on original observation sheets (Vogler et al. 2006).

During the IGY, a global total ozone network was initiated and operated. Therefore, much more data are available after that year (most of the data are stored at the World Ozone and Ultraviolet Data Center (WOUDC)). The network included a station in Halley Bay, Antarctica, which is famous for the discovery of the ozone hole in the 1980s (Farman et al. 1985). At the same time observation techniques and calibration were improved and the procedures were standardised. This allows re-evaluating many total ozone series back to 1957 based on meta-information and calibration information.

Prior to around 1970, all Dobson instruments were calibrated in Dobson's laboratory before they were shipped to the stations. After that time the WMO took on the responsibility over the ground-based total ozone observations. The Dobson and Brewer networks are based on primary world standard instruments, which are regularly calibrated by the Langley plot method at the Mauna Loa observatory at Hawaii. Side-by-side calibrations with a transfer standard and regional standards allow the calibration of the primary instrument to be transferred to the other instruments. This results in a high-quality standard for well-maintained instruments. Today, including some carefully re-evaluated series, trend-quality total ozone data from various parts of the world reach back to around 1964.

Near-global coverage of total ozone measurements was only reached with satellite data. Figure 2 (right) shows BUV data for the location of Table Mountain, from 1970 to 1972. The BUV do not have trend quality, but can provide useful information for analysing total ozone variability. Much better data were obtained with TOMS and SBUV instruments since 1978. TOMS (Version 8) data for Table Mountain (1978 to 1981) are shown in Fig. 2 (right). The agreement between these data and both the Dobson data from the 1920s as well as the BUV data is very good.

Satellite instruments can suffer from degradation that must be corrected in order to use the data for trend analyses. For TOMS and SBUV, the main degradation issue was that of the solar diffuser plate used to reflect sunlight into the detector to make solar flux measurements. The ozone measurement is a ratio of the UV-scattered sunlight from the Earth to the direct UV irradiance from the sun. Thus, any degradation of the reflective characteristics of the diffuser plate must be taken into account. This has been done with the TOMS and SBUV instruments by using redundant measurements at multiple wavelengths (Herman et al. 1991).

To create a time series over the entire satellite data record, it is necessary to combine the measurements from multiple satellites. Stolarski and Frith (2006) combined the data from the Nimbus 7 TOMS, Earth Probe TOMS, and 4 SBUV instruments (Nimbus 7, NOAA 9, NOAA 11, and NOAA 16) to produce a single time series over the globe (Fig. 4). They used overlapping periods of measurement to establish calibrations relative to the Earth Probe TOMS and evaluated the potential for drift uncertainty in this record and found that an instrument drift uncertainty of about 1% per decade needed to be included in determing the signficance of trend

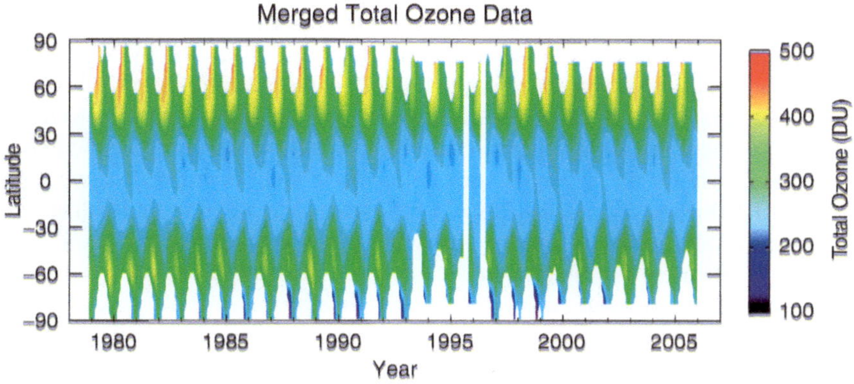

Fig. 4 Total ozone measurements as a function of latitude and time from 1978 through 2005. Data are taken from two TOMS instruments and four SBUV instruments and put on a common calibration scale (see Stolarski and Frith 2006 for details).

results from the satellite data. Similar combined data sets were also constructed using TOMS, SBUV, GOME, and ground-based records by Bodeker et al. (2001); using TOMS, SBUV, and ground-based records by Fioletov et al. (2002); and using SBUV records by Miller et al. (2002). Fioletov et al. (2002) compared these records and found them to agree with one another to within error bars. Fioletov et al. (2002) also found that the satellite data agreed with ground data within the estimated drift uncertainties.

The agreement between satellite and ground-based data results from an extensive effort to calibrate and intercompare systems. For instance, well-calibrated ground-based data can be used to test the drift corrections in long-term satellite measurements. The satellite data have also proven useful for finding problems with individual ground-station measurements. The TOMS instrument on Nimbus 7 has been used as a transfer standard to identify anomalies in ground-station data. If a station had a sudden jump with respect to the TOMS measurements during the satellite's overpass of the station, examination of the station record would usually identify some change that had been made (either a change in location, procedure, or instrument). This screening procedure proves valuable for "cleaning up" the ground-based data records.

4 Historical and Present Data from Polar Regions

Since the discovery of the ozone hole over Antarctica in the mid-1980s (Farman et al. 1985), its first mapping by satellite (Stolarski et al. 1986), and the determination of Arctic ozone depletion in the late 1980s (Hofmann et al. 1989), ozone in polar regions has become an important research topic. How high were typical total ozone levels over the Arctic and Antarctica in the chemically undisturbed stratosphere? Historical total ozone data can provide a background for comparisons (and for constraining and validating climate models).

The first measurements in the high Arctic (Spitsbergen) were performed during expeditions in the 1920s and 1930s (Brönnimann et al. 2003). Continuous observations started in 1950 and were recently digitised and re-evaluated by Vogler et al. (2006). Data from Antarctica date back to the IGY and can be found on the WOUDC data base. Interestingly, during these early years, obervations at both sites were also taken during the polar night using moon light, thus allowing a year-round comparison. Figure 5 shows the averaged seasonal cycle of total ozone at Longyearbyen (79°N) in 1950–1962 and Halley Bay (75.5°S) from 1957 to 1963 (individual years are shown in thin lines, the averaged seasonal cycle in thick lines). Note that the Halley Bay data are offset by 6 months compared to those of Longyearbyen.

The agreement between total ozone at the two stations is good in summer and fall, when ozone levels are similar (and interannual variability is small). However, total ozone develops in completely different ways over the winter season in the polar regions of each hemisphere. Ozone increases rapidly over the Arctic and reaches a maximum in spring, with a high interannual variability. In contrast, ozone over Antarctica is almost constant until late spring and then increases, with relatively

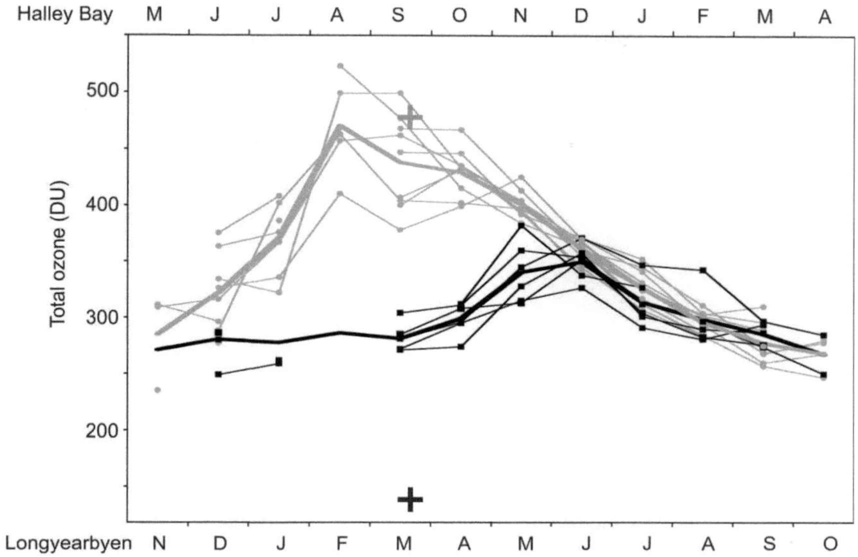

Fig. 5 Total ozone at Longyearbyen (grey lines, Vogler et al. 2006), 1950–1962 and Halley Bay (black lines, WOUDC data), 1957–1963. Dots and thin lines denote individual years (at least three values must be available per month). Thick lines denote the averaged seasonal cycle over the period. The crosses indicate the corresponding values from Fig. 6.

little interannual variability. These differences are due to a weaker (and more variable) stratospheric polar vortex and more poleward transport in the Arctic compared to Antarctica. Hence, for the spring season, we expect to see differences of 100–200 DU between total ozone over Antarctica and the Arctic that are not caused by chemical destruction but are due to different dynamics.

Current measurements of total ozone must be analysed in context of this difference. Figure 6 shows total ozone (OMI) over Antarctica and the Arctic in the respective spring months in 2006. Note the larger spatial variability of total ozone in the Arctic spring compared to Antarctic spring. The Antarctic ozone hole was particularly large in this year and once again received a lot of media attention. Total ozone dropped below 150 DU over a large area. In contrast, total ozone in the Arctic on 19 March 2006 was everywhere above 400 DU and exceeded 500 DU over a wide area. Even though there was some ozone depletion in January, temperatures in the Arctic stratosphere rose above the threshold under which polar stratospheric clouds are formed (to which ozone depletion is related) and were high throughout the rest of the winter (see bulletin at http://www.cpc.noaa.gov/products/stratosphere/winter_bulletins/nh_05-06/, http://www.wmo.int/web/arep/gaw/arctic_ozobull.html).

Comparing these differences with those noted in Fig. 5 (for comparison, the 2006 values are marked with crosses in Fig. 5), we find an additional difference of about 150 DU between low values in the 1950s and the current values, which is due to chemical destruction. Instead of a 100–200 DU difference (Fig. 5), we find a 250–350 DU difference (Fig. 6).

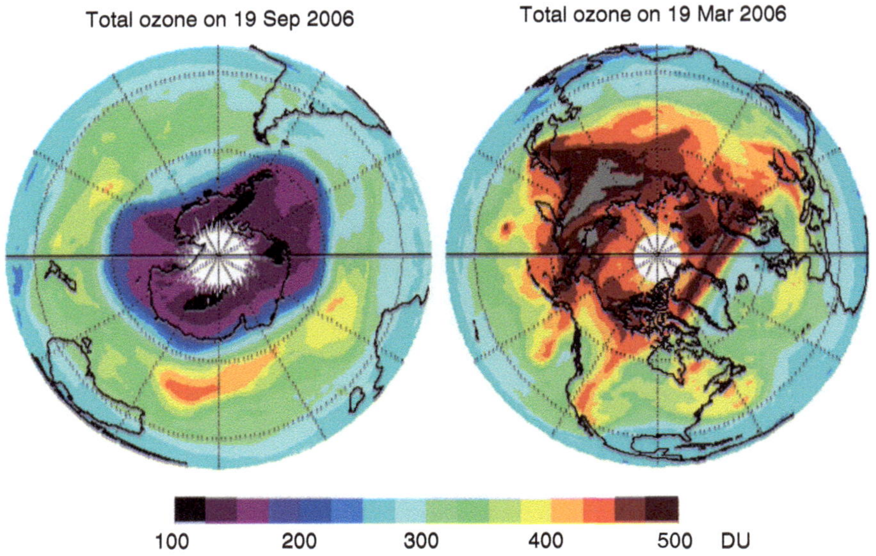

Fig. 6 OMI total ozone data for the South Polar Region on 19 September (left) and for the North Polar Region on 19 March (right) 2006 allowing a snapshot at the current state of the ozone layer during the respective spring season.

5 Conclusions and Outlook

Total ozone is probably the stratospheric variable with the longest measurement history, reaching back more than 80 years. Historically these data have played a major role in shaping our understanding of stratospheric processes, both chemical and dynamical. Their role in understanding stratosheric chemistry is summarised by Brasseur (this volume). Through the discovery of the Antarctic ozone hole and global ozone depletion, monitoring total ozone became of vital interest and the "old" data records became again important by providing an "undisturbed" background climatology of ozone against which recent data can be compared. Historically, total ozone data were also important for understanding dynamical processes in the stratosphere and the tropopause region. In fact, this focus was one of the main drivers behind the development of total ozone measurements in the 1920s–1940s (Simmons 1990). Analyses of total ozone in a dynamical context were (and still are) important contributors to our current understanding of stratospheric dynamics ranging from the large-scale meridional circulation of the stratosphere (the Brewer-Dobson circulation) to small scale wave-breaking events (e.g., ozone mini holes).

Historical total ozone data could again become important in the context of analysing climate variability and extremes during the 20th century. However, with few exceptions, not much information is currently available for the first half of the

20th century. Recently, the total ozone series from Oxford, UK, measured by G. M. B. Dobson himself and covering the period 1924–1975 (though with gaps) was re-evaluated (Vogler et al. 2007). Further historical total ozone series still await digitising and re-evaluation, and the rich sources of stellar spectra have not yet been fully exploited. Such efforts will provide important data for studying daily-to-interannual variability of ozone (and hence stratospheric processes) and might be useful for model validation. However, there is little hope to obtain another long, trend-quality ozone series. The situation is much better for the period since 1957, when a global network was established, and after 1978, when satellite data of sufficient quality became available.

Continuation of a high-quality total ozone record requires maintaining both space and ground-based capabilities. The space observing system gives global coverage and quality control of individual station measure the observing system through new space missions and bringing the ground-based network further ahead by removing systematic differences between Brewer and Dobson instruments.

Acknowledgments SB and CV were funded by the Swiss National Science Foundation. RS was funded by the NASA TOMS Science Team and the US OMI Science Team.

References

Angione, R. J. and R. G. Roosen, 1983: Baseline ozone results from 1923 to 1955. *J. Clim. Appl. Meteorol.*, **22**, 1377–1383.

Bodeker, G. E., J. C. Scott, K. Kreher, and R. L. McKenzie, 2001: Global ozone trends in potential vorticity coordinates using TOMS and GOME intercompared against the Dobson network, 1978–1998. *J. Geophys. Res.*, **106**, 23029–23042.

Bojkov, R. D., V. E. Fioletov, and A. M. Shalamjansky, 1994: Total ozone changes over Eurasia since 1973 based on reevaluated filter ozonometer data. *J. Geophys. Res.*, **99**, 22985–22999.

Brönnimann, S., 2005: Interactive comment on "Detection and measurement of total ozone from stellar spectra: paper 2. Historic data from 1935–1942" by R. E. M. Griffin. *Atmos. Chem. Phys. Disc.*, **5**, S4045–S4048.

Brönnimann, S., J. Staehelin, S. F. G. Farmer, J. C. Cain, T. M. Svendby, and T. Svenøe, 2003: Total ozone observations prior to the IGY. I: A history. *Q. J. Roy. Meteorol. Soc.*, **129**, 2797–2817.

Brönnimann, S., J. Luterbacher, J. Staehelin, T. M. Svendby, G. Hansen, and T. Svenøe, 2004: Extreme climate of the global troposphere and stratosphere in 1940–1942 related to El Niño. *Nature*, **431**, 971–974.

Dobson, G. M. B. and D. N. Harrison, 1926: Observations of the amount of ozone in the Earth's atmosphere and its relation to other geophysical conditions. *Proc. R. Soc. A.*, **110**, 660–693.

Dobson, G. M. B., 1968: Forty years' research on atmospheric ozone at Oxford: a history. *Appl. Optics*, **7**, 387–405.

Fabry, C., 1950: L'ozone atmosphérique. Paris.

Farman, J. C., B. G. Gardiner, and J. D. Shanklin, 1985: Large losses of total ozone in Antarctica reveal seasonal ClO_x/NO_x interaction. *Nature*, **315**, 207–210.

Fioletov, V., G. Bodeker, A. Miller, R. McPeters, and R. Stolarski, 2002: Global and zonal total ozone variations estimated from ground-based and satellite measurements: 1964–2000. *J. Geophys. Res.*, **107**, doi 10.1029/2001JD001143.

Griffin, R. E. M., 2006: Detection and measurement of total ozone from stellar spectra: paper 2. Historic data from 1935–1942. *Atmos. Chem. Phys.*, **6**, 2231–2240.

Hansen, G. and T. Svenøe, 2005: Multilinear regression analysis of the 65-year Tromsø total ozone series. *J. Geophys. Res.*, **110**, doi:10.1029/2004JD005387.

Heath, D. F., C. L. Mateer, and A. J. Krueger, 1973: The Nimbus-4 BUV atmospheric ozone experiment – two year's operation. *Pure Appl. Geophys.*, **106–108**, 1238–1253.

Herman, J. R., R. McPeters, R. Stolarski, D. Larko, and R. Hudson, 1991: Global average ozone change from November 1978 to May 1990. *J. Geophys. Res.*, **96**, 17279–17305.

Hofmann, D. F. et al., 1989: Stratospheric clouds and ozone depletion in the Arctic during January 1989. *Nature*, **340**, 117–121.

Kerr, J. B., C. T. McElroy, D. I. Wardle, R. A. Olafson, and W. P. J. Evans, 1985: The automated Brewer Spectrophotometer. In: *Atmospheric Ozone*, C. S. Zerefos and A. Ghazi, eds., 543–546.

Miller, A. J. et al., 2002: A cohesive total ozone data set from the SBUV(/2) satellite system. *J. Geophys. Res.*, **107**, doi 10.1029/2001JD000853.

Pommereau, J. P. and F. Goutail, 1988: O_3 and NO_2 ground-based measurements by visible spectrometry during arctic winter and spring 1988. *Geophys. Res. Lett.*, **15**, 891–894.

Prabhakara, C., V. V. Salomonson, B. J. Conrath, J. Sterania, and L. J. Allison, 1971: Nimbus 3 IRIS ozone measurements over southeast Asia and Africa during June and July 1969. *J. Atmos. Sci.*, **28**, 828–831.

Stolarski, R. S., 2001: History of the study of atmospheric ozone. *Ozone-sci. Eng.*, **23**, 421–428.

Stolarski, R. S. and S. M. Frith, 2006: Search for evidence of trend slow-down in the long-term TOMS/SBUV total ozone data record: the importance of instrument drift uncertainty. *Atmos. Chem. Phys.*, **6**, 4057–4065.

Stolarski, R. S., A. J., Krueger, M. R. Schoeberl, R. D. McPeters, P. A. Newman, and J. C. Alpert, 1986: Nimbus-7 satellite measurements of the springtime Antarctic ozone decrease. *Nature*, **322**, 808–811.

Simmons, E. L., 1990: Sixty-four years of regular ozone measurements. *Meteorol Mag.*, **119**, 53–57.

Solomon, S., 1999: Stratospheric ozone depletion: a review of concepts and history. *Rev. Geophys.*, **37**, 275–316.

Staehelin, J., N. R. P. Harris, C. Appenzeller, and J. Eberhard, 2001: Ozone trends: a review. *Rev. Geophys.*, **39**, 231–290.

Staehelin, J. et al., 1998: Total ozone series at Arosa (Switzerland): homogenisation and data comparison. *J. Geophys. Res.*, **103**, 5827–5841.

Vogler, C., S. Brönnimann, and G. Hansen, 2006: Re-evaluation of the 1950–1962 total ozone record from Longyearbyen, Svalbard. *Atmos. Chem. Phys.*, **6**, 4763–4773.

Vogler, C., S. Brönnimann, J. Staehelin, R. E. M. Griffin, 2007: The Dobson total ozone series of Oxford: re-evaluation and applications. *J. Geophys. Res.*, 112, D20116, doi:10.1029/2007 JD008894.

Section B
Climate Trends: Forcings and Processes

Arctic Sea Ice Variability During the Last Half Century

J. Stroeve[1] and W. Maslowski[2]

Abstract Observational data and modeling results are analyzed to describe changes in the Arctic sea ice cover during the last half century. Accelerated melt of sea ice cover is reported during the late 1990s and 2000s both based on satellite observations of sea ice extent and model simulations of sea ice thickness. The observed and modeled changes are in qualitative agreement but model results imply higher rate of ice thickness decline compared to sea ice extent. Possible causes of variability in sea ice cover include increased surface air temperatures, changes in atmospheric circulation and changes in the absorption of incoming radiative flux. However, atmospheric forcings, such as the Arctic Oscillation (AO), explain less than half of the total variance in Arctic sea ice cover. Model results analyzed in the Greenland Sea as well as observations in the western Arctic Ocean indicate that oceanic forcing might be an important overlooked factor in driving recent sea ice melt. The main oceanic processes relevant to variability of sea ice cover include advection of heat and melting of sea ice in marginal ice zones and at the ice–ocean interface downstream of the warm water paths. Such changes have potential significant ramifications to the entire pan-Arctic region, including the physical environment, regional ecosystems, native communities, and use of the region for commercial exploration and transportation. Continued studies including in situ and remote sensing observations and modeling are critical to advancing the knowledge of Arctic climate change and predicting scenarios of future change.

1 Introduction

The Arctic is an integral part of the Earth's climate system, through its influence on surface energy and moisture fluxes, atmospheric and oceanic circulation. Studies confirm the importance of the Arctic in climate, including

[1] University of Colorado, CIRES, Campus Box 449, Boulder, CO 80309, USA
stroeve@kodiak.colorado.edu

[2] Naval Postgraduate School, Department of Oceanography, 833 Dyer Road, Monterey CA 93943, USA
maslowsk@nps.edu

mechanisms that could cause abrupt climate change (e.g., Overpeck et al. 2005; Lindsay and Zhang 2005; Foley 2005). One of the factors that is among the most important in strongly coupling the Arctic with the rest of the Earth system is the ice-albedo feedback. Changes in land, lake and sea ice cover, in addition to changes in seasonal snow cover, impart a strong albedo feedback that can be quickly transmitted to the global atmosphere. Arctic sea ice is currently undergoing profound changes (e.g., Stroeve et al. 2005; Meier et al. 2005; Rothrock et al. 2003). However, basin-wide observations of Arctic sea ice conditions over the last several decades are limited primarily to ice extent measurements from satellites (obtained by summing the area covered by all pixels that have 15% or greater ice concentration). Ice concentration data exist over the same period but include larger uncertainties during summer due to large percentage of melt ponds (Cavalieri et al. 1992). Finally, sea ice thickness data are available from submarines, ice drifting stations and upward looking sonars (ULS) since the 1950s but they are spatially and temporally limited (e.g., Rothrock et al. 1999).

Recently the summer sea ice extent has shown large reductions (Comiso 2002; Stroeve et al. 2005) with smaller, yet statistically significant declines observed in winter (Meier et al. 2005). Observations of ice thickness suggest the perennial sea ice could have thinned by as much as 40% between the late 1950s/1970s and 1990s (Rothrock et al. 1999), while changes in seasonal ice thickness are not statistically significant (Melling et al. 2005; Haas 2004; Polyakov et al. 2003). The observed changes in extent and thickness are related to a variety of factors, including increased ocean and atmospheric temperatures, changes in atmospheric and oceanic circulation and the role of feedbacks, such as the ice-albedo feedback. Some studies suggest the Arctic may have recently reached a threshold where the thick perennial ice cover is increasingly replaced by thinner, first-year ice that is more easily melted each summer (Stroeve et al. 2005; Lindsay and Zhang 2005).

A common feature of climate model predictions is the loss of seasonal Arctic sea ice from the effects of greenhouse gas (GHG) loading (ACIA 2005), with some models predicting the Arctic could be ice-free during summer by 2040 (Holland et al. 2006). Such changes in sea ice cover potentially have many significant ramifications, including changes in the global ocean circulation and heat budget, regional ecosystems and wildlife, the indigenous human population, and commercial exploration and transportation. To accurately predict and effectively plan for the impact of these changes it is necessary to improve model simulations of Arctic sea ice and climate as well as continue monitoring Arctic sea ice extent and employ modern advanced technologies to improve knowledge of ice thickness. In addition, further research is needed to obtain the best possible understanding of all the factors that influence past, present and future ice cover. This paper reports on observed changes in Arctic sea ice extent and modeled changes in ice thickness and discusses impacts of varying atmospheric and oceanic conditions on changes of sea ice cover during the last 50 years.

2 Summary of Ice Extent Observations

Passive microwave satellite observations provide the longest and most consistent estimates of sea ice cover and many studies have reported on the decline of the Arctic sea ice using this data (e.g., Bjørgo et al. 1997; Cavalieri et al. 1997; Comiso 2002; Johannessen et al. 1999; Parkinson et al. 1999; Serreze et al. 2003; Stroeve et al. 2005). Figure 1 shows updated (e.g., from November 1978 through September 2006) standardized monthly mean anomalies of Arctic ice extent over the satellite era (determined by dividing the monthly mean anomalies by the monthly standard deviations relative to 1979–2000 mean) together with a 12-month running mean. Since the late 1970s, the extent of sea ice in the Arctic has continued to decline and since about 1995, the anomalies have mostly been negative, suggesting something quite different is happening now to the Arctic ice pack. The annual decline in Arctic sea ice is currently at a rate of –4.3% per decade.

Serreze et al. (2003) first reported on the record low summer sea ice in 2002. In September 2002, ice extent was nearly 15% lower than the satellite-era long-term mean (1979–2000). Before 2002, the trend in September sea ice extent was just over –6.5% per decade. After summer 2002, this value jumped to –7.3% per decade. Although the ice extent in the winter that followed was similar to that observed in previous years, September 2003 was nearly as low as that in 2002, with an ice cover 12% below the mean. The situation repeated itself yet again in 2004, with a September ice cover that was 13% below the mean, resulting in a decline of September sea ice of –7.8% per decade (Stroeve et al. 2005). In 2005, the September sea ice retreated

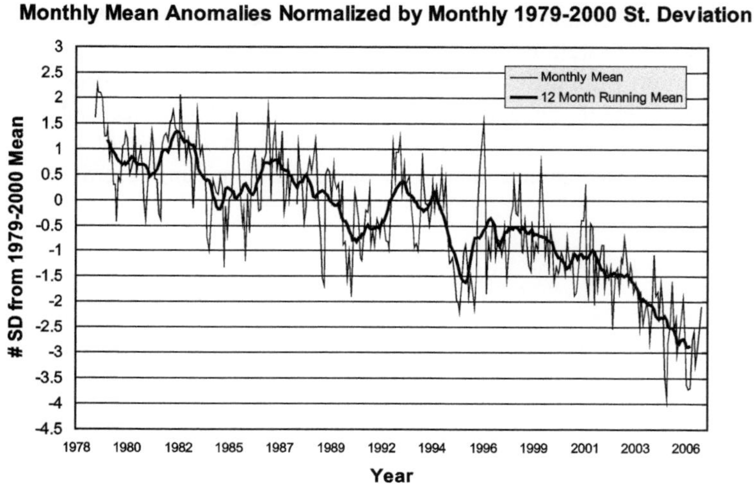

Fig. 1 Standardized monthly mean sea ice extent anomalies relative to 1979–2000 mean for the Arctic from November 1978 through September 2006. The thick black line is the 12-month running mean. Ice extent is derived from NASA Team sea ice concentrations (Cavalieri et al. 1996).

to its smallest area since 1979, increasing the overall decline to a rate of approximately −8.6% per decade. After yet another record low sea ice year in 2006, the rate of decline in September now stands at −8.8% per decade.

Figure 2 shows the location of ice losses in September for 2002–2006. In 2002–2004, reductions in sea ice concentration were particularly pronounced off the shores of Alaska and Siberia and over the east Greenland shelf. In fact, 2002 was the first year during the satellite era where there was essentially no sea ice in the Greenland Sea at the end of summer. In 2005, the largest ice losses occurred primarily in the Chukchi, East Siberian, Laptev, Kara and Barents seas. Ice extent from January through the middle of July 2006 was well below 2005 conditions, with large ice losses in the Chukchi and East Siberian seas. However, air temperatures dropped in August which slowed the ice losses, resulting in 2006 being the second lowest ice year on record.

Although the decline during summer has been the most dramatic, statistically significant (at the 99% level) negative trends have occurred during every month since 2002, and the last three winters (2004/2005–2006/2007) have seen ice extents that were at least 6% below the 1979–2000 mean. This may indicate the recent large ice losses during summer are beginning to affect the winter ice cover. While increased absorbed shortwave energy from the sun resulting from less summer sea ice will delay ice formation in the autumn, this will mainly affect the ice thickness of the subsequent winter ice cover, rather than the ice extent. However, with continued warming of the system, reductions in winter ice extent are likely. The decline of sea ice in winter (e.g., March) is now at a rate of −2.9% per decade since 1979.

Although the satellite era provides the most consistent and accurate record of sea ice extent, there are earlier observations from a variety of sources such as aircraft and ship reports that allow us to put the current decline from the satellite record into longer-term perspective. Monthly mean sea ice concentration fields from the Hadley Centre sea ice and sea surface temperature data set (HadISST) provide for an extension of the satellite record back in time to 1870 (Rayner et al. 2003). Sea ice concentrations from different sources (e.g., satellites, ship and aircraft observations) were analyzed to make them as consistent as possible throughout the record. However, large observational gaps occur, resulting in the use of climatologies derived from calibrated passive microwave data to fill in missing records, particularly

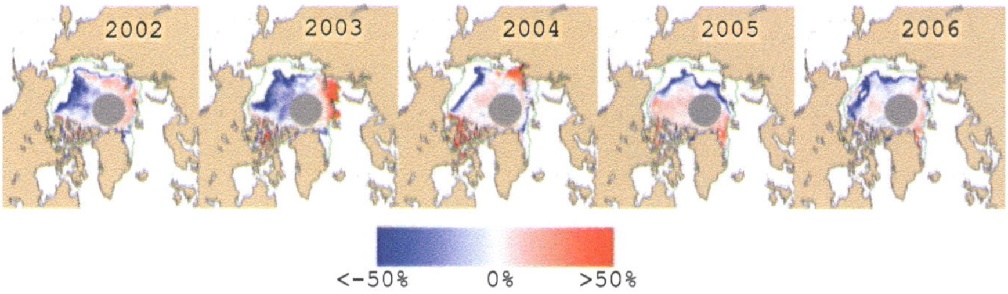

Fig. 2 Monthly mean September ice concentration anomalies (blue = negative, red = positive) from 2002 through 2006. The median ice extent (1979–2000) is given by the green line.

prior to 1953. Recently, Meier et al. (2007) reprocessed the Had1SST data to correct for a significant inconsistency found between 1996 and 1997 when the Had1SST data set switched to a different source/algorithm for sea ice concentration. Using this "adjusted" Had1SST data set, September sea ice has declined at a rate of −7.8% per decade from 1953 through 2006 (see Fig. 5), while the March ice extent has retreated at a rate of −1.8% per decade. Relative to 1953–2000, ice extent during the last 5 years represents a 24% decrease in the area of the ocean covered by sea ice during September.

3 Modeling of Ice Thickness

Rothrock et al. (1999) suggest Arctic sea ice thickness declined by more than 40% between the 1990s and the 1950/1970s. However, these ice thickness measurements are sparse in space and time and do not cover the time period of recent large losses in ice extent. Therefore, we turn to analysis of model output from a high-resolution coupled ice-ocean model of the pan-Arctic region (Maslowski et al. 2004; Maslowski and Lipscomb 2003) forced with realistic atmospheric data for 1979–2004 to address questions related to recent changes in Arctic ice thickness. The sea ice model includes a parallel version of Zhang and Hibler (1997) dynamic model with viscous-plastic rheology and the thermodynamic model based on Parkinson and Washington (1979) with the zero-layer approximation for heat conduction through the ice (Semtner 1976). It is coupled to a regional adaptation of the Los Alamos National Laboratory Parallel Ocean Program (POP) with the free-surface approach (Dukowicz and Smith 1994) and unsmoothed realistic bathymetry. The model domain is pan-Arctic, including all northern hemisphere sea ice-covered oceans and marginal seas. It is configured on a 1/12° (or ~9 km) and 45 vertical level grid in a rotated spherical coordinate system to eliminate the singularity at the North Pole. The interannual integration started at the end of a 48-year spin up run including climatological forcing from the European Centre for Medium Range Weather Forecasts (ECMWF) for the first 26 years and followed with repeated 1979–1981 ECMWF forcing for 12 more years. Additional details about the coupled model, including its set up and forcing are discussed in Maslowski et al. (2004) and Clement et al. (2005). Model output includes monthly mean results from the sea ice model for 1979–2004, which are analyzed below.

Model results suggest (Fig. 3) that between 1979 and 1996 the ice thickness mode (as defined by the highest percentage of the total model grid cells per ice thickness bin) was in thickness range of 2.5–3.5 m. Beginning in 1997 through 2003 the thickness mode has shifted toward significantly thinner ice in the range of 1.0–2.5 m. The reduction of the modal ice thickness in the late 1990s represents the thinning of the multiyear pack ice toward the ice thickness range representative of the first-year ice. The modeled temporal change of ice thickness coincides with the observed September ice extent decrease during the last decade (e.g., Fig. 1). This implies that changes in sea ice extent are of the same sign as those of ice thickness, hence

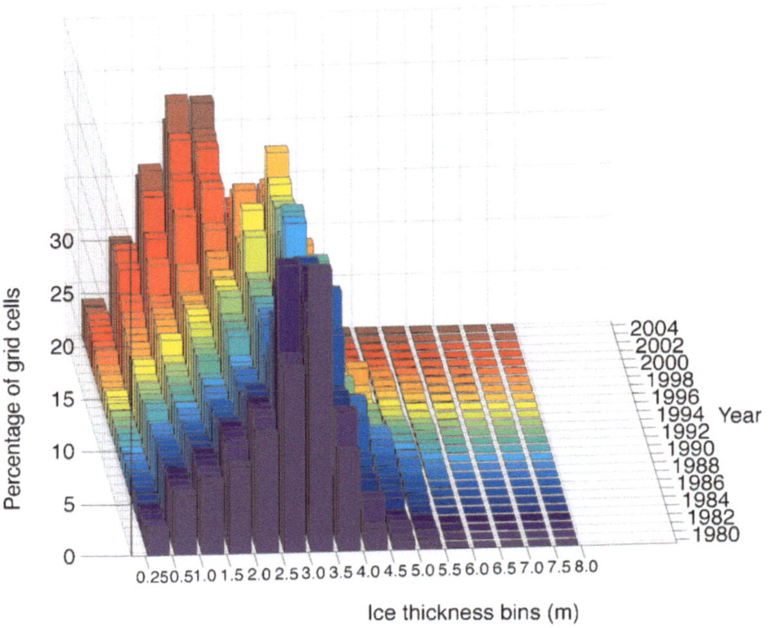

Fig. 3 Probability Distribution Function (PDF) of modeled annual mean binned ice thickness. The z-axis shows percentage of the total model grid cells per thickness bin defined along the x-axis. The y-axis is time in years from 1979 to 2004 with a different color histogram for each year.

also volume. However, according to Fig. 3 the rate of ice thickness reduction might be greater than that of the ice extent, given that the thickness of maximum PDF decreased by over 30% between 1979 and 2004.

4 Possible Causes of the Sea Ice Reduction

4.1 Impact of Atmospheric Temperatures on Sea Ice

Sea ice reduction can be attributed to warming and changes in atmospheric and oceanic circulation. Globally air temperatures have warmed by 0.8°C in the last century, but the warming has been especially pronounced in the Arctic, with a warming of 1.2°C since 1875 (Polyakov et al. 2003). In 2005, the Arctic was particularly warm, causing 2005 to surpass 1998 as the warmest year on record since 1880. Although each of the last 5 years had different atmospheric patterns that contributed to the large ice losses, anomalously high temperatures have been consistent throughout the Arctic since 2002 (Stroeve et al. 2004). Figure 4 shows annual NCEP/NCAR air temperature anomalies (Kalnay et al. 1996) as a function of height and latitude for 2000–2005 relative to 1979–1999. Air temperatures near

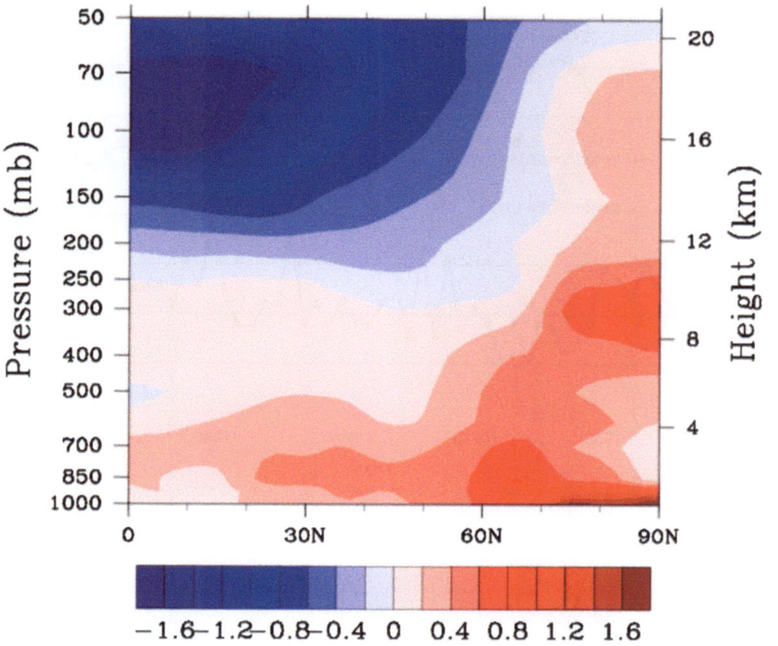

Fig. 4 NCEP/NCAR annual air temperature anomalies from 2000 to 2005 relative to 1979–1999 as a function of height and latitude. The scale goes from red for temperatures strongly above average to blue for temperatures strongly below average.

the surface were more than 1°C warmer over the Arctic, consistent with the large ice losses observed during these years.

The connection between rising air temperatures and decreasing summer ice extent holds for earlier years as well, and may help explain why ice extent was less in the 1950s than it was in the late 1960s and early 1970s. Figure 5 shows September ice extent is inversely correlated ($r = -0.6$) with the annual NCEP/NCAR 925 mbar air temperatures averaged from 60°N to 90°N. Using 12-month running mean temperature and sea ice anomalies increases the correlation to −0.74 (Meier et al. 2007). Since a warmer Arctic should manifest itself as changes in the melt season that favor less total ice cover, it is reasonable to suspect that the length of the melt season has increased in recent years, leading to earlier disappearance of the sea ice and later freeze-up. Stroeve et al. (2006) reported on changes in the Arctic melt season using estimates of melt onset dates and sea ice freeze-up dates derived from satellite passive microwave data. Results from this analysis show the Arctic is experiencing an overall lengthening of the melt season at a rate of about 2 weeks per decade. Melt is beginning on average 8 days/decade earlier than normal in the eastern Arctic (e.g., Kara and Barents Seas) and 4 days/decade earlier than normal in the western Arctic (e.g., Beaufort and Chukchi Seas) based on the satellite record (1979–2005). Freeze-up occurs about a week per decade later throughout most of the Arctic.

These observations are consistent with the hypothesis that warmer temperatures are a dominant factor in the continued decline of the Arctic ice cover. However, while

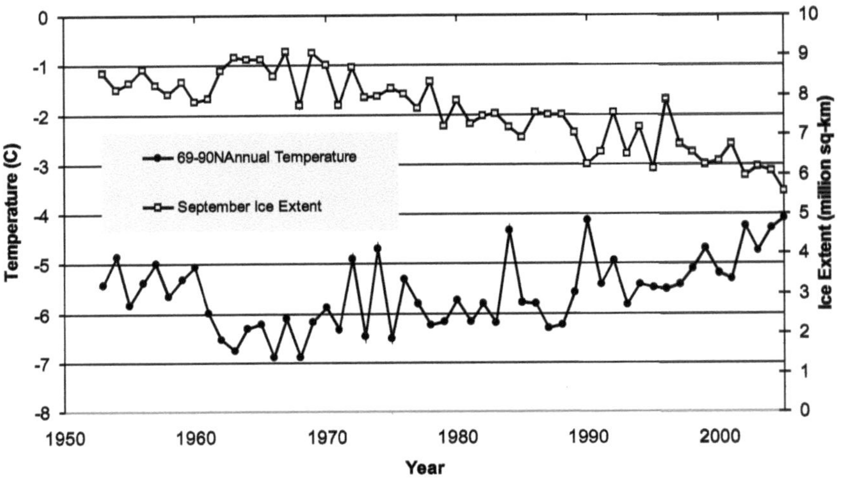

Fig. 5 Annual Arctic (60–90°N) 925 mbar NCEP/NCAR air temperatures and September sea ice extent from 1953–2005.

unusually warm air temperatures may help explain the decrease of the ice cover by causing the ice to melt earlier and form later, the higher air temperatures of recent years may also be a reflection of the loss of sea ice exposing the warmer ocean that transfers the energy back to the atmosphere (as sensible and latent heat).

4.2 Impact of the Arctic Oscillation on Sea Ice

In many studies, the positive mode of the Arctic Oscillation (AO) has been associated with decreasing Arctic sea ice cover in the 1990s (e.g., Rigor et al. 2002; Rigor and Wallace 2004; Zhang et al. 2003). Rigor et al. (2002) showed that when the AO is positive in winter, altered wind patterns result in more offshore ice motion and ice divergence along the Siberian and Alaskan coastal areas, leading to the production of more first-year ice in spring that is thinner and therefore requires less energy to melt in summer. Zhang et al. (2003) explained the lower summer ice cover during positive AO phases as a result of the spread of warmer air temperatures from Eurasia into the ocean. However, while this may have been the case through the mid-1990s, the AO has since become more neutral in recent winters, and yet the summer sea ice continues to decline. It is worth noting that anthropogenic forcing (i.e., warming) and the AO are likely not independent from each other. Some studies suggest anthropogenic warming may lead to increased amplitude and frequency of the positive AO phase, which would in turn cause the AO to strengthen the impact of the warming on the loss of sea ice. Rigor and Wallace (2004) suggested the strongly positive phase of the winter AO in the early to mid-1990s led to the export of thick, multiyear ice out of the Arctic basin, leaving behind thinner ice that is more easily melted. However, they estimate that the combined winter and summer AO-indices can explain less than 20% of the variance in summer sea ice extent in the western Arctic, where most of the recent reduction in sea ice cover occurred. A separate analysis of the combined effect of winds, radiative fluxes and advected heat, suggests that atmospheric forcing can account for about one half of the total variance in summer sea ice extent in the western Arctic (Francis et al. 2005). This implies that other factors must be involved in forcing variability of the Arctic sea ice cover.

4.3 Impact of Ocean Temperatures on Sea Ice

The role of oceanic forcing on the sea ice cover is explored using coupled ice-ocean pan-Arctic model results in the Greenland Sea (Maslowski et al. 2004), where sea ice has disappeared from the Greenland shelf for the first time in the summer of 2002 over the entire satellite record. Figure 6a shows the modeled anomalies of area-averaged monthly mean net ice melt/growth and total oceanic heat flux into the area indicated in Fig. 6b.

Oceanic heat fluxes associated with Atlantic Water recirculating at and to the south of Fram Strait can explain over 60% of the total variance in sea ice cover over the Greenland shelf on the annual basis (based on correlation of the 13-month running means – smoothed data). Recent observations across the West Spitsbergen Current (Walczowski and Piechura 2006) show significant warming and widening of Atlantic Water flowing to the north, which according to the model should be accompanied by an increased heat flux of Atlantic Water recirculating into the east

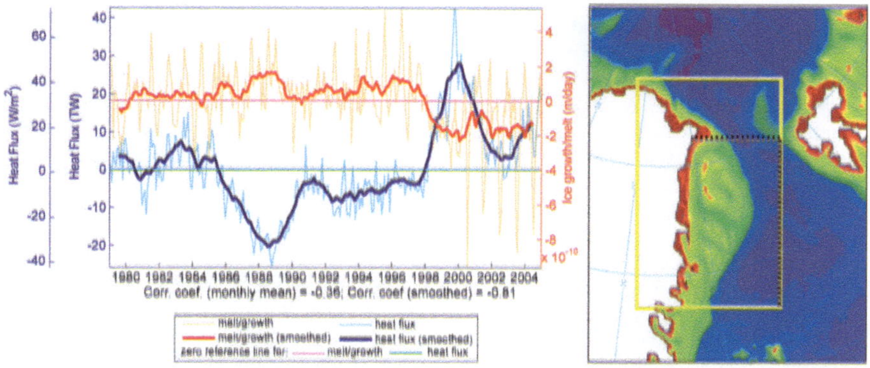

Fig. 6 (a) Total monthly mean heat flux onto the Greenland shelf (blue) and area-averaged monthly mean anomalies of ice melt/growth (red) after removing 24-year mean annual cycle. The blue curve also represents the flux per unit area (W/m^2; leftmost axis), in addition to the total heat flux into that area (TW; second left axis). The smoothed lines represent a 13-month running mean. (b) The area for ice melt/growth calculation (yellow rectangle) and two oceanic sections for the total on-shelf heat flux (black dotted lines).

Greenland current and onto the shelf. The decreasing ice cover, through positive ice-albedo feedback, may lead to further warming of the upper ocean and subsequent reductions of sea ice. Similar impacts of warm summer Pacific Water on sea ice cover in the western Arctic Ocean are discussed by Shimada et al. (2006) and are seen in model results.

5 Summary

The Arctic is currently experiencing a dramatic reduction in its sea ice cover. Taken together, the decline in ice thickness and extent imply a precipitous decline in total ice volume over the past 50 years. The decline in sea ice has been attributed to several factors such as rising air temperatures and variations in the Arctic Oscillation. However, the oceanic thermodynamic control of sea ice through under-ice ablation and lateral melt along marginal ice zones appears to be largely overlooked. Similarly, the effects of ice-free and warmer oceanic surface on the atmosphere need to be properly accounted for in the Arctic Ocean, especially in regional and global atmospheric model simulations, which commonly prescribe climatological sea ice and surface ocean conditions instead (Rinke et al. 2006). Those ocean-ice and ocean-atmosphere interactions may act to melt sea ice, as well as to de-correlate AO forcing, which helps explain the timing issues between AO/atmospheric forcing and sea ice variability in the 2000s.

References

ACIA (2005) *Arctic Climate Impact Assessment*. Cambridge University Press, 1042 p.

Bjørgo, E., O. M. Johannessen, and M. W. Miles, 1997: Analysis of merged SSMR-SSMI time series of Arctic and Antarctic sea ice parameters 1978–1995. *Geophys. Res. Lett.*, **24**(4), 413–416.

Cavalieri, D. J., J. Crawford, M. Drinkwater, W. J. Emery, D. T. Eppler, L. D. Farmer, M. Goodberlet, R. Jentz, A. Milman, C. Morris, R. Onstott, A. Schweiger, R. Shuchman, K. Steffen, C. T. Swift, C. Wackerman, and R. L. Weaver, 1992: *NASA Sea Ice Validation Program for the DMSP SSM/I: Final Report*. NASA Technical Memorandum 104559. National Aeronautics and Space Administration, Washington, DC, 126 p.

Cavalieri, D., C. Parkinson, P. Gloersen, H. J. Zwally, 1996, updated 2006: *Sea ice Concentrations from Nimbus-7 SMMR and DMSP SSMI Passive Microwave Data* (list dates used). National Snow and Ice Data Center, Digital Media, Boulder, CO.

Cavalieri, D. J., P. Gloersen, C. L. Parkinson, J. C. Comiso, and H. J. Zwally, 1997: Observed hemispheric asymmetry in global sea ice changes. *Science*, **272**, 1104–1106.

Clement, J. C., W. Maslowski, L. W. Cooper, J. M. Grebmeier, and W. Walczowski, 2005: Ocean circulation and exchanges through the northern Bering Sea – 1979–2001 model results. *Deep Sea Res. II*, **52**(24–26), 3509–3540.

Clement, J. L., W. Maslowski, L. Cooper, J. Grebmeier, and W. Walczowski, 2005: Ocean circulation and exchanges through the northern Bering Sea – 1979–2001 model results. *Deep Sea Res. II*, **52**, 3509–3540, doi:10.1016/j.dsr2.2005.09.010.

Comiso, J. C., 2002: A rapidly declining perennial sea ice cover in the arctic. *Geophys. Res. Lett.*, **29**(20), 1956, doi:10.1029/2002GL015650.

Dukowicz, J. K. and R. D. Smith, 1994: Implicit free-surface method for the Bryan-Cox-Semtner ocean model. *J. Geophys. Res.*, **99**(C4), 7991–8014.

Foley, J. A., 2005: Tipping points in the Tundra. *Science*, **310**, 627–28, doi:10.1126/science.1120104.

Francis, J. A., E. Hunter, J. R. Key, and X. Wang, 2005: Clues to variability in Arctic minimum sea ice extent. *Geophys. Res. Lett.*, **32**, 21501, doi:10.1029/2005GL024376.

Haas, C., 2004: Late-summer sea ice thickness variability in the Arctic transpolar drift 1991–2001 derived from ground-based electromagnetic sounding. *Geophys. Res. Lett.*, **31**, L09402, doi:10.1029/2003GL019394.

Holland, M. M., C. M. Bitz, and B. Tremblay, 2006: Future abrupt reductions in the summer Arctic sea ice. *Geophys. Res. Lett.*, **33**, L23503, doi:10.1029/2006GL028024.

Johannessen, O., E. V. Shalina, and M. W. Miles, 1999: Satellite evidence for an arctic sea ice cover in transformation. *Science*, **286**, 1937–1939.

Kalnay, E. et al., 1996: The NCEP/NCAR reanalysis 40-year project. *Bull. Am. Meteorol. Soc.*, **77**, 437–471.

Lindsay, R. W. and J. Zhang, 2005: The thinning of Arctic Sea ice, 1988–2003: have we passed a tipping point? *J. Climate*, **18**, 4879–94, doi:10.1175/JCLI3587.1.

Maslowski, W. and W. H. Lipscomb, 2003: High-resolution simulations of Arctic Sea ice during 1979–1993. *Polar Res.*, **22**, 67–74.

Maslowski, W., D. Marble, W. Walczowski, U. Schauer, J. L. Clement, and A. J. Semtner, 2004: On climatological mass, heat, and salt transports through the Barents Sea and Fram Strait from a pan-Arctic coupled ice-ocean model simulation. *J. Geophys. Res.*, **109**, C03032, doi:10.1029/2001JC001039.

Meier, W. N., J. Stroeve, F. Fetterer, and K. Knowles 2005: Reductions in Arctic Sea ice cover no longer limited to summer. Eos, *Trans. Amer. Geophys. Union*, **86**(36), 326.

Meier W. N., J. Stroeve, and F. Fetterer, 2007: The declining Arctic Sea ice: how much of an indicator of change is it? *Ann. Glaciol.*, **46**.

Melling, H. D., Riedel, and Z. Gedalof, 2005: Trends in the droft and extent of seasonal pack ice, Canadian Beaufort Sea. *Geophys. Res. Lett.*, **32**, doi:10.1029/2005GLO24483.

Overpeck, J. T. et al., 2005: Arctic system on trajectory to new, seasonally ice-free state. Eos, *Trans. Amer. Geophys. Union*, **86**, **309**, 312–313.

Parkinson, C. L. and W. M. Washington, 1979: A large-scale numerical model of sea ice. *J. Geophys. Res.*, **84**, 311–337.

Parkinson, C. L., D. J. Cavalieri, P. Gloersen, H. J. Zwally, and J. C. Comiso, 1999: Arctic Sea ice extents, areas, and trends, 1978–1996. *J. Geophys. Res.*, **104**(C9), 20837–20856.

Polyakov, I. V., R. V. Bekryaev, G. V. Alekseev, U. Bhatt, R. L. Colony, M. A. Jonson, A. P. Makshtas, and D. Walsh, 2003: Variability and trends of air temperature and pressure in the maritime Arctic, 1875–2000. *J. Climate.*, **16**(12), 2067–2077.

Rayner, N. A., D. E. Parker, E. B. Horton, C. K. Folland, L. V. Alexander, D. P. Rowell, E. C. Kent, and A. Kaplan, 2003: Global analyses of sea surface temperature, sea ice, and night marine air temperature since the late nineteenth century. *J. Geophys. Res.*, **108**(D14), 4407, doi:10.1029/2002JD002670.

Rigor, I. G. and J. M. Wallace, 2004: Variations in the age of sea ice and summer sea ice extent. *Geophys. Res. Lett.*, **31**, doi:10.1029 /2004GL019492.

Rigor, I. G., J. M. Wallace, and R. L.Colony, 2002: Response of sea ice to the Arctic oscillation. *J. Climate*, **15**(18), 2648–2668.

Rinke, A., W. Maslowski, K. Dethloff, and J. Clement, 2006: Influence of sea ice on the atmosphere: a study with an Arctic atmospheric regional climate model. *J. Geophys. Res.*, **111**, D16103, doi:10.1029/2005JD006957.

Rothrock, D. A., Y. Yu, and G. A. Maykut, 1999: Thinning of the Arctic Sea ice cover. *Geophys. Res. Lett.*, **26**, 3469–3472.

Rothrock, D. A., J. Zhang, and Y. Yu, 2003: The Arctic ice thickness anomaly of the 1990s: a consistent view from observations and models. *J. Geophys. Res.*, **108**, C3, doi:10.1029/2001 JC001208.

Semtner, A. J., 1976: A model for the thermodynamic growth of sea ice in numerical investigations of climate. *J. Phys. Oceanogr.*, **6**, 379–389.

Serreze, M. C., J. A. Maslanik, T. A. Scambos, F. Fetterer, J. Stroeve, K. Knowles, C. Fowler, S. Drobot, R. G. Barry, and T. M. Haran, 2003: A record minimum Arctic Sea ice extent and area in 2002. *Geophys. Res. Lett.*, **30**(3), 1110, doi:10.1029/2002GL016406.

Shimada, K., T. Kamoshida, , M. Itoh, S. Nishino, E. Carmack, F. McLaughlin, S. Zimmermann, and A. Proshutinsky, 2006: Pacific ocean inflow: influence on catastrophic reduction of sea ice cover in the Arctic Ocean. *Geophys. Res. Lett.*, **33**, L08605, doi:10.1029/2005GL025624.

Stroeve, J. C., M. C. Serreze, F. Fetterer, T. Arbetter, W. Meier, J. Maslanik, and K. Knowles, 2005: Tracking the Arctic's shrinking ice cover: another extreme September minimum in 2004. *Geophys. Res. Lett.*, **32**, L04501, doi:10.1029/2004GL021810.

Stroeve, J., T. Markus, W. N. Meier, and J. Miller, 2006: Recent changes in the Arctic melt season, *Ann. Glaciol.*, **44**, 367–374.

Walczowski, W. and J. Piechura, 2006: New evidence of warming propagating toward the Arctic Ocean. *Geophys. Res. Lett.*, **33**, L12601, doi:10.1029/2006GL025872.

Zhang, J. and W. D. Hibler, 1997: On an efficient numerical method for modeling sea ice dynamics. *J. Geophys. Res.*, **102**, 8691–8702.

Zhang, J., D. Thomas, D. A. Rothrock, R. Lindsay, Y. Yu, and R. Kwok, 2003: Assimilation of ice motion observations and comparisons with submarine ice thickness data. *J. Geophys. Res.*, **108**(C6), 3170, doi:10.1029/2001JC001041.

Decadal Changes in Surface Radiative Fluxes and Their Role in Global Climate Change

M. Wild

Abstract The major anthropogenic impact on climate over the 20th century occurred through a modification of the earth radiation balance by changing the amount of greenhouse gases and aerosol in the atmosphere. Radiative energy reaching the ground is particularly important for mankind as it is a key determinant of the climate of our environments and strongly influences the thermal and hydrological conditions at the Earth surface. Recent evidence suggests that significant anthropogenic-induced variations occurred in both surface solar and thermal radiation over the past decades, related to anthropogenic air pollution and greenhouse gas emissions, respectively. Observed solar radiation incident at the surface showed a continuous decrease ("global dimming" or "surface solar dimming") since the beginning of worldwide measurements in the mid-20th century up to the 1980s, when a widespread trend reversal towards an increase ("global brightening" or "surface solar brightening") occurred. This trend reversal was favoured by an increasing transparency of the cloud-free atmosphere, due to air pollution regulations and the breakdown of the economy in former communist countries. In the thermal spectrum of radiation, which is directly modified by changes in atmospheric greenhouse gas concentrations, a gradual increase in surface downwelling thermal radiation over recent years can be seen, in line with our expectations from an increasing greenhouse effect. This increasing greenhouse effect has become only fully apparent after the decline of solar dimming, which effectively masked greenhouse warming prior to the 1980s. The present article discusses the variations in surface radiation and their impact on various aspects of the climate system over the past decades.

1 Introduction

The climate of the 20th century has undergone significant decadal changes of natural and anthropogenic origin. Many studies have documented the changes in surface temperature and global warming over the past decades. Yet only few studies investigated

Institute for Atmospheric and Climate Science, ETH Zurich, Universitätstr. 16, CH-8092 Zurich, Switzerland
wild@env.ethz.ch

S. Brönnimann et al. (eds.), *Climate Variability and Extremes during the Past 100 Years.*

the underlying surface energy balance components, which are the key processes governing the long-term evolution of surface temperature and climate. The largest components in terms of energy exchanges are thereby the surface radiative fluxes. Changes in these fluxes have therefore the potential to significantly alter climate. In the present study we summarize the current knowledge on decadal variations in the surface radiative fluxes and discuss how these variations relate to global warming and how they affect other elements of the climate system, such as the water cycle, soil moisture or glaciers.

2 Observational Data on Surface Radiation

Widespread measurements of surface radiative fluxes began in the late 1950s. Many of these historic radiation measurements have been collected in the Global Energy Balance Archive (GEBA, Ohmura et al. 1989) located at ETH Zurich, and in the World Radiation Data Centre (WRDC) of the Main Geophysical Observatory in St. Petersburg. In addition, more recently, high quality surface radiation measurements from the Baseline Surface Radiation Network (BSRN, Ohmura et al. 1998) of the World Climate Research Program (WCRP) and the Atmospheric Radiation Program (ARM, Ackerman and Stokes 2003) became available. These networks measure surface radiative fluxes at the highest possible accuracy with well-defined and calibrated state-of-the-art instrumentation at selected worldwide distributed sites. A further network of high-accuracy measurements at five remote sites is maintained by the Global Monitoring Division of NOAA (Dutton et al. 2006).

3 Observed Changes in Surface Radiation

3.1 The Era of "Global Dimming" from the 1960s to the 1980s

Trends in solar radiation from the beginning of worldwide measurements in the early 1960s until 1990 have been determined in numerous studies (e.g., Ohmura and Lang 1989; Gilgen et al. 1998; Stanhill and Cohen 2001 and references therein; Liepert 2002; Wild et al. 2004). These studies report a general decrease of sunlight at widespread locations over land surfaces between 1960 and 1990. This phenomenon is now popularly known as "global dimming", an expression coined by Stanhill and Cohen (2001). This expression may be misleading since it implies a globally unequivocal occurrence of the "dimming" which is difficult to prove due to the lack of observations from many parts of the world, particularly from ocean and remote land areas. Note that satellite data, which would provide a global coverage, only date back to the early 1980s and do not provide stable and reliable time series due to calibration problems and difficulties in inferring surface fluxes from their top-of-atmosphere measurements.

Increasing air pollution and associated increase in aerosol concentration are considered major causes of the measured decline of surface solar radiation, although changes in cloud amount and optical properties may also contribute. A direct assessment of the origins of the decline of surface solar radiation is complicated by the lack of adequate long-term observational data sets of major influencing factors such as clouds and aerosol. An attempt has been made in Norris and Wild (2007) to differentiate between aerosol and cloud impact on radiative changes over Europe. They show that changes in cloud amount cannot explain the changes in surface insolation, pointing to aerosol direct and indirect effects as major cause of these variations. Alpert et al. (2005) found that the decline in surface solar radiation in the 1960–1990 period is largest in areas with dense population, which also suggest a significant anthropogenic influence through air pollution and aerosol. Their results further suggest that dimming, of smaller magnitude, occurred also at sparsely populated sites. This is in line with other studies showing a dimming over the 1960–1990 period also at remote sites (e.g., Dutton et al. 2006), suggesting that the dimming is not of purely local nature and air pollution may have large-scale climatic effects.

Modelling approaches can give additional insight into the possible origins, distribution and magnitude of surface solar dimming as will be discussed in Sect. 4. In a study by Wild et al. (2004), estimates of changes in the different radiation components over land surfaces were brought together for the period 1960–1990. These estimates indicate that the reported surface solar dimming over global land surfaces, if real and representative, may have exceeded the greenhouse induced increase in downward thermal radiation. This would imply a slight reduction rather than an increase in radiative heating between 1960 and 1990 (Wild et al. 2004). This is in line with an independent analysis of long-term records of 66 surface net radiation stations over the same period, which also suggests a slight decrease rather than an increase in radiative energy available at the surface between 1960 and 1990 (Wild et al. 2004). This may have had significant impacts on the thermal and hydrological conditions of the climate system as outlined in Sect. 5.

3.2 The Era of "Global Brightening" From the 1980s to 2000

The studies on global dimming were all limited to data prior to 1990, since no data after 1990 were available at the time of the completion of these studies. ETH Zurich recently undertook a major effort to update the worldwide measured surface radiation data in GEBA for the period from 1990 to present. Wild et al. (2005) evaluated the newly available surface observational records in GEBA and BSRN to investigate the trends in surface solar radiation in the more recent years. This analysis showed that the decline in solar radiation at land surfaces seen in earlier data is no longer visible in the 1990s. Instead, the decline leveled off or even turned into a brightening since the late 1980s at the majority of observation sites (Fig. 1). The brightening during the 1990s is not just found under all sky conditions, but also under clear skies, pointing to aerosol as major cause of this trend reversal (Wild et al. 2005;

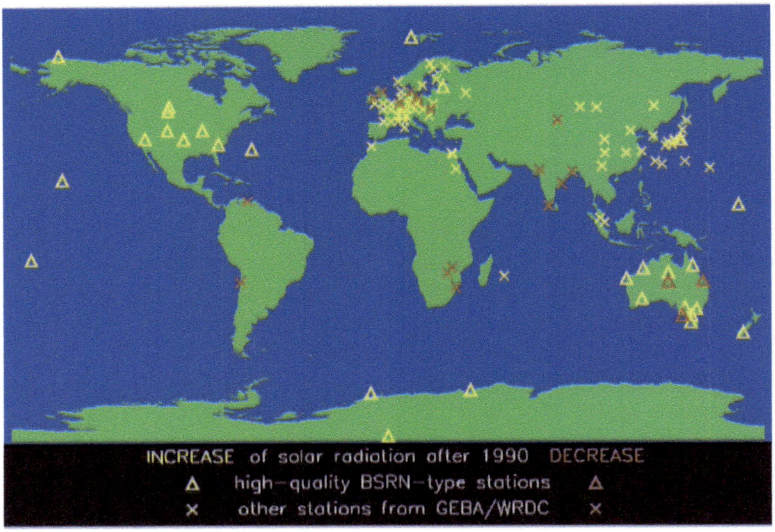

Fig. 1 Global distribution of surface radiation sites with data covering the 1990s. Sites measuring an increase in surface solar radiation after 1990 are marked in yellow; sites measuring a decrease are shown in brown. High-quality observation sites fulfilling the BSRN standards are shown as triangles; other sites from the updated GEBA as crosses. Information from 300 sites over Europe and 45 sites over Japan are displayed as aggregated regional means. The majority of the sites show an increase in surface solar radiation after 1990 (from Wild et al. 2005).

Norris and Wild 2007). The trend reversal is reconcilable with recently estimated radiation trends from satellites and from the Earthshine method (Pallé et al. 2005), as well as with trends in aerosol emission and atmospheric transmission, which also show a distinct trend reversal during the 1980s (Streets et al. 2006; Stern 2006; Wild et al. 2005). The documented trend reversal in aerosol emission towards a reduction and the associated increasing atmospheric transmission since the mid-1980s may be related to air pollution regulations and the breakdown of the economy in former communist countries (Wild et al. 2005).

Climate-relevant surface radiative forcing in recent decades has not only been caused by surface solar variations, but has also been strongly governed by the reduced thermal surface cooling with enhanced greenhouse capacity of the atmosphere, manifest in increased downward thermal radiation from the atmosphere to the surface. Evidence for increasing downward thermal radiation is obtained from the surface measurements of the BSRN. Downward thermal radiation measured at 12 worldwide distributed sites from BSRN shows on average an increase of 0.26 W m^{-2} per year since the beginning of the measurements in 1992 (Wild and Ohmura 2004), in line with our expectations from greenhouse theory and models (Wild et al. 1997, Fig. 1). A significant increase in downward thermal radiation over recent years was also found by Philipona et al. (2004) in the Swiss Alps. Therefore, unlike in the preceding decades, where a reduction in downwelling surface solar radiation compensated for the greenhouse-induced increase in downwelling surface thermal radiation, both downwelling solar and thermal radiation show signs of increase since the mid-1980s. Therefore, also the surface net radiation (the available radiative energy at the surface) likely increased over the 1990s, in contrast to previous decades, which rather showed a decrease as noted in Wild et al. (2004). This is critical for the global water cycle, since the surface net radiation is the principal driver behind the hydrological cycle (see Sect. 5).

4 Modelling of Global Dimming and Brightening

Complementary to the observational data analyses presented so far, General Circulation Models (GCMs) can be used to get more insight into the variations in surface radiative fluxes and their origins. To interpret the role of aerosol in the context of surface solar dimming and brightening, transient GCM simulations with an interactive time-dependent representation of aerosol are particularly useful. However, so far most transient GCM simulations used either prescribed aerosol climatologies, or restricted their prognostic aerosol treatment to sulphate, as for example, in the majority of the simulations carried out for the IPCC AR4 report. This is a considerable limitation, since not only scattering aerosol such as sulphates, but particularly also absorbing aerosols not considered in these experiment have shown significant variations over the past decades (Streets et al. 2006) and are believed to be crucial for the understanding of the variations in surface solar radiation. We addressed this issue using a version of the Max Planck Institute GCM, which includes a microphysical formulation of all major global aerosol components, with prognostic treatment of their composition, size-distribution and mixing state (ECHAM5-HAM, Stier et al. 2005). We investigated the temporal evolution of surface solar clear sky fluxes simulated in a transient experiment with ECHAM5-HAM which uses time-dependent aerosol emissions (Stier et al. 2006). The model successfully simulates a general decrease in clear sky surface solar radiation from the 1950s to the 1980s and a more recent recovery over large parts of the globe (Fig. 2), in line with observational evidence (cf. Fig. 2b) with Fig.1 (Wild et al., in preparation). This suggests that direct aerosol effects may have played a significant role in the observed changes in surface solar radiation and in the recent transition from dimming to brightening. This supports the observational evidence for the importance of aerosol in this context found in Wild et al. (2005) and Norris and Wild (2007). A further inspection of Fig. 2 suggests a distinct latitudinal dependence of surface solar dimming and brightening (Wild et al., in preparation). While most of the extratropics show a reversal from dimming to brightening during the 1980s in this model simulation, dimming persists up to the present day in many low latitude areas. This is favoured by a transition from increasing to decreasing sulphur and black carbon emissions in industrialized countries since the 1980s (Streets et al. 2006), which are mostly situated in the extratropics. Some of the developing countries in the low latitudes, on the other hand, show a continuing increase in aerosol

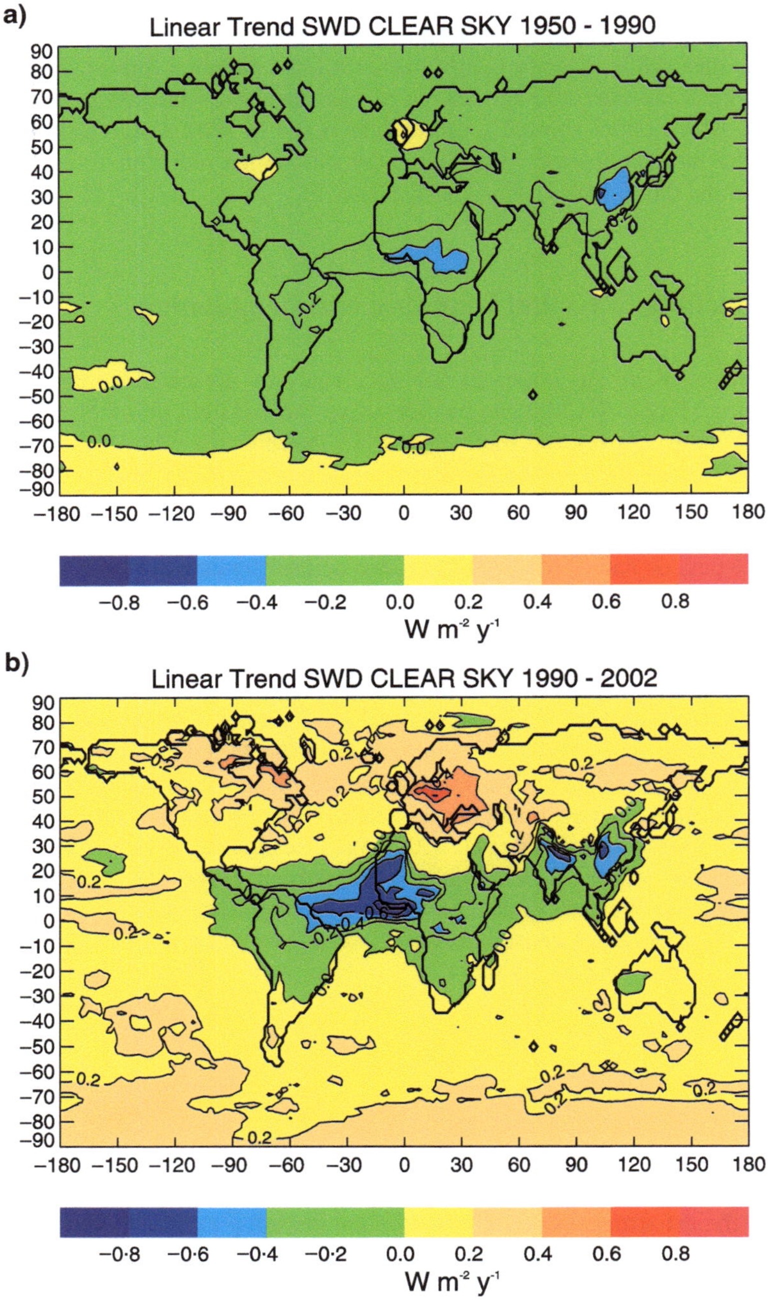

Fig. 2 Simulated trends in surface solar radiation under cloud-free conditions for the "surface solar dimming" period 1950–1990 (a) and the "surface solar brightening" period 1990–2002 (b). Simulations done with the aerosol-climate modelling system ECHAM5 HAM. Units are W m^{-2}/y.

emissions, contributing to the continuing dimming in these areas. There are not enough direct observations in the tropics to strictly verify this latitudinal dependence of solar dimming and brightening (cf. Fig. 1). However, the available observations are not in conflict with this hypothesis.

5 Role of Global Dimming and Brightening in the Context of Global Climate Change

5.1 Impact on Global Warming

Solar and thermal radiative fluxes reaching the ground are key determinants of surface temperature. Concerns have been raised that increases in aerosol from air pollution and associated dimming of surface solar radiation (see Sect. 3) could have masked to a large extent the temperature rise induced by increasing greenhouse gases, so that the observed temperature records would not reflect the entire dimension of greenhouse warming (Andreae et al. 2005). This would imply that we underestimate the sensitivity of the climate system to increased levels of greenhouse gases, which has potentially major implications for predictions of future climate. On the other hand, the above discussed evidence fore a widespread decline of solar dimming during the 1980s and reversal to a brightening thereafter may give raise to speculations that recent global warming could be due to surface solar brightening rather than the greenhouse effect. In Wild et al. (2007), we investigated this issue by using, in addition to the surface radiation data from GEBA and BSRN, a gridded surface temperature data set from the Climate Research Unit (CRU), University of East Anglia (Mitchell and Jones 2005). We focused on land surfaces and on the period 1958–2002, where our knowledge of the common variation in surface radiation and temperature is best. Annual mean temperature changes averaged over land surfaces from 1958 to 2002 as deviations from 1960 are determined from the CRU data set in Fig. 3

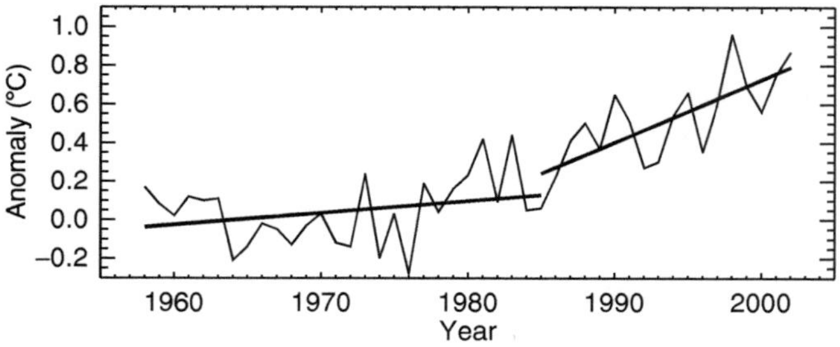

Fig. 3 Temperature evolution over the global land surfaces from 1958 to 2002 with respect to 1960. While the temperature rise during the period of surface solar dimming from the 1960s to the 1980s is moderate, temperature rise is more rapid in the last two decades where dimming was no more present. Derived from CRU temperature data set (Mitchell and Jones 2005). Units are °C (from Wild et al. 2007).

(Wild et al. 2007). It is evident that the temperature rise was small in the first half of the period, but became significant during the second half. In the timeframe 1958–1985, representing the period of "global dimming", only a marginal linear temperature increase is found (0.0036°C/year or 0.036°C/decade). The linear temperature increase during the period 1985–2002, where "global dimming" was no longer effective, is an order of magnitude larger (0.038°C/year or 0.38°C/decade). This indicates that changes in surface solar radiation may have effectively influenced the temperature evolution over the past decades.

To separate the influence of surface solar and thermal radiation on global warming in Wild et al. (2007) we use the fact that solar and thermal radiations have different effects on the daily temperature cycle. Since the solar flux is only in effect during daylight, it affects the daily maximum temperature (TMAX) more than daily minimum temperature (TMIN). TMIN, on the other hand, is mainly affected by the thermal radiative exchanges. An analysis of TMIN and TMAX therefore holds the potential to separate the influence of solar and thermal radiation on surface temperature. In Wild et al. (2007) we found that during the period with prevailing solar dimming, TMAX is in fact is slightly declining, at 0.04°C/decade between 1958 and 1985. This decline is consistent with the decreasing availability of solar energy at the surface and supports the assumption that solar dimming had a discernible influence on surface temperature in this period. TMIN on the other hand, less affected by solar dimming, shows an increase over the same period, at 0.11°C/decade, indicative of an increasing greenhouse forcing. On the other hand, the period 1985–2002, during which surface solar dimming reversed into brightening, exhibits a significant increase in both TMAX and TMIN, at 0.37°C/decade and 0.40°C/decade, respectively (Wild et al. 2007). This implies that the increase in TMAX has almost caught up to the increase in TMIN since 1985, and is in line with the hypothesis that solar dimming was not present anymore to prevent TMAX from keeping pace with TMIN. These tendencies are also evident in changes in the diurnal temperature range (DTR), defined as difference between TMAX and TMIN. DTR variations correlate very well with variations in surface solar radiations, as for example, shown by Bristow and Campbell (1984) and Liu et al. (2004a). In Fig. 4 we determined the change in annual land mean DTR for the same period 1958–2002 from the CRU data set, with respect to the 1960–1990 mean. Several studies reported a decrease in this quantity over much of the 20th century (Karl et al. 1993; Easterling et al. 1997; Dai et al. 1999). Figure 4 shows that the decrease of DTR over the global land surfaces only lasted into the mid-1980s, but then levelled off. This has also been noted in Vose et al. (2005), but has not been interpreted further. In Wild et al. (2007) we point to the striking similarity between the distinct change in the DTR regime during the 1980s in Fig. 4, and the simultaneous change in the surface solar radiation regime from dimming to brightening (Sect. 3). Thus, the evolution of DTR provides independent evidence for a large-scale change in the surface radiative forcing regime during the 1980s and its major impact on temperature. An interesting feature is that the levelling off in the DTR is more pronounced in higher than in low latitudes, supporting the hypothesis of a latitudinal dependence of the strength of the brightening presented in Sect. 4. The close correlation between DTR and surface insolation changes may

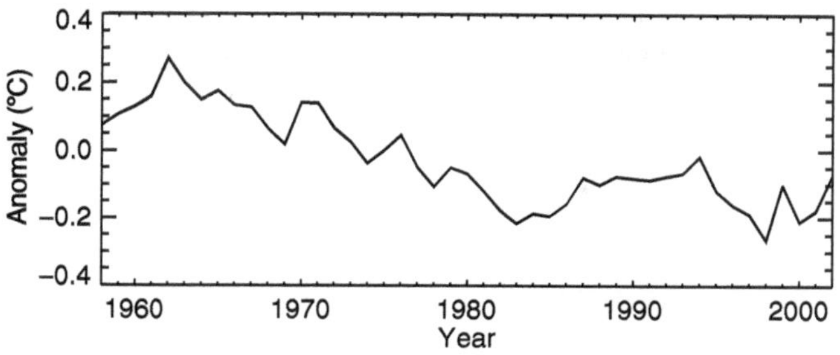

Fig. 4 Anomalies in diurnal temperature range averaged over the global land surfaces from 1958 to 2002, shown as deviations from the 1961–1990 mean. Diurnal temperature range decreases until the mid-1980 and levels off afterwards. Derived from CRU temperature data set (Mitchell and Jones 2005). Units are °C (from Wild et al. 2007).

also allow a reconstruction of surface insolation trends back to the first part of the 20th century, when radiation measurements were lacking, but DTR observations were already abundant. An analysis of the global land mean DTR record from the CRU data set over the whole 20th century shows that the decline in DTR noted above in the 1958–1985 period was not present in the first half of the 20th century. This suggests that global dimming may not have been present the first half of the 20th century, but rather a form of an "early brightening".

Note that despite the widespread trend reversal from dimming to brightening, the amount of solar radiation incident at the Earth surface at the turn of the millennium has not gone back to the 1960 levels. Nevertheless, land surface temperatures have increased by 0.8°C over the period 1960–2000 (Fig. 3). This suggests that the net effect of surface solar forcing over the past decades cannot be the principal driver behind the overall temperature increase, since over the past 40 years, cooling from solar dimming still outweighs warming from solar brightening. Rather, the overall temperature increase since the 1960s can be attributed to (thermal) greenhouse forcing as also evident in the thermal BSRN radiation data outlined above. Thus, speculations that solar brightening rather than the greenhouse effect could have been the main cause of the overall global warming over the past decades appear unfounded (Wild et al. 2007). Rather, it implies a substantial increase in thermal (greenhouse) surface radiative forcing since the 1960s, in order to enhance land surface temperatures by 0.8°C despite indications for a reduced heating due to surface solar changes over the past 40 years.

5.2 Impact on the Global Hydrological Cycle, Soil Moisture, and Glaciers

Radiative energy available at the earth's surface is the principal driver of the hydrological cycle (Ramanathan et al. 2001). Variations in the surface radiation balance induce changes in evaporation and associated precipitation, and thereby govern the intensity of the global hydrological cycle. Potential additional energy from surface solar brightening noted during the 1990s may not only have gone into heating of the surface, but also into additional evaporation. This is supported by pan-evaporation measurements in energy-limited environments which partially indicate, after decades of decrease (Roderick and Farquhar 2002; Ohmura and Wild 2002), a recent recovery (e.g., Liu et al. 2004b), in line with the changes in surface solar and net radiation discussed in Sect. 3. Also, there is evidence for an acceleration of the hydrological cycle over land surfaces during the 1990s, with increasing land precipitation after decreasing tendencies in the 1970s and 1980s (Beck et al. 2005). Here, I want to point out that that this is consistent with the increasing availability of surface radiative energy from both increasing downwelling solar and thermal radiation during the 1990s which fuelled the hydrological cycle (Wild and Grieser, in preparation). On the other hand, radiative energy available at the surface was rather decreasing prior to the 1990s, when global dimming outweighed the greenhouse-induced increase in downwelling thermal radiation, thereby attenuating the hydrological cycle (Wild et al. 2004) (see Sect. 3). This implies that variations in surface solar and thermal radiation, related to anthropogenic air pollution and greenhouse gas changes, respectively, indeed had a discernible impact on the variations of the global hydrological cycle over the past decades.

In a modelling study, Rotstayn and Lohmann (2002) present evidence that enhanced northern hemispheric anthropogenic aerosol emissions and associated radiative forcing may have favoured the severe droughts and starvation in the Sahel region in the 1970s and 1980s. It is interesting to note here that in the 1990s, when northern hemispheric aerosol forcing showed a tendency to decrease, precipitation in the Sahel started to recover. It can be speculated that the transition from global dimming to brightening may have helped to restore vital precipitation in the Sahel.

The imprint of global dimming and brightening can also be seen in long-term observations in soil moisture. A 45-year record of soil moisture measurements in the Ukraine covering the period 1958–2002 shows an increase in soil moisture in the first half of the record, followed by a slight decrease in the latter half (Robock et al. 2005). These changes cannot be explained by changes in precipitation alone. Rather, they suggest that the decrease in surface solar radiation caused a decrease in surface evaporation up to the 1980s, which lead to increasing soil moisture, while the more recent surface solar brightening induced a recovery of surface evaporation, and thereby a depletion of soil moisture (Li et al. 2007).

Another consequence of the global dimming/brightening phenomenon can be seen in the behaviour of mountain glaciers in Switzerland. Between 1973 and 1985, no significant changes in glacier area were found (1%). However, from 1985–2000 glacier area extent in Switzerland was reduced by 18% (Paul et al. 2004). This is also in line with the hypothesis portrayed here, that the greenhouse effect was masked by surface solar dimming up to the 1980s, so that no major changes in glacier area extent occurred, and only showed its full dimension and impact after dimming has declined in the mid-1980s. Since then no dimming was around anymore to counterbalance the enhanced greenhouse effect, resulting in a strong reduction of glacierized area.

6 Summary

Available observational data enable to trace the amount of solar radiation incident at the Earth surface back to the mid-20th century at worldwide distributed sites. Analyses of these records suggest that surface solar radiation declined in the first decades of the second half of the 20th century, a phenomenon known as "global dimming" or "solar dimming". A major contributor to this decline is assumed to be the increasing air pollution. Evidence has been presented that the relative cooling due to the dimming was able to outweigh greenhouse warming up to the mid-1980s. Since then recent studies suggest that global dimming faded and reversed into a brightening, partially because of reduced aerosol content in the atmosphere related to air pollution measures and the breakdown of the communist economy in Eurasia. With the fade of the dimming in the 1980s, the greenhouse effect became finally visible at its full dimension. We showed that this had major consequences for various elements of the climate system. For example, surface temperature rise accelerated over recent decades when the damping effect of solar dimming was no longer present. The increase in available surface energy from both increasing downwelling solar and thermal radiation may have been at the origin of the recent acceleration of the hydrological cycle. It may also have contributed to the significant retreat of Swiss glaciers in the same period. It is therefore evident that the variations in surface radiative fluxes have largely influenced the evolution of climate over the 20th century.

Acknowledgements Atsumu Ohmura is highly acknowledged for his valuable input to this study and his leadership in the establishment of the GEBA and BSRN databases. Hans Gilgen and Andreas Roesch made a tremendous effort to manage the GEBA and BSRN databases. This study is supported by the National Center of Competence in Climate Research (NCCR Climate) sponsored by the Swiss National Science Foundation. Climate modelling at ETH is generously funded with computational resources by the Swiss Scientific Computing Center CSCS.

References

Ackerman, T. P. and G. M. Stokes, 2003: The atmospheric radiation measurement program. *Phys. Today*, **56**, 38–44.

Alpert, P., P. Kishcha, J. Y. Kaufman, and R. Schwarzbard 2005: Global dimming or local dimming? Effect of urbanization on sunlight availability. *Geophys. Res. Lett.*, **32**, L17802, doi:10.1029/2005GL023320.

Andreae, M. O., C. D. Jones, and P. M. Cox, 2005: Strong present day aerosol cooling implies a hot future. *Nature*, **435**, 1187–1190.

Beck, C., J. Grieser, and B. Rudolf, 2005: A new monthly precipitation climatology for the global land areas for the period 1951 to 2000. *Klimastatusbericht 2004*, DWD, 181–190.

Bristow, K. L. and S. Campbell, 1984: On the relationship between in coming solar radiation and daily maximum and minimum temperature. *Agr. Forest Meteorol.*, **31**, 159–166.

Dai, A., K. E. Trenberth, and T. R. Karl, 1999: Effects of clouds, soil moisture, precipitation, and water vapour on diurnal temperature range. *J. Climate*, **12**, 2451–2473.

Dutton, E. G., D. W. Nelson, R. S. Stone, D. Longenecker, G. Carbaugh, J. M. Harris, and J. Wendell, 2006: Decadal variations in surface solar irradiance as observed in a globally remote network. *J. Geophys. Res.*, **111**, D19101, doi:10.1029/2005JD006901.

Easterling, D. R. et al., 1997: Maximum and minimum temperature trends for the globe. *Science*, **277**, 364–367.

Gilgen, H., M. Wild, and A.Ohmura, 1998: Means and trends of shortwave irradiance at the surface estimated from global energy balance archive data. *J. Climate*, **11**, 2042–2061.

Karl, T. R. et al., 1993: Asymmetric trends of daily maximum and minimum temperature. *Bull. Am. Meteorol. Soc.*, **74**, 1007–1023.

Li, H., A. Robock, and M. Wild, 2007: Evaluation of Intergovernmental Panel on climate change fourth assessment soil moisture simulations for the second half of the twentieth century. *J. Geophys. Res.*, **112**, D06106, doi:10,1029/2006JD007455.

Liu, B., M. Xu, M. Henderson, Y. Qi, and Y. Li, 2004a: Taking China's temperature: daily range, warming trends, and regional variations, 1955–2000. *J. Climate*, **17**, 4453–4462.

Liu, B., M. Xu, M. Henderson, and W. Gong, 2004b: A spatial analysis of pan evaporation trends in China, 1955–2000. *J. Geophys. Res.*, **109**, doi:10.1029/2004JD004511.

Liepert, B. G., 2002: Observed reductions of surface solar radiation at sites in the United States and worldwide from 1961 to 1990. *Geophys. Res. Lett.*, **29**, doi:10.1029/2002GL014910.

Mitchell, T. D. and P. D. Jones, 2005: An improved method of constructing a database of monthly climate observations and associated high-resolution grids. *Int. J. Climateatol.*, **25**, 693–712.

Norris, J. R. and M. Wild, 2007: Trends in direct and indirect aerosol radiative effects over Europe inferred from observed solar "dimming" and "brightening". *J. Geophys. Res.* (in press).

Ohmura, A. and H. Lang, 1989: Secular variation of global radiation over Europe. In: *Current Problems in Atmospheric Radiation*, J. Lenoble, J. F. Geleyn, eds., Deepak, Hampton, VA, 98–301.

Ohmura, A., H. Gilgen, and M. Wild, 1989: Global Energy Balance Archive GEBA, World Climate Program – Water Project A7, Zürcher Geografische Schriften No. 34, Verlag der Fachvereine, Zürich.

Ohmura, A. et al., 1998: Baseline Surface Radiation Network (BSRN/WCRP), a new precision radiometry for climate research. *Bull. Am. Meteorol. Soc.*, **79**, 2115–2136.

Ohmura, A. and M. Wild, 2002: Is the hydrological cycle accelerating? *Science*, **298**, 1345–1346.

Pallé, E., P. Montanes-Rodriguez, P. R. Goode, S. E. Koonin, M. Wild, and S. Casadio, 2005: A multi-data comparison of shortwave climate forcing changes. *Geophys. Res. Lett.*, **32**, L21702, doi:10.1029/2005GL023847.

Paul, F., A. Kaab, M. Maisch, T. Kellenberger, and W. Haeberli, 2004: Rapid disintegration of Alpine glaciers observed with satellite data. *Geophys. Res. Lett.*, **31**, L21402.

Philipona, R., B. Dürr, C. Marty, A. Ohmura, and M. Wild, 2004: Radiative forcing – measured at Earth's surface – corroborate the increasing greenhouse effect. *Geophys. Res. Lett.*, **31**, L03202, doi:10.1029/2003GL01876.

Ramanathan, V., P. J. Crutzen, J. T. Kiehl, and D. Rosenfeld, 2001: Aerosol, climate and the hydrological cycle. *Science*, **294**, 2119–2124.

Robock, A., M. Mu, K. Vinnikov, I. V. Trofimova, and T. I. Adamenko, 2005: Forty-five years of observed soil moisture in the Ukraine: no summer desiccation (yet). *Geophys. Res. Lett.*, **32**, L 03401, doi:10.1029/2004GL021914.

Roderick, M. L. G. D. Farquhar, 2002: The cause of decreased pan evaporation over the past 50 years. *Science*, **298**, 1410–1411.

Rotstayn, L. D. U. Lohmann, 2002: Simulation of the tropospheric sulfur cycle in a global model with a physically based cloud scheme. *J. Geophys. Res.*, **107**, 4592.

Stanhill, G. and S. Cohen, 2001: Global dimming: a review of the evidence for a widespread and significant reduction in global radiation with discussion of its probable causes and possible agricultural consequences. *Agr. Forest Meteorol.*, **107**, 255–278.

Stern, D. I., 2006: Reversal of the trend in global anthropogenic sulfur emissions. *Glob. Environ. Change*, **16**, 207–220.

Stier, P., J. Feichter, S. Kinne, S. Kloster, E. Vignatti, J. Wilson, L. Ganzeveld, I. Tegen, M. Werner, Y. Balkanski, M. Schulz, and O. Boucher, 2005: The aerosol-climate model ECHAM5-HAM. *Atm. Chem. Phys.*, **5**, 1125–1156.

Stier, P., J. Feichter, E. Roeckner, S. Kloster, and M. Esch, 2006: The evolution of the global aerosol system in a transient climate simulation from 1860 to 2100. *Atmos. Chem. Phys.*, **6**, 3059–3076.

Streets, D. G., Y. Wu, and M. Chin, 2006: Two-decadal aerosol trends as a likely explanatio nof the global dimming/brightening transition. *Geophys. Res. Lett.*, **33**, L15806, doi:10.1029/2006GL026471.

Vose, R. S., D. R. Easterling, and B. Gleason, 2005: Maximum and minimum temperature trends for the globe: an update through 2004. *Geophys. Res. Lett.*, **32**, L23822, doi:10.1029/2005GL024379.

Wild, M., A. Ohmura, and U. Cubasch, 1997: GCM simulated surface energy fluxes in climate change experiments. *J. Climate*, **10**, 3093–3110.

Wild, M. and A. Ohmura, 2004: BSRN longwave downward radiation measurements combined with GCMs show promise for greenhouse detection studies. *GEWEX News*, **14**(4), 9–10.

Wild, M., A. Ohmura, H. Gilgen, and D. Rosenfeld, 2004: On the consistency of trends in radiation and temperature records and implications for the global hydrological cycle. *Geophys. Res. Lett.*, **31**, L11201, doi:10.1029/2003GL019188.

Wild, M. et al., 2005: From dimming to brightening: decadal changes in surface solar radiation. *Science*, **308**, 847–850.

Wild, M., A. Ohmura, and K. Makowski, 2007: Impact of global dimming and brightening on global warming. *Geophys. Res. Lett.*, **34**, L04702, doi:10.1029/2006GL028031.

Observed Interdecadal Changes in Cloudiness: Real or Spurious?

J. R. Norris

Abstract This study documents multidecadal variations in low-level, upper-level, and total cloud cover over land and ocean independently obtained from surface synoptic observations and from satellite data produced by the International Satellite Cloud Climatology Project. Substantial agreement exists between global mean time series of surface- and satellite-observed upper-level cloud cover, indicating that the reported variations in this cloud type are likely to be real. Upper-level cloud cover has decreased over almost all land regions since 1971 and has decreased over most ocean regions since 1952. Global mean time series of surface- and satellite-observed low-level and total cloud cover exhibit very large discrepancies, however, implying that artifacts exist in one or both data sets. The global mean satellite total cloud cover time series appears spurious because the spatial pattern of correlations between grid box time series and the global mean time series closely resembles the fields of view of geostationary satellites rather than geophysical phenomena. The surface-observed low-level cloud cover time series averaged over the global ocean appears suspicious because it reports a very large 5%-sky-cover increase between 1952 and 1997. Unless low-level cloud albedo substantially decreased during this time period, the reduced solar absorption caused by the reported enhancement of cloud cover would have resulted in cooling of the climate system that is inconsistent with the observed temperature record.

1 Introduction

Clouds have a large impact on the Earth's radiation budget due to their reflection of solar or shortwave (SW) radiation back to space and their restriction of the emission of thermal or longwave (LW) radiation to space. The difference between actual radiation flux and what it would be were clouds absent is called cloud radiative forcing (CRF). Clouds at upper levels in the troposphere make the largest contribution to LW

Scripps Institution of Oceanography, University of California, San Diego,
9500 Gilman Drive 0224, La Jolla, CA 92093-0224, USA
jnorris@ucsd.edu

S. Brönnimann et al. (eds.), *Climate Variability and Extremes during the Past 100 Years.*

CRF because they are cold and thus emit less radiation than the surface under clear skies, and clouds that are optically thick make the largest contribution to SW CRF because they have greater albedo than the surface under clear skies. In the global annual average, the magnitude of top-of-atmosphere (TOA) SW CRF is greater than the magnitude of TOA LW CRF, corresponding to a loss of energy and consequently an overall cloud cooling effect in the present climate state.

Despite the greater sensitivity of the Earth's radiation budget to variations in cloudiness than to equivalent percentage variations in CO_2 concentration (Slingo 1990), we currently do not know whether clouds are changing so as to mitigate or exacerbate anthropogenic greenhouse warming (Moore et al. 2001). One leading reason for this uncertainty is that global climate models (GCMs) do not correctly or consistently represent clouds and their radiative effects (e.g., Bony et al. 2006). The severe deficiencies in GCM simulations of cloudiness motivate the investigation of how observed clouds have changed over the past several decades, a period of rapidly rising global temperature.

The present study documents changes in low-level, upper-level, and total cloud cover since 1971 over land and since 1952 over ocean. Global mean time series averaged from gridded surface and satellite observations are compared during their period of overlap to assess how well the two independent data sets agree. Substantial similarity indicates which cloud type variations are reliable whereas disagreement implies the existence of artifacts in one or both data sets. The particular regions of the world that contribute most to the global mean time series are discerned by creating maps of correlation values between grid box time series and global mean time series. Maps of linear trends at each grid box are created for upper-level cloud cover, and a concluding section describes our current knowledge of the likely impact of cloud trends on the Earth's radiation budget.

2 Data

Individual surface synoptic cloud reports were obtained from the Extended Edited Cloud Report Archive (Hahn and Warren 1999) globally over land during 1971–1996 and globally over ocean during 1952–1997. The land cloud reports came from stations assigned official numbers by the World Meteorological Organization (WMO), and the ocean cloud reports, primarily made by Volunteer Observing Ships, came from the Comprehensive Ocean-Atmosphere Data Set (Woodruff et al. 1987). A synoptic cloud report includes the fractions of sky dome covered by all clouds and by the lowest cloud layer, as seen by a human observer at the surface (WMO 1987).

Upper-level cloud cover, defined in this study as the coverage by mid-level and high-level clouds, was inferred by assuming random overlap with obscuring lower-level clouds using the method described in Norris (2005a). Surface-observed low-level cloud cover values were also adjusted in some cases to represent the "satellite view" by assuming random overlap and removing the portion of low-level cloud cover overlapped by higher clouds (Norris 2005a).

Cumulonimbus clouds were included in the low-level category even though they have high tops because that is the usual practice for surface observations. This choice has little impact on the results of this analysis since sky-dome coverage by cumulonimbus (Warren et al. 1988) is usually less than one third of sky-dome coverage by upper-level clouds (Norris 2005b). Moreover, horizontal coverage of cumulonimbus clouds is actually smaller than that reported by surface observers, who include cloud sides as part of cloud cover.

Only daytime observations are used to avoid observer biases resulting from poor nighttime illumination (Hahn et al. 1995). Individual synoptic reports were averaged to monthly 5° × 5° values using a special procedure that minimized geographical and temporal sampling biases (Norris 2005a). Subtraction of long-term monthly means from each grid box yielded 5° × 5° monthly anomalies. Although the surface cloud data lack complete global sampling due to the scarcity of stations and ships in remote land and ocean regions, the occurrence of missing data has little impact on global mean values.

Monthly mean cloud fraction as seen by geostationary and polar-orbiting weather satellites was obtained from the International Satellite Cloud Climatology Project (ISCCP) at 2.5° × 2.5° grid spacing for July 1983 onwards (Rossow et al. 1996; Rossow and Schiffer 1999). Pixels are classified as cloudy if they are brighter in the visible channel (VIS) and/or colder in the infrared window channel (IR) than clear-sky pixels beyond a specified threshold. ISCCP high clouds have tops above the 440 hPa level; ISCCP mid-level clouds have tops between 680 hPa and 440 hPa; and ISCCP low-level clouds have tops below the 680 hPa level. For this study, ISCCP high-level and mid-level clouds were combined to provide upper-level cloud fraction. Note that ISCCP low-level cloud fraction takes into account only those low clouds that are not obscured by higher clouds. Only ISCCP daytime (VIS+IR) data were examined since ISCCP may have trouble correctly detecting transmissive cirrus or low-level clouds using IR alone (Rossow and Schiffer 1999). As done for the surface data, the seasonal cycle was removed by subtracting long-term monthly means from each grid box.

3 Results

Global mean anomalies of surface- and satellite-observed upper-level, low-level, and total cloud cover for land only, ocean only, and land+ocean were created by averaging monthly anomalies over all grid boxes with weighting by grid box area and land/ocean fraction. A 13-point Lanczos 1-year low-pass filter was applied to the resulting time series, displayed in Fig. 1, in order to improve readability and emphasize low-frequency variability. The surface and satellite observations exhibit agreement only for upper-level cloud cover, and linear correlations between the upper-level cloud cover time series are 0.58, 0.65, and 0.68 for land only, ocean only, and land+ocean, respectively. Assuming ten degrees of freedom (from 12 years of overlapping data), the coefficients are significant at the 95% level. The similarity

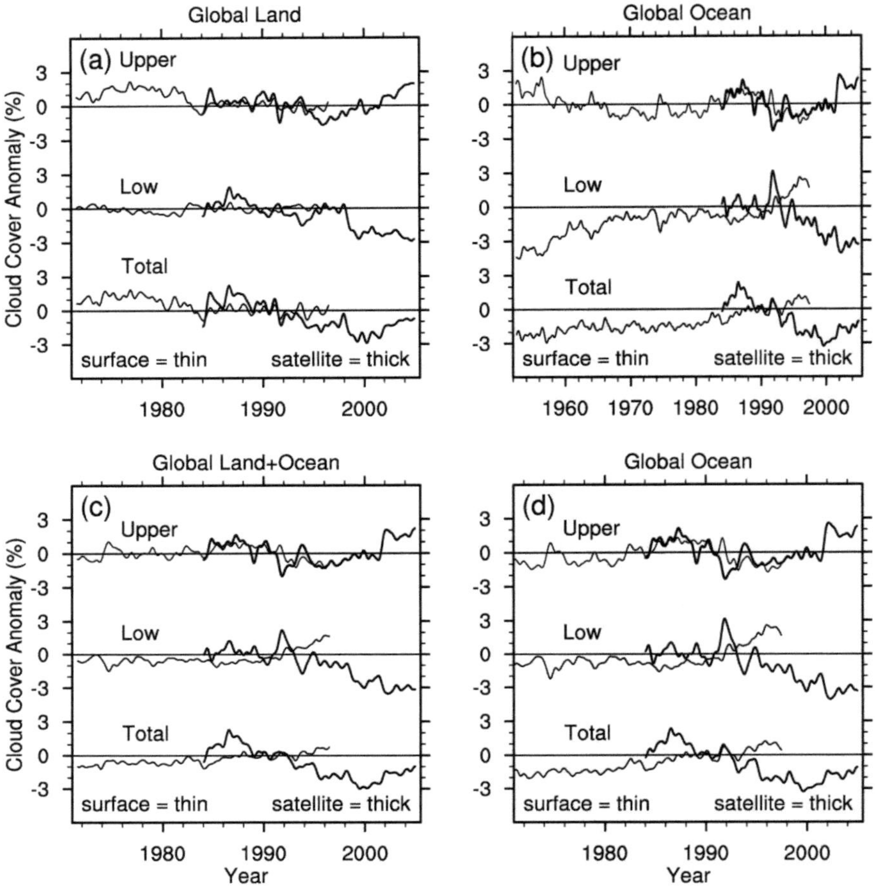

Fig. 1 Monthly anomalies of surface-observed (thin) and satellite-observed (thick) upper-level (top), low-level (middle), and total (bottom) cloud cover globally averaged over: (a) land during 1971–2005, (b) ocean during 1952–2005, (c) land and ocean during 1971–2005, and (d) ocean during 1971–2005). Surface-observed low-level cloud cover was adjusted for overlapping higher clouds to correspond to the satellite view, and the time series were smoothed with a 13-point 1-year low-pass filter.

of these two independent data sets strongly suggests that the reported upper-level cloud variations are real during their period of overlap and furthermore likely to be reliable before and after that interval. One exception to this conclusion, however, is the June 1991–1993 time period when the presence of Mt. Pinatubo volcanic aerosol caused some upper-level clouds to be misdiagnosed as low-level clouds in the satellite data over the ocean but not over land (Luo et al. 2002). Another exception is the period of October 2001 onwards when there appears to be a large discontinuity in the satellite time series.

A very large discrepancy occurs between the low-level cloud cover time series; the surface data exhibit a generally increasing trend over ocean whereas the satellite data exhibit a generally decreasing trend over land and ocean (Fig. 1). This leads to a corresponding disagreement between the total cloud cover time series since total cloud cover is the sum of upper-level cloud cover and low-level cloud cover as seen from a satellite. Although perfect correspondence should not be expected due to the differing measurement methods, the disparity between the time series is so great that it is very likely that one or both of the datasets is contaminated by spurious variability.

The time series of ocean-only surface-observed low-level cloud cover appears suspicious because a least-squares linear fit reveals that marine low-level cloudiness strongly increased by about 5%-sky-cover between 1952 and 1997. Such a change would enhance the amount of SW radiation reflected back to space with little change in outgoing LW radiation due to the low altitude of the clouds, thus acting to cool the Earth. An examination of Earth Radiation Budget Experiment data (Barkstrom et al. 1989) indicates that the ratio of net CRF divided by low-level cloud cover is on the order of 1 W m^{-2} per 1%-sky-cover in regions with few upper-level clouds (e.g., Klein and Hartmann 1993). Assuming that this value applies to low-level clouds over all ocean regions and that all other cloud properties remain constant, this implies that the reported increase in low-level cloud cover caused an approximate 5 W m^{-2} decline in energy absorbed by ocean areas between 1952 and 1997. This estimated change in TOA radiation flux does not match the record reported by the Earth Radiation Budget Satellite during 1985–1997 (Wielicki et al. 2002; Norris 2005a). Moreover, considering that the increase in anthropogenic greenhouse radiative forcing since pre-industrial times (1750) is only about 2.5 W m^{-2} (Ramaswamy et al. 2001) and that global temperature increased between 1952 and 1997, it seems that the large low-level cloud cover increase reported by ship observers must either be spurious or accompanied by a compensating substantial decrease in cloud albedo. Unfortunately, synoptic cloud reports do not provide quantitative information on cloud optical properties that could be used to examine whether cloud albedo has experienced a long-term decline.

The nature of global mean cloud variability was explored by calculating correlation coefficients between the global mean time series and the time series at each grid box. High positive correlation values identified those regions that contributed most to the global mean cloud changes and thus to the apparent artifacts. The top plot of Fig. 2 presents the correlation map for satellite-observed total cloud cover, which was analyzed instead of low-level cloud cover to avoid ambiguities related to overlap by higher cloud layers. As previously noted in Norris (2000a), the spatial correlation pattern of satellite total cloud cover consists of circular regions filled with positive values that closely correspond to the fields of view of the geostationary satellites contributing to ISCCP. These obviously artificial features strongly suggest that low-frequency variations in global mean satellite total cloud cover are not real. Positive correlations are also highest near the east/west boundaries of the fields of view, which Campbell (2004) attributes to biases related to systematic decreases in satellite view angle over time. The correlation map for satellite-observed upper-level cloud cover does not exhibit the pattern seen in top plot of Fig. 2 (not shown).

The bottom plot of Fig. 2 presents the correlation map for ocean-only surface-observed low-level cloud cover without any adjustment to "satellite view." Values

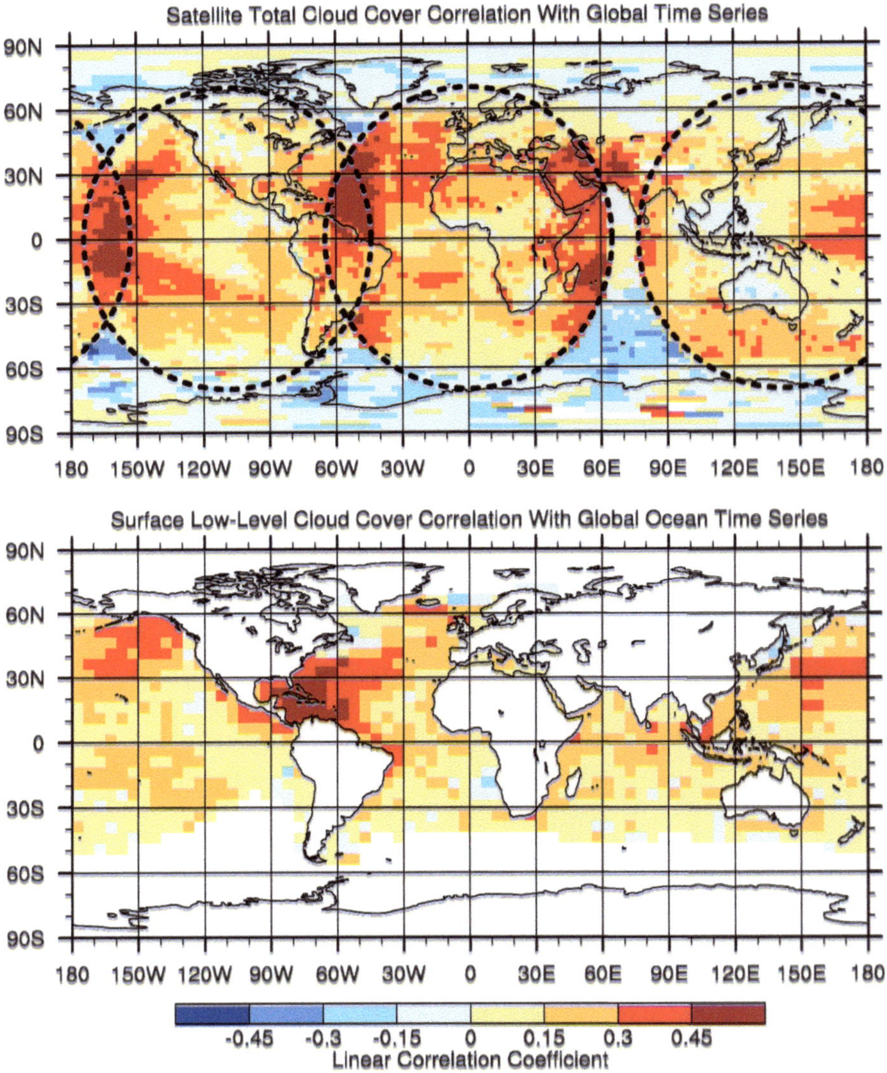

Fig. 2 Map of correlation values between grid box anomaly time series and 1-year low-pass global mean anomaly time series for: (top) satellite-observed total cloud cover and (bottom) surface-observed low-level cloud cover not adjusted to "satellite view." Grid boxes with data missing more than half the time are not plotted. Dashed circles in the top plot show the approximate fields of view of the US, European, and Japanese geostationary satellites.

are not plotted for the high latitude oceans where sampling is sparse. Almost all of the remaining grid boxes are positively correlated with the global ocean mean time series, a feature that is consistent with the finding of Norris (1999) that low-level cloud cover reported by ships increased from 1952 to 1995 not only in the global mean but also in every latitude band. Since individual ships travel on many different routes, it is possible that some unidentified artifact would thereby be spread around the global ocean. Low-level cloud trends should therefore be deemed reliable only when they are corroborated by physically consistent trends in related meteorological parameters, as is the case for the midlatitude North Atlantic and North Pacific (e.g., Norris and Leovy 1994; Norris 2000b). The origin of high positive correlations in the western subtropical Atlantic has not yet been identified. Land grid boxes do not exhibit a large number of positive correlations with the global mean time series of low-level cloud cover (not shown).

Only 12% of the coefficients in the correlation map for ocean-only surface-observed upper-level cloud cover are greater than 0.15 (not shown), unlike 47% for low-level cloud cover. This suggests the absence of a globally coherent artifact in surface-observed upper-level cloud cover over the ocean. It is currently not known how a potential artifact could be absent from upper-level clouds yet presumably present in low-level clouds that are directly seen by surface observers.

Least-squares linear trends in upper-level cloud cover were calculated between 1952 and September 2001 over ocean and between 1971 and September 2001 over land. Surface and satellite cloud anomaly records were concatenated in order to have the longest time period possible with averaging of the two data sets during the interval in common (1984–1996). Furthermore, the anomalies were calculated with respect to the 1984–1996 period in both data sets so as to avoid discontinuities with the pre-1984 surface-only period and post-1996 satellite-only period. Satellite observations over ocean during June 1991–1993 were not used to avoid the volcanic aerosol problem. Assuming that each year is independent, the median value for 95% statistical significance is 0.2%-sky-cover per decade over ocean and 0.4%-sky-cover per decade over land.

The top plot of Fig. 3 shows that upper-level cloud cover has decreased over most of the North Pacific, the western tropical Pacific, the subtropical regions of the Atlantic Ocean and the South Pacific, and the equatorial Indian Ocean. The only areas that have experienced widespread increases in upper-level cloud cover are the equatorial South Pacific, the tropical Atlantic Ocean, and the midlatitude North Atlantic. The study of Norris (2005b) demonstrated that the positive and negative cloud trends over the tropical Indo-Pacific region were physically consistent with shifts in surface wind convergence and precipitation. Moreover, the long-term changes in upper-level cloud cover resembled the spatial pattern of the interannual cloud response to El Niño, although the cloud trends were larger than would be expected from a linear relationship to the trend in El Niño SST anomalies. The distribution of positive and negative cloud trends over the North Atlantic is similar to the interannual upper-level cloud response to the North Atlantic Oscillation (not shown), and the directions of the long-term cloud changes are consistent with the previously reported trend in the North Atlantic Oscillation between the 1960s and

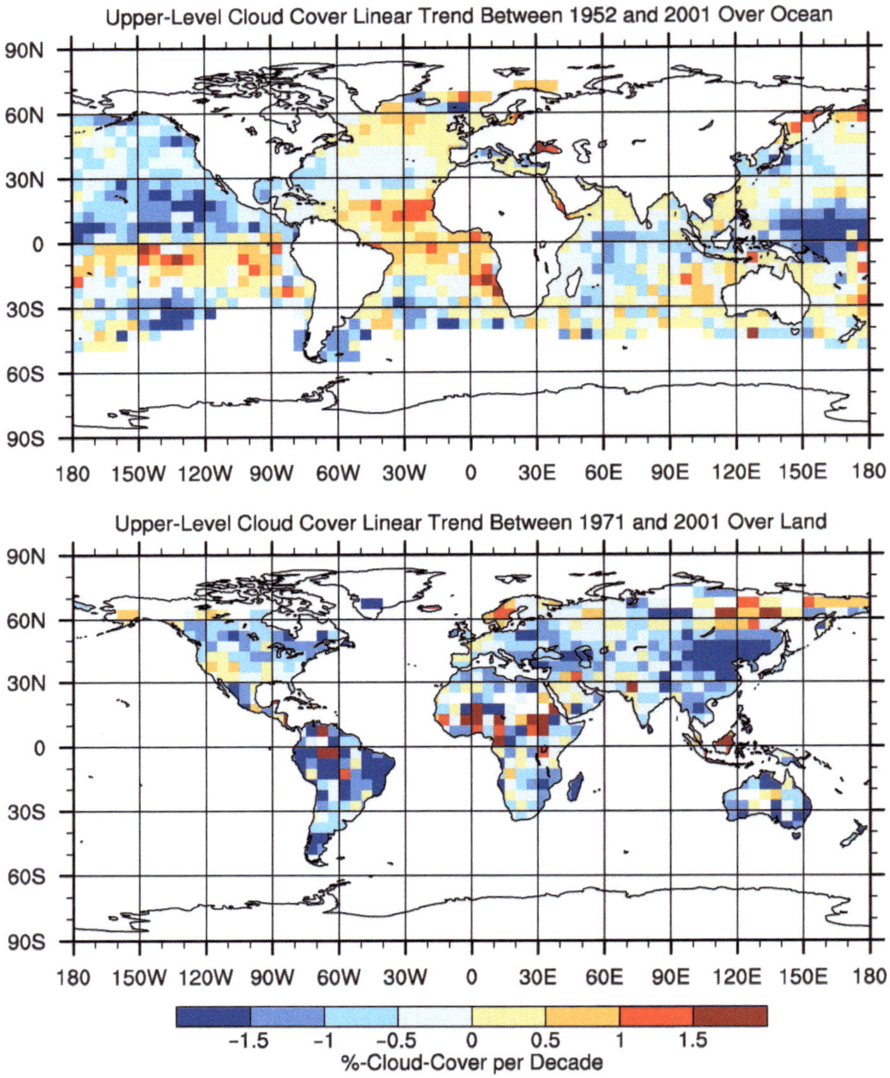

Fig. 3 Map of linear trends in combined surface- and satellite-observed upper-level cloud cover at each grid box over: (top) ocean between 1952 and September 2001 and (bottom) land between 1971 and September 2001. Grid boxes with data missing more than half the time are not plotted, and ocean satellite data during June 1991–1993 were not included. Assuming that each year is independent, the median value for 95% statistical significance is 0.2%-sky-cover per decade over ocean and 0.4%-sky-cover per decade over land.

1990s (e.g., Hurrell and van Loon 1997). The origin of the large increase in upper-level cloud cover off the coast of southwest Africa is currently unknown.

The bottom plot of Fig. 3 shows reductions in upper-level cloud cover over almost all land regions since 1971. Although the causes of these trends have not yet been identified, the trends appear to be real since surface and satellite cloud variations exhibit close agreement in several different regions (not shown).

4 Conclusions

Multidecadal cloud cover variability over land and ocean was examined using surface and satellite observations. Global mean time series of low-level and total cloud cover obtained from surface and satellite data substantially disagree, implying that low-frequency variability in one or both data sets is spurious. Further evidence of an artifact in satellite total cloud cover is the pattern of correlations between grid box time series and the global mean time series that closely resembles the fields of view of geostationary satellites rather than meteorological features. Contrastingly, global mean time series of surface- and satellite-observed upper-level cloud cover are similar, providing confidence that the reported cloud variations are real. Based on the calculation of least-squares linear trends at each grid box, upper-level cloud cover decreased over most ocean regions between 1952 and 2001 and decreased over almost all land regions between 1971 and 2001.

The reduction of upper-level cloud cover during the past several decades will have allowed more LW radiation to escape to space (all else held constant), thus keeping the Earth cooler than it otherwise would have been. Diminished upper-level cloud cover will also have reflected less SW radiation back to space, thus acting to warm the Earth at the same time, but the entire cloud impact on net radiation unfortunately cannot be determined without reliable knowledge of trends in low-level cloud cover. Surface observers report that low-level cloud cover has increased by about 5%-sky-cover between 1952 and 1997 over the ocean, corresponding to an approximate $5\,W\ m^{-2}$ decrease in energy absorbed by the climate system if cloud albedo remained constant during that time period. Such a large change in net CRF is physically implausible and indicates that there has either been a substantial decline in cloud albedo or that the surface-observed low-level cloud cover trend is spurious. Consequently, it is currently not possible to ascertain whether recent multidecadal variations in clouds have mitigated or exacerbated anthropogenic global warming. More research needs to be done to identify and remove apparent artifacts from the satellite and surface cloud datasets.

Acknowledgment An NSF CAREER award, ATM02-38527, supported this work.

References

Barkstrom B et al (1989) Earth Radiation Budget Experiment (ERBE) archival and April 1985 results. *B. Am. Meteorol. Soc.*, **70**, 1254–1262.

Bony S et al. (2006) How well do we understand and evaluate climate change feedback processes? *J. Climate*, **19**, 3445–3482.

Campbell GG (2004) View angle dependence of cloudiness and the trend in ISCCP cloudiness. Paper presented at the 13th Conference on Satellite Meteorology and Oceanography, Norfolk, VA, September 2004. http://ams.confex.com/ams/pdfpapers/79041.pdf

Hahn CJ, Warren SG (1999) *Extended Edited Synoptic Cloud Reports from Ships and Land Stations over the Globe, 1952–1996, NDP026C*. Carbon dioxide Information Analysis Center, Oak Ridge National Laboratory, Oak Ridge, TN. http://cdiac.ornl.gov/ftp/ndp026c/ndp026c.pdf.

Hahn CJ, Warren SG, London J (1995) The effect of moonlight on observation of cloud cover at night, and application to cloud climatology. *J. Climate*, **8**, 1429–1446.

Hurrell JW, van Loon H (1997) Decadal variations in climate associated with the North Atlantic Oscillation. *Clim. Chang*., **36**, 301–326.

Klein SA, Hartmann DL (1993) The seasonal cycle of low stratiform clouds. *J Climate*, **6**, 1587–1606.

Luo Z, Rossow WB, Inoue T, Stubenrauch CJ (2002) Did the eruption of Mt. Pinatubo volcano affect cirrus properties? *J. Climate*, **15**, 2806–2820.

Moore B et al. (2001) Advancing our understanding. In: *Climate Change 2001: The Scientific Basis*, Houghton JT et al., eds., Cambridge University Press, New York.

Norris JR (1999) On trends and possible artifacts in global ocean cloud cover between 1952 and 1995. *J. Climate*, **12**, 1864–1870.

Norris JR (2000a) What can cloud observations tell us about climate variability. *Space Sci. Rev*., **94**, 375–380.

Norris JR (2000b) Interannual and interdecadal variability in the storm track, cloudiness, and sea surface temperature over the summertime North Pacific. *J. Climate*, **13**, 422–430.

Norris JR (2005a) Multidecadal changes in near-global cloud cover and estimated cloud cover radiative forcing. *J. Geophys. Res*., **110**, D08206, doi:10.1029/2004JD005600.

Norris JR (2005b) Trends in upper-level cloud cover and surface divergence over the tropical Indo-Pacific Ocean between 1952 and 1997. *J. Geophys. Res*., **110**, D21110, doi:10.1029/2005JD006183.

Norris JR, Leovy CB (1994) Interannual variability in stratiform cloudiness and sea surface temperature. *J. Climate*, **7**, 1915–1925.

Ramaswamy V et al. (2001) Radiative forcing of climate change. In: *Climate Change 2001: The Scientific Basis*, Houghton JT et al., eds., Cambridge University Press, New York.

Rossow WB, Schiffer RA (1999) Advances in understanding clouds from ISCCP. *B. Am. Meteorol. Soc*., **80**, 2261–2287.

Rossow WB, Walker AW, Beuschel DE, Roiter MD (1996) *International Satellite Cloud Climatology Project (ISCCP) Documentation of New Cloud Datasets, WMO/TD-No. 737*. World Meteorological Organization. http://isccp.giss.nasa.gov/docs/documents.html.

Slingo A (1990) Sensitivity of the earth's radiation budget to changes in low clouds. *Nature*, **343**, 49–51.

Warren SG, Hahn CJ, London J, Chervin RM, Jenne RL (1988) *Global Distribution of Total Cloud Cover and Cloud Type Amounts over the Ocean, NCAR/TN-317+STR*. National Center for Atmospheric Research, Boulder, CO.

Wielicki BA et al. (2002) Evidence for large decadal variability in the tropical mean radiative energy budget. *Science*, **295**, 841–844.

Woodruff SD, Slutz RJ, Jenne RL, Steurer PM (1987) A comprehensive ocean-atmosphere data set. *B. Am. Meteorol. Soc*., **68**, 1239–1250.

World Meteorological Organization (WMO) (1987) *International Cloud Atlas*, vol. 1. WMO Publication 407.

The Role of Land–Atmosphere Interactions for Climate Variability in Europe

S. I. Seneviratne[1] and R. Stöckli[1,2]

Abstract We provide here a brief review on the role of land–atmosphere interactions for climate variability, with a special focus on the European continent. First, an overview of the land energy and water balances and of the underlying physical, biophysical, and biogeochemical soil–vegetation–atmosphere processes is presented. Further, we highlight how land–atmosphere feedbacks can impact seasonal to interannual climate variability in transitional climate zones and midlatitude regions along three main paths: Soil moisture–temperature interactions, soil moisture–precipitation interactions, and vegetation–climate interactions. In this context, we discuss recent results based on findings from terrestrial observational networks, satellite observations, and numerical climate models across a number of spatial and temporal scales. These results illustrate the extent to which land-surface processes, land–atmosphere interactions, and associated memory effects can modulate the dynamics of the climate system. Finally, the concluding section addresses current areas of uncertainty and open questions for research in this field.

1 Introduction

The importance of land–atmosphere interactions and all processes they involve for the climate system is increasingly being recognized. Similar to the oceans, land areas provide the lower boundary for the atmosphere, with which they exchange energy, water and chemical compounds such as CO_2 (Fig. 1). Storage of water on land (e.g., as soil moisture, groundwater, snow, surface water or ice) constitutes a significant memory component within the climate system, similar in many ways to heat storage in the oceans. Moreover, anomalies of soil moisture (positive or negative)

[1] Institute for Atmospheric and Climate Science, ETH Zurich, Zurich, Switzerland
sonia.seneviratne@env.ethz.ch

[2] Department of Atmospheric Science, Colorado State University, Fort Collins, CO, USA
reto.stoeckli@env.ethz.ch

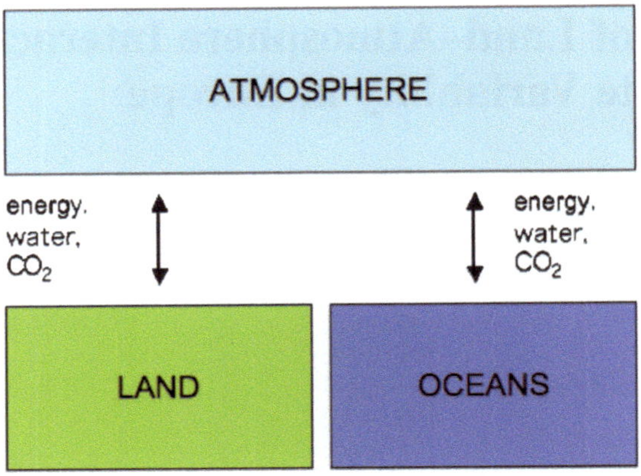

Fig. 1 Land–atmosphere and oceans–atmosphere interactions.

have strong effects on the land energy and water balances in regions where evapotranspiration is limited by soil moisture availability. Consequently, also vegetation processes are critical in constraining the uptake of soil moisture for evapotranspiration.

While the role of land for climate variability has often been neglected in the past, recent studies have highlighted how land–atmosphere interactions can be critical in modulating variations in climate on a range of temporal (seasonal to centennial) and spatial (local to global) scales. In particular, the role of soil moisture for precipitation (e.g., Betts et al. 1996; Beljaars et al. 1996; Eltahir 1998; Schär et al. 1999; Betts 2004; Koster et al. 2004a) and temperature (e.g., Seneviratne et al. 2006a) in midlatitude and transitional climate zones has been highlighted in several investigations. Soil moisture is also an important memory component of the climate system (e.g., Koster and Suarez 2001; Seneviratne et al. 2006b) and thus a useful source of persistence for seasonal forecasting (e.g., Koster et al. 2004b; Ferranti and Viterbo 2006). Further relevant land–atmosphere interactions in the framework of climate change involve interactions with the carbon cycle and in particular links between CO_2 assimilation and water use in plants (e.g., Field et al. 1995; Körner 2000; Gedney et al. 2006).

In this review, we focus more particularly on the role of land–atmosphere interactions for the seasonal-to-interannual variability of the European summer climate. However, land–atmosphere interactions have been shown to be relevant for several other regions and time scales. The recent Global Land-Atmosphere Climate Experiment (GLACE, Koster et al. 2004a, 2006) pinpointed that land–atmosphere interactions tend to be particularly important in transitional zones between dry and wet climates. For present climate, this applies, e.g., to the Great Plains of North America, the Sahel, equatorial Africa, and Northern India (Koster et al. 2004a), but also to the Mediterranean region (Seneviratne et al. 2006a). Moreover, these "hot spots" of land–atmosphere coupling are also inherently modified with shifts in climate regimes, for instance due to climate change (Seneviratne et al. 2006a). They can thus be displaced on longer time scales. Finally, long-term vegetation dynamics and human-induced land use changes can also interact with the rest of the climate system on decadal-to-centennial time scales (Cramer et al. 2001; Claussen et al. 2004; Pielke 2005). These longer-term feedbacks will not be treated in detail as part of the present review.

The structure of this chapter is as follows. Section 2 provides a short overview of the processes governing land energy and water balances, and of their interconnections; Sect. 3 presents soil moisture–temperature interactions; Sect. 4, soil moisture–precipitation interactions; and Sect. 5, vegetation–climate interactions. Finally, we present a summary and an outlook in Sect. 6.

2 Land Energy and Water Balances

The land energy balance for a surface soil layer (including possible snow or ice cover) can be expressed as:

$$\frac{dH}{dt} = R_n - \lambda E - SH - G \tag{1}$$

where dH/dt is the change of energy within the considered surface layer (e.g., temperature change, phase changes), R_n is the net radiation, λE is the latent heat flux (latent heat of vaporization λ times the evapotranspiration E), SH is the sensible heat flux and G is the ground heat flux to deeper layers (Fig. 2, left).

Similarly, the land water balance for a surface soil layer (including a possible snow or ice cover) is expressed as:

$$\frac{dS}{dt} = P - E - R_s - R_g \tag{2}$$

where dS/dt is the change of water content within the considered layer (e.g., changes in soil moisture, snow content, ice content, surface water, groundwater), P is the precipitation, E is the evapotranspiration, R_s is the surface runoff, and R_g is (depending on the soil depth considered) the drainage or groundwater runoff (Fig. 2, right).

Equations (1) and (2) show that the land energy and water balances are coupled through the evapotranspiration term. If soil moisture is lacking, then no evapotranspiration can take place and most of the incoming energy (net radiation) goes into sensible heat flux, thus strongly enhancing air temperature. Reversely, if water is available in ample supply (moist surface or water body), then a large amount of energy will be used for evapo(transpi)ration thus effecting a net cooling compared to dry surfaces. These effects are, however, only important in regions where soil moisture is the main controlling factor for evapotranspiration. In high-latitude regions, for instance, evapotranspiration is limited by net radiation and the length of the growing season.

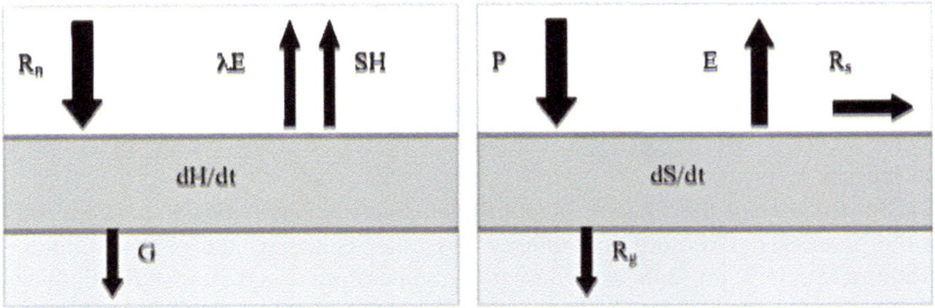

Fig. 2 Land energy (left) and water (right) balances.

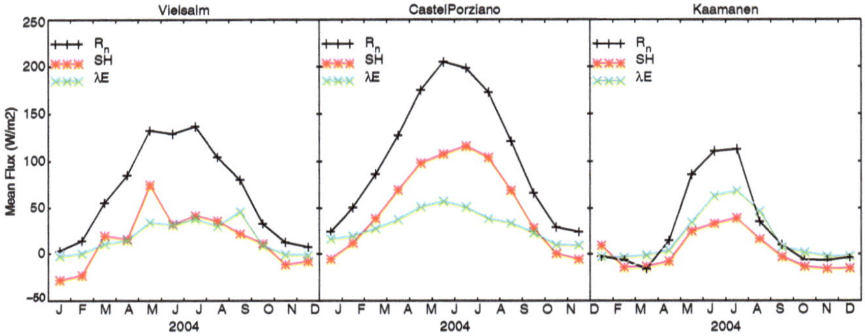

Fig. 3 Monthly net radiation (R_n), latent heat flux (λE) and sensible heat flux (SH) during 2004 at three CarboEurope flux tower sites covering a wide range of climate zones: Vielsalm, Belgium (temperate mixed forest, Aubinet et al. 2001); Castelporziano, Italy (summer-dry Mediterranean evergreen forest, Reichstein et al. 2002); and Kaamanen, Finland (Arctic tundra, Aurela et al. 2002). Fluxes were averaged from the original Level 2 CarboEurope data set.

Surface energy and water exchanges can for instance be analyzed from flux tower measurements provided by CarboEurope (http://www.carboeurope.org/). As an example, we display in Fig. 3 surface fluxes from three sites encompassing a wide range of climate zones: Vielsalm, Belgium (temperate mixed forest); Castelporziano, Italy (summer-dry Mediterranean evergreen forest); and Kaamanen, Finland (Arctic tundra). The temperate forest (left) has a well-balanced distribution between the monthly λE and SH fluxes. At the Mediterranean site (middle), lack of precipitation resulting in low soil moisture severely constrains vegetation activity; this limits the λE flux, thus inducing an enhanced SH flux. The arctic Tundra site's (right) seasonal courses of SH and λE are controlled by the short growing season and the low magnitude of available energy R_n.

Finally, the land energy and water balances are themselves linked with the terrestrial carbon as carbon assimilation and evapotranspiration are tightly coupled. These aspects will be discussed in more detail in Sect. 5.

3 Soil Moisture–Temperature Interactions

The main mechanism by which soil moisture can impact air temperature has been discussed in the previous section. Namely, soil moisture exerts a strong control on the partitioning of incoming surface energy (net radiation) in the latent and sensible heat fluxes in any region where it is the limiting factor for evapotranspiration. The fact that these regions are often transitional zones between dry and wet climates (e.g., Koster et al. 2004a) can be understood in the following way: In constantly wet regions, soil moisture is not a limiting factor for evapotranspiration and thus will not have a strong impact on the land energy balance; in constantly dry regions (deserts), there is too little soil moisture to allow significant evapotranspiration, independent of the season or of interannual variations in further climate variables. However, in regions where soil moisture can vary seasonally and interannually between dry and wet conditions, it will necessarily be an important factor impacting temperature variability.

For Europe, Seneviratne et al. (2006a) recently investigated how land–atmosphere coupling impacts summer temperature variability in present- and future-climate Regional Climate Model (RCM) simulations. For present climate, land–atmosphere coupling is found to have a significant impact in the Mediterranean region, where ca. 60% of the simulated interannual variability of summer temperature is due to interannual variations in soil moisture content (Fig. 4). Observational evidence for a link between spring precipitation deficits and summer temperature in the Mediterranean region for present climate (Della-Marta et al. 2007) lends support to these modeling results. Interestingly, the GLACE study (Koster et al. 2004a, 2006) did not identify a strong impact of soil moisture for temperature or precipitation in this region. This may be due to the setup of this study, for instance, because the impact of interannual variations in sea surface temperatures was not considered (see also discussion in Seneviratne et al. 2006a).

In future climate, the same study finds strong soil moisture–temperature coupling in most of Central and Eastern Europe (Fig. 5). This is due to a gradual shift of climatic regimes within the continent, whereby the transitional climate zone is shifted northward from the Mediterranean region to Central and Eastern Europe. This shift also appears responsible for the very large increase in summer temperature variability projected in this region (e.g., Schär et al. 2004). A further analysis of the correlation between evapotranspiration and temperature (which can be seen as an indirect measure of coupling) in the RCM experiments and in Global Climate Model (GCM) simulations from the Intergovernmental Panel on Climate Change (IPCC) 4th Assessment Report (AR4) confirmed this result (Seneviratne et al. 2006a).

4 Soil Moisture–Precipitation Interactions

The possible existence of strong soil moisture–precipitation feedbacks has been the topic of several investigations, based either on observations (e.g., Betts et al. 1996) or modeling studies (e.g., Beljaars et al. 1996; Schär et al. 1999; Pal and Eltahir

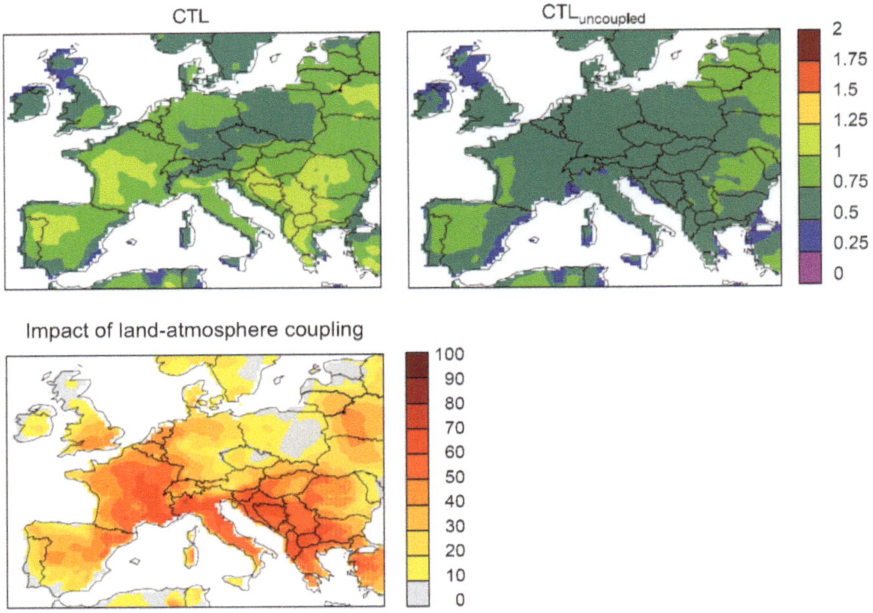

Fig. 4 Impact of land–atmosphere coupling for interannual summer (JJA) temperature variability in simulations for the time period 1970–1989. (Top) Standard deviation of the JJA 2 m temperature in the "coupled" (left) and "uncoupled" (right) experiment (K). (Bottom) Percentage of interannual JJA temperature variance due to land–atmosphere coupling (%) (from Seneviratne et al. 2006a).

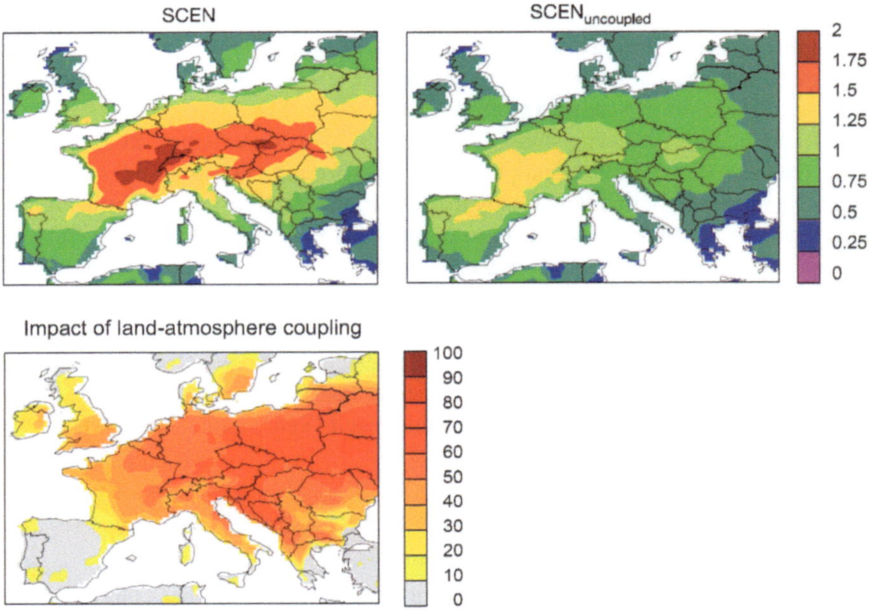

Fig. 5 Same as Fig. 4 for the time period 2080–2099 (from Seneviratne et al. 2006a).

2001; Koster et al. 2004a). The processes involved are more complex than a simple recycling mechanism by which additional moisture input from evapotranspiration to the atmosphere would lead automatically to additional precipitation. Rather, it appears that the suite of processes leading to a positive feedback loop between soil moisture content and subsequent precipitation involves modifications of the boundary layer structure and of the atmospheric stability profile (e.g., Eltahir 1998; Schär et al. 1999; Betts 2004).

As for temperature, soil moisture mostly appears to impact subsequent precipitation in transitional regions between dry and wet climate (Koster et al. 2004a). Again, this is due to the fact that soil moisture is not a limiting factor for evapotranspiration in wet climate, and that evapotranspiration is too limited in dry climate to significantly impact the regional climate system. For Europe, the GLACE study did not identify strong soil moisture–precipitation coupling (Koster et al. 2004a) in this region. However, as for soil moisture–temperature coupling, this could possibly be due to the setup of the numerical experiments (see preceding section). While investigations of possible soil moisture–precipitation coupling from observations is difficult in Europe due to the lack of soil moisture observations, modeling studies did find some impact of soil moisture for subsequent precipitation (e.g., Schär et al. 1999; Fischer et al. 2007). For instance, Schär et al. (1999) investigated the impact of initial soil moisture in 2-month-long RCM simulations for the months of July 1990 and July 1993. Their findings suggest that soil moisture anomalies can have strong impact on subsequent precipitation in Spain, France and Central Europe (Fig. 6). Soil moisture–precipitation feedbacks also appear relevant for future increases in precipitation variability in Europe, mostly in the Alpine region (Seneviratne et al. 2006a).

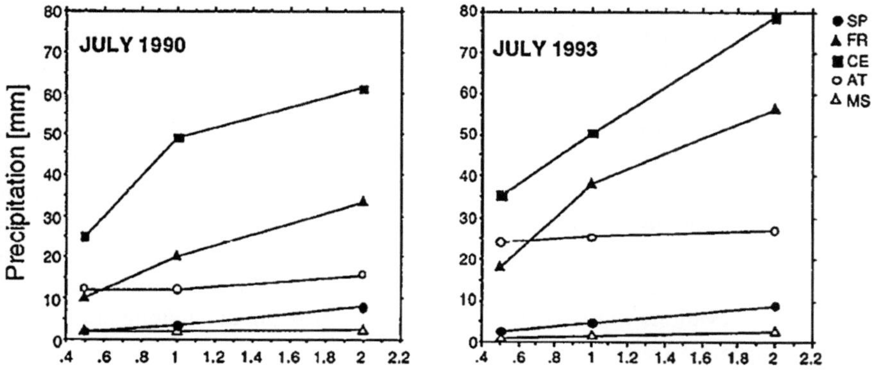

Fig. 6 Impact of initial soil moisture content on subsequent precipitation in various European regions (SP: Spain, FR: France, CE: Central Europe, AT: Atlantic, MS: Mediterranean Sea). Displayed is the simulated total precipitation in month-long regional climate model experiments for July 1990 and 1993, as function of the factor applied to the initial moisture content (from Schär et al. 1999. Copyright AMS).

5 Vegetation–Climate Interactions

In vegetated landscape water flux from the soil to the atmosphere mostly occurs through leaf stomates during the growing season (Sellers et al. 1997; Bonan 2002) and to a lesser extent through soil evaporation. During photosynthesis plants open stomates to take up CO_2. This chemical process is primarily driven by the light energy but its rate depends on various environmental conditions and the plant's biophysical and biochemical state. While stomates are open, water leaks from the saturated leaf interior (Farquhar et al. 1980; Ball et al. 1987) and has to be redrawn from the soil through the plant's root system in order to avoid dessication. Thus, evapotranspiration from vegetated surfaces occurs mostly as a by-product of photosynthesis and is constrained by this process. Sunlight, soil moisture, atmospheric vapor pressure, temperature and carbon dioxide concentration are the main physical environmental regulators for photosynthesis (Dickinson 2001), but it is also modulated by a number of biotic and abiotic factors such as tree age, nutrient availability, pests and the phenological state of the plant.

Hence, while vegetation cover, plant growth and photosynthesis are obviously strongly constrained by regional climate, they can also have a significant impact on climate on seasonal to interannual time scales. Many observational and modeling studies document such effects in Europe. For instance, in a study for the summer 2003 heat wave in France, Zaitchik et al. (2006) show how air temperature can be sensitive to vegetation cover in this region: Using satellite data, they identify that temperature differences between August 10, 2003 (at the peak of the heat wave) and a normal August day in 2000 (August 1, 2000) are much higher over pasture/active crops (+20°C) than forest (+11°C) areas. This complex spatial pattern is even more evident in Fig. 7 where the highest anomalies of MODIS radiative land surface temperatures for summer 2003 are concentrated in the predominantly agricultural areas of central France.

A better resilience of the forest areas to the heat wave conditions similarly shows up in Normalized Difference Vegetation Index (NDVI) measurements (a measure for vegetation activity): For the pasture/active crops areas, NDVI on August 10, 2003 corresponds to 50% of the value in 2000, while forested areas show no differences (Zaitchik et al. 2006). Ciais et al. (2005) found that in general drought-tolerant ecosystems in the Mediterranean had a lower response to the unusual 2003 conditions than more drought-susceptible temperate vegetation in central Europe. A complex pattern of impacts was seen for Alpine ecosystems (Jolly et al. 2005): Longer-growing seasons were observed at high elevations due to a longer snow-free period while lower elevations were experiencing a shorter growing due to temperature and moisture stress.

These temporal variations in vegetation greenness feed back on the hydrological cycle over land through modifications of the surface heat, water (Guillevic et al. 2002) and carbon balances (Schaefer et al. 2005). Start and length of the growing season over Europe derived from 20 years of NDVI data reveal a significant inter-annual variability in European phenology (Stöckli and Vidale 2004; Studer et al.

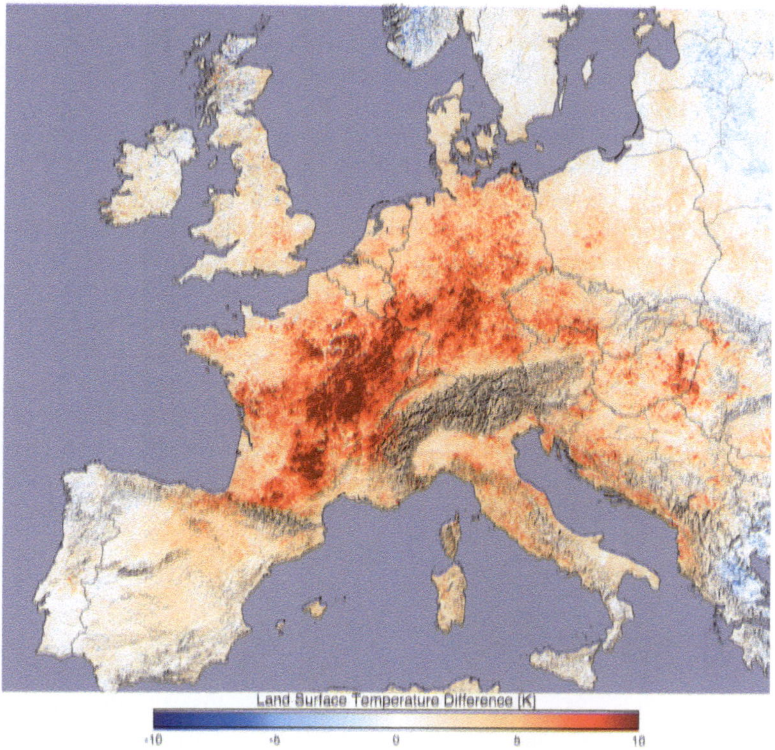

Fig. 7 MODIS (MODerate resolution Imaging Spectroradiometer) radiative land surface temperature (LST) anomaly over Europe for 2003: LST difference (all cloud-free pixels during July 20–August 20) for the years 2000, 2001, 2002, 2004, 2005 and 2006 subtracted from 2003. (Visualization by R. Stöckli as published in Allen and Lord 2004 but with updated data from 2005, 2006).

2007). These patterns of vegetation phenological states act in concert with climatic drivers such as temperature and precipitation (Los et al. 2001; Zhang et al. 2004). Accounting for interannual variability of phenology in numerical weather prediction (Chapeaux et al. 2000) and climate modeling (Bounoua et al. 2000; Lu and Shuttleworth 2002) results in differences of up to 0.9 K in air temperature in Mediterranean and Central Europe and can modify precipitation rates up to 9 mm/month in northern latitudes.

Climate change might further alter these vegetation–climate feedback processes. Twentieth-century warming already resulted in earlier springs and generally longer growing seasons over Europe (Menzel 2000; Defila and Clot 2001; Studer et al. 2005; Menzel et al. 2007). These trends are likely linked to increases in temperature and decreases in snow duration in temperate and alpine ecosystems during the last century. Furthermore, higher atmospheric CO_2 levels can possibly lead to enhanced water-use efficiency in plants (e.g., Field et al. 1995). In a recent modeling study,

Gedney et al. (2006) suggest for instance that observed positive runoff trends during the 20th century may be due to this effect. Also ground observations of CO_2-enriched trees document such water saving effects, though the absolute response is highly species dependent (Leuzinger et al. 2005).

Reversely, potential future modifications of the hydrological cycle, such as more frequent drought conditions, could also have an impact on the carbon cycle. In their study for the 2003 heat wave, Ciais et al. (2005) performed a thorough analysis of 15 CarboEurope tower sites, and found that 2003 switched Central Europe from a net carbon sink to a net carbon source. Gross Primary Production (GPP, i.e., the carbon uptake of plants during photosynthesis) decreased by 30% as a result of rainfall deficit and heat. Despite the high soil temperatures, the carbon loss through ecosystem respiration also decreased slightly, heterotrophic decomposition being inhibited by the prevailing dry soil conditions. A decreased vegetation carbon sink would ultimately result in a further enhancement of atmospheric CO_2 concentrations and consequent impacts on the climate system. Finally, direct CO_2 effects on carbon assimilation may also be possible (enhanced carbon assimilation in enhanced CO_2 conditions), though these may be more limited than previously assumed (Körner et al. 2005).

6 Conclusions and Outlook

We have seen in this review that the interactions between land and the atmosphere are manifold and can impact climate variability along various paths. We have focused here only on Europe and on seasonal-to-interannual climate variability, but land-atmosphere interactions can impact climate in many other regions (e.g., Koster et al. 2004a) and on a much wider range of temporal scales (Claussen et al. 2004).

While the study of land–atmosphere interactions offers promising perspectives for future research, there are also some open issues impeding progress in this field. The main limitation is the lack of ground observations of key variables such as soil moisture or evapotranspiration. The Global Soil Moisture Data Bank (Robock et al. 2000) provides access to soil moisture observations from several measurement networks around the globe, but data is lacking in many regions and in particular in Europe. The Fluxnet network (Baldocchi et al. 2001) provides flux measurements of energy, water (evapotranspiration) and CO_2 around the world, and especially in Europe (through CarboEurope). Some studies have shown the usefulness of these measurements for the process-based assessment of climate models (Stöckli and Vidale 2005; Teuling et al. 2006). Nonetheless, for certain applications the data set lacks spatial and temporal continuity.

In this light, approaches that allow to obtain indirect estimates of relevant land surface quantities such as soil moisture or evapotranspiration are very promising and could significantly advance research in this field. For instance, combined atmospheric–terrestrial water balance estimates using reanalysis data and runoff observations (Seneviratne et al. 2004; Hirschi et al. 2006a) have been shown to provide useful information on basin-scale variations in terrestrial water storage and have been employed in several applications (e.g., Andersen et al. 2005; Hirschi et al. 2006b, 2007; Jacob et al. 2007; Seneviratne et al. 2006b; van den Hurk et al. 2005). Moreover, several satellite data products show some promising results such as the Gravity Recovery and Climate Experiment Mission (e.g., Tapley et al. 2004; Rodell et al. 2004a), microwave remote sensing products (e.g., Reichle and Koster 2005), and radiometrically derived biophysical vegetation products, e.g., land cover maps, NDVI or LAI (Tucker et al. 1985; Champeaux et al. 2000; Justice et al. 2002; Stöckli and Vidale 2004; Running et al. 2004). Finally, approaches combining observations and model data such as the Global Soil Wetness Project (Dirmeyer et al. 1999, 2002), the Global Land Data Assimilation System (Rodell et al. 2004b), and other land data assimilation products might ultimately help to obtain reliable global estimates of the relevant climate variables.

In the area of ecosystem fluxes and vegetation–climate interactions, improving our process-based understanding at different scales will require the integration of ground and space-based observational networks and numerical modeling initiatives, combined with the complementary collaboration of different research communities (Canadell et al. 2000; Running et al. 1999; Turner et al. 2004). For instance, bottom-up studies from flux towers allow ecosystem researchers to document the seasonal-to-interannual biospheric functioning in response to climate variability. They are valuable both for the development and the validation of empirical and process-based ecosystem models (Running et al. 1999; Stockli and Vidale 2005). These data sets have nonetheless some limitations, in terms of availability and consistency (Houghton 2003), as well as with regard to their applicability to regional-to-global scale processes. Inverse modeling through data assimilation can for instance provide effective means to generate regional flux estimates of the global hydrological and biogeochemical cycle based on a heterogeneous and incomplete distribution of local measurements (Gurney et al. 2002). By linking such top-down measurements with local-scale bottom-up ecosystem measurements, the mentioned scale gap can possibly be overcome (e.g., Denning et al. 2003).

In conclusion, the investigation of land–atmosphere interactions and their role for seasonal to interannual climate variability is a growing interdisciplinary research field offering significant promises for climate research. Land–atmosphere interactions are relevant for climate predictability on several time and spatial scales, and their better understanding could advance many climate applications important for society such as seasonal forecasting and climate-change modeling. Finally, we have shown that they are relevant in many ways for climate in Europe and should therefore be better investigated on this continent, especially in terms of improved soil moisture networks and the integrated analysis of the already existing models and data.

Acknowledgments We would like to acknowledge the CarboEurope site investigators Marc Aubinet, Mika Aurela, Bernard Heinesch, Tuomas Laurila, Dario Papale, and Riccardo Valentini for providing the flux tower measurements for Fig. 3.

References

Allen, M. R. and R. Lord, 2004: Climate change: the spectre of liability. *Nature*, **432**, 551–552.

Andersen, O. B., S. I. Seneviratne, J. Hinderer, and P. Viterbo, 2005: GRACE-derived terrestrial water storage depletion associated with the 2003 European heat wave. *Geophys. Res. Lett.*, **32**, L18405, doi:10.1029/2005GL023574.

Aubinet, M., B. Chermanne, M. Vandenhaute, B. Longdoz, M. Yernaux, and E. Laitat, 2001: Long term carbon dioxide exchange above a mixed forest in the Belgian ardennes, *Agr. Forest Meteorol.*, **108**, 293–315.

Aurela, M., T. Laurila, and J. P. Tuovinen, 2002: Annual CO2 balance of a subarctic fen in northern Europe: importance of the wintertime efflux. *J. Geophys. Res.*, **107**, doi:10.1029/2002JD002055.

Baldocchi, D. et al., 2001: FLUXNET: a new tool to study the temporal and spatial variability of ecosystem-scale carbon dioxide, water vapor, and energy flux densities. *B. Am. Meteorol. Soc.*, **82**, 2415–2433.

Beljaars, A. C. M., P. Viterbo, M. J. Miller, and A. K. Betts, 1996: The anomalous rainfall over the United States during July 1993: sensitivity to land surface parameterization and soil moisture. *Mon. Wea. Rev.*, **124**, 362–383.

Ball, J. T., I. E. Woodrow, and J. A. Berry, 1987: A model predicting stomatal conductance and its contribution to the control of photosynthesis under different environmental conditions. In: *Progress in Photosynthesis Research*, J. Biggins, ed., Nihjoff, Dordrecht, 221–224.

Betts, A. K., J. H. Ball, A. C. M. Beljaars, M. J. Miller, and P. Viterbo, 1996: The land surface-atmosphere interaction: a review based on observational and global modeling perspectives. *J. Geophys. Res.*, **101**, 7209–7225.

Betts, A. K., 2004: Understanding hydrometeorology using global models. *B. Am. Meteorol. Soc.*, **85**, 1673–1688.

Bonan, G. B., 2002: *Ecological Climatology*. Cambridge University Press, Cambridge.

Bounoua, L., G. J. Collatz, S. O., Los, P. J. Sellers, D. A. Dazlich, C. J. Tucker, and D. A. Randall, 2000: Sensitivity of climate to changes in NDVI. *J. Climate*, **13**, 2277–2292.

Champeaux, J.-L., D. Arcos, E. Bazile, D. Giard, J.-P. Goutorbe, F. Habets, J. Noilhan, and J.-L. Roujean, 2000: AHVRR-derived vegetation mapping over Western Europe for use in numerical weather prediction models. *Int. J. Remote Sens.*, **21**, 1183–1199.

Canadell, J. G., H. A. Mooney, D. D. Baldocchi, J. A. Berry, J. R. Ehleringer, C. B. Field, S. T. Gower, D. Y. Hollinger, J. E. Hunt, R. B. Jackson, S. W. Running, G. R. Shaver, W. Steffen, S. E. Trumbore, R. Valentini, and B. Y. Bond, 2000: Carbon metabolism of the terrestrial biosphere: a multitechnique approach for improved understanding. *Ecosystems*, **3**, 115–130.

Ciais, P. et al., 2005: Europe-wide reduction in primary productivity caused by the heat and drought in 2003. *Nature*, **437**, 529–533.

Claussen, M., P. M. Cox, X. Zeng, P. Viterbo, A. C. M. Beljaars, R. A. Betts, H.-J. Bolle, T. Chase, and R. Koster, 2004: The global climate. In: *Vegetation, Water, Humans and the Climate: A New Perspective on an Interactive System*, P. Kabat, M. Claussen, P. A. Dirmeyer, J. H. C. Gash, Guenni L. Bravo de, M. Meybeck, R. Pielke Sr, C. J. Vörösmarty, R. W. A. Hutjes, and S. Lütkemeier, eds., Springer, Berlin.

Cramer, W., A. Bondeau, F. I. Woodward, I. C. Prentice, R. A. Betts, V. Brovkin, P. M. Cox, V. Fisher, J. A. Foley, A. D. Friend, C. Kucharik, M. R. Lomas, N. Ramankutty, S. Sitch, B. Smith, A. White, and C.Young-Molling, 2001: Global response of terrestrial ecosystem structure and function to CO_2 and climate change: results from six dynamic global vegetation models. *Glob. Change Biol.*, **7**, 357–373.

Defila, C. and B. Clot, 2001: Phytophenological trends in Switzerland. *Int. J. Biometeorol.*, **45**, 203–207.

Della-Marta, P. M., J. Luterbacher, H. von Weissenfluh, E. Xoplaki, M. Brunet, and H. Wanner, 2007: Summer heat waves over western Europe 1880–2003, their relationship to large scale forcings and predictability. *Clim. Dyn.*, **29**, 251–279.

Denning, A. S., M. Nicholls, L. Prihodko, I. Baker, P. L. Vidale, K. Davis, and P. Bakwin, 2003: Simulated variations in atmospheric CO_2 over a Wisconsin forest using a coupled ecosystem-atmosphere model. *Glob. Change Biol.*, **9**, 1241–1250.

Dickinson, R. E., 2001: Linking ground hydrology to ecosystems and carbon cycle in a climate model. In: *Present and Future of Modeling Global Environmental Change: Toward Integrated Modeling*, T. Matsuno and H. Kida, eds., TERRAPUB, 137–144.

Dirmeyer, P. A., A. J. Dolman, and N. Sato, 1999: The pilot phase of the Global Soil Wetness Project. *Bull. Am. Meteorol. Soc.*, **80**, 851–875.

Dirmeyer, P. A., X. Gao, and T. Oki, 2002: GSWP-2: the second global soil wetness project science and implementation plan. *IGPO Publication Series 37*, International GEWEX Project Office.

Eltahir, E. A. B., 1998: A soil moisture-rainfall feedback mechanism. 1. Theory and observations. *Water Res. Res.*, **34**, 765–776.

Farquhar, G. D., S. V. Caemmerer, and J. A. Berry, 1980: A biochemical-model of photosynthetic CO_2 assimilation in leaves of C-3 species. *Planta*, **149**, 78–90.

Ferranti, L. and P. Viterbo, 2006: Sensitivity to soil water initial conditions. *J. Climate*, **19**, 3659–3680.

Field, C. B., R. B. Jackson, and H. A. Mooney, 1995: Stomatal responses to increased CO2 – implications from the plant to the global scale. *Plant Cell and Environment*, **18**, 1214–1225.

Fischer, E., S. I. Seneviratne, P. L. Vidale, D. Lüthi, and C. Schär, 2007: Soil moisture – atmosphere interactions during the 2003 European summer heatwave. *J. Climate* (in press).

Gedney, N., P. M. Cox, R. A. Betts, O. Boucher, C. Huntingford, and P. A. Stott, 2006: Detection of a direct carbon dioxide effect in continental river runoff records. *Nature*, **439**, 835–838.

Guillevic, P., R. D. Koster, M. J. Suarez, L. Bounoua, G. J. Collatz, S. O. Los, and S. P. P. Mahanama, 2002: Influence of the interannual variability of vegetation on the surface energy balance – a global sensitivity study. *J. Hydrometeor*, **3**, 617–629.

Gurney, K. R., R. M. Law, A. S. Denning, P. J. Rayner, D. Baker, P. Bousquet, L. Bruhwiler, Y. H. Chen, P. Ciais, S. Fan, I. Y. Fung, M. Gloor, M. Heimann, K. Higuchi, J. John, T. Maki, S. Maksyutov, K. Masarie, P. Peylin, M. Prather, B. C. Pak, J. Randerson, J. Sarmiento, S. Taguchi, T. Takahashi, and C. W. Yuen, 2002: Towards robust regional estimates of CO_2 sources and sinks using atmospheric transport models. *Nature*, **415**, 626–630.

Hirschi, M., S. I. Seneviratne, and C. Schär, 2006a: Seasonal variations in terrestrial water storage for major mid-latitude river basins. *J. Hydrometeor*, **7**, 39–60.

Hirschi, M., P. Viterbo, and S. I. Seneviratne, 2006b: Basin-scale water-balance estimates of terrestrial water storage from ECMWF operational forecast analysis data. *Geophys. Res. Lett.*, **33**, L21401, doi:10.1029/2006GL027659.

Hirschi, M., S. I. Seneviratne, S. Hagemann, and C. Schär, 2007: Analysis of seasonal terrestrial water storage variations in regional climate simulations over Europe. *J. Geophys. Res.* (in press).

Houghton, R. A., 2003: Why are estimates of the terrestrial carbon balance so different? *Glob. Change Biol.*, **9**, 500–509.

Jacob, D., L. Barring, O. B. Christensen, J. H. Christensen, S. Hagemann, M. Hirschi, E. Kjellström, G. Lenderink, B. Rockel, C. Schär, S. I. Seneviratne, S. Somot, A. van Ulden, and B. van den Hurk, 2007: An inter-comparison of regional climate models for Europe: design of the experiments and model performance. *Clim. Change*, **81**(1), 31–52.

Jolly, W. M., M. Dobberthin, N. E. Zimmermann, and M. Reichstein, 2005: Divergent vegetation growth responses to the 2003 heat wave in the Swiss alps. *Geophys. Res. Lett.*, **32**, L18409, doi:10.1029/2005GL023252.

Justice, C. O., J. R. G. Townshend, E. F. Vermote, E. Masuoka, R. E. Wolfe, N. Saleous, D. P. Roy, and J. T. Morisette, 2002: An overview of MODIS land data processing and product status. *Remote Sens. Environ.*, **83**, 3–15.

Körner, C., 2000:. Biosphere responses to CO_2 enrichment. *Ecol. Appl.*, **10**, 1590–1619.

Körner, C., R. Asshoff, O. Bignucolo, S. Hattenschwiler, S. G. Keel, S. Pelaez-Riedl, S. Pepin, R. T. W. Siegwolf, and G. Zotz, 2005: Carbon flux and growth in mature deciduous forest trees exposed to elevated CO_2. *Science*, **309**, 1360–1362.

Koster, R. D. et al., 2006: GLACE: the Global Land – atmosphere coupling experiment. Part 1. Overview. *J. Hydrometeor.*, **7**, 590–610.
Koster, R. D. et al., 2004a: Regions of strong coupling between soil moisture and precipitation. *Science*, **305**, 1138–1140.
Koster, R. D., M. J. Suarez, P. Liu, U. Jambor, A. Berg, M. Kistler, R. Reichle, M. Rodell, and J. Famiglietti, 2004b: Realistic initialization of land surface states: impacts on subseasonal forecast skill. *J. Hydrometeor.*, **5**, 1049–1063.
Koster, R. D. and M. J. Suarez, 2001: Soil moisture memory in climate models. *J. Hydrometeor.*, **2**, 558–570.
Leuzinger, S., G. Zotz, R. Asshoff, and C. Körner, 2005: Responses of deciduous forest trees to severe drought in Central Europe. *Tree Physiology*, **25**, 641–650.
Los, S. O., G. J. Collatz, L. Bounoua, P. J. Sellers, and C. J. Tucker, 2001: Global interannual variations in sea surface temperature and land surface vegetation, air temperature, and precipitation. *J. Climate*, **14**, 1535–1549.
Lu, L. and W. J. Shuttleworth, 2002: Incorporating NDVI-derived LAI into the climate version of RAMS and its impacts on regional climate. *J. Hydrometeor.*, **3**, 347–362.
Menzel, A., 2000: Trends in phenological phases in Europe between 1951 and 1996. *Int. J. Biometeorol.*, **44**, 76–81.
Menzel, A., N. Estrella, and C. Schleip, 2007: *Climate Variability, Trends and the NAO – As Assessed by Impacts on European Plant Phenology Across the 20th Century* (this volume).
Pal, J. S. and E. A. B. Eltahir, 2001: Pathways relating soil moisture conditions to future summer rainfall within a model of the land-atmosphere system. *J. Climate*, **14**, 1227–1242.
Pielke Sr, R. A., 2005: Land use and climate change. *Science*, **310**, 1625–1626.
Reichle, R. H. and R. D. Koster, 2005: Global assimilation of satellite surface soil moisture retrievals into the NASA Catchment land surface model. *Geophys. Res. Lett.*, **32**, L02404.
Robock, A. et al., 2000: The global soil moisture data bank. *Bull. Am. Meteorol. Soc.*, **81**, 1281–1299.
Rodell, M., J. S. Famiglietti, J. Chen, S. I. Seneviratne, P. Viterbo, S. Holl, and C. R. Wilson, 2004a: Basin scale estimates of evapotranspiration using GRACE and other observations. *Geophys. Res. Lett.*, **31**, L20504, doi:10.1029/2004GL020873.
Rodell, M. et al., 2004b: The global land data assimilation system. *Bull. Am. Meteorol. Soc.*, **85**, 381–394.
Reichstein, M., J. D. Tenhunen, O. Roupsard, J. M. Ourcival, S. Rambal, F. Miglietta, A. Peressotti, M. Pecchiari, G. Tirone, and R. Valentini, 2002: Severe drought effects on ecosystem CO2 and H2O fluxes at three Mediterranean evergreen sites: revision of current hypotheses? *Global Change Biol.*, **8**, 999–1017.
Running, S. W., D. D. Baldocchi, D. P. Turner, S. T. Gower, P. S. Bakwin, and K. A. Hibbard, 1999: A global terrestrial monitoring network integrating tower fluxes, flask sampling, ecosystem modeling and EOS satellite data. *Remote Sens. Environ.*, **70**, 108–127.
Running, S. W., R. R. Nemani, F. A. Heinsch, M. S. Zhao, M. Reeves, and H. Hashimoto, 2004: A continuous satellite-derived measure of global terrestrial primary production. *Bioscience*, **54**, 547–560.
Schaefer, K., A. S. Denning, and O. Leonard, 2005: The winter arctic oscillation, the timing of spring, and carbon fluxes in the northern hemisphere. *Global Biochemical Cycles*, **19**, doi:10.1029/2004GB002336.
Schär, C., P. L. Vidale, D. Lüthi, C. Frei, C. Häberli, M. Liniger, and C. Appenzeller, 2004: The role of increasing temperature variability in European summer heat waves. *Nature,* **427**, 332–336.
Schär, C., D. Lüthi, U. Beyerle, E. and Heise, 1999: The soil-precipitation feedback: a process study with a regional climate model. *J. Climate*, **12**, 722–741.
Seneviratne, S. I., D. Lüthi, M. Litschi, and C. Schär,2006a: Land-atmosphere coupling and climate change in Europe. *Nature*, **443**, 205–209.
Seneviratne, S. I., R. D. Koster, Z. Guo, P. A. Dirmeyer, E. Kowalczyk, D. Lawrence, P. Liu, C.-H. Lu, D. Mocko, K. W. Oleson, and D. Verseghy, 2006b: Soil moisture memory in AGCM

simulations: analysis of global land-atmosphere coupling experiment (GLACE) data. *J. Hydrometeor.*, **7**, 1090–1112.

Seneviratne, S. I., P. Viterbo, D. Lüthi, and C. Schär, 2004: Inferring changes in terrestrial water storage using ERA-40 reanalysis data: the Mississippi River basin. *J. Climate*, **17**, 2039–2057.

Sellers, P. J., R. E. Dickinson, D. A. Randall, A. K. Betts, F. Hall, J. Berry, G. Collatz, A. Denning, H. Mooney, C. Nobre, N. Sato, C. Field, A. Henderson-Sellers, 1997: Modeling the exchanges of energy, water, and carbon between continents and the atmosphere. *Science*, **275**, 502–509.

Stöckli, R. and P. L. Vidale, 2004: European plant phenology and climate as seen in a 20-year AVHRR land-surface parameter dataset. *Int. J. Remote Sens.*, **25**, 3303–3330.

Stöckli, R. and P. L. Vidale, 2005: Modeling diurnal to seasonal water and heat exchanges at European Fluxnet sites. *Theor. Appl. Climatol.*, **80**, 229–243.

Studer S, Appenzeller C, Defila C (2005) Inter-annual variability and decadal trends in alpine spring phenology: a multivariate analysis approach. *Climatic Change*, **73**, 395–414.

Studer, S., R. Stöckli, C. Appenzeller, and P. L. Vidale, 2007: A comparative study of satellite and ground-based phenology. *Int. J. Biometeorol.*, **51**, 405–414, doi:10.1007/s00484–006–0080–5.

Tapley, B. D., S. Bettadpur, J. C. Ries, P. F. Thompson, and M. M. Watkins, 2004: GRACE measurements of mass variability in the earth system. *Science*, **305**, 503–505.

Teuling, A. J., S. I. Seneviratne, C. Williams, and P. A. Troch, 2006: Soil moisture persistence from evapotranspiration time series. *Geophys. Res. Lett.*, **33**, L023403, doi:10.1029/2006GL028178.

Tucker, C. J., J. R. G. Townshend, and T. E. Goff, 1985: African land-cover classification using satellite data. *Science*, **227**, 369–375.

Turner, D. P., S. V. Ollinger, and J. S. Kimball, 2004: Integrating remote sensing and ecosystem process models for landscape- to regional-scale analysis of the carbon cycle. *Bioscience*, **54**, 573–584.

van den Hurk, B., M. Hirschi, C. Schär, G. Lenderink, E. van Meijgaard, A. van Ulden, B. Rockel, S. Hagemann, P. Graham, E. Kjellström, and R. Jones, 2005: Soil control on runoff response to climate change in regional climate model simulations. *J. Climate*, **18**, 3536–3551.

Zaitchik, B. F., A. K. Macaldy, L. R. Bonneau, and R. B. Smith, 2006: Europe's 2003 heat wave: a satellite view of impacts and land atmosphere feedbacks. *Int. J. Climateatol.*, **26**, 743–769.

Zhang, X. Y., M. A. Friedl, C. B. Schaaf, and A. H. Strahler 2004: Climate controls on vegetation phenological patterns in northern mid- and high latitudes inferred from MODIS data. *Glob. Change Biol.*, **10**, 1133–1145.

Aerosol Effects on Precipitation Locally and Globally

U. Lohmann

Abstract The question of whether anthropogenic emissions of aerosols or their precursors can contribute to droughts and heavy precipitation events is still an open one. While there is a microphysical link between an increase in aerosols and an increase in cloud albedo, a direct link to surface precipitation is less straight forward because it involves interactions between cloud microphysics and dynamics.

Locally it has been suggested that in some instances increases in aerosols can increase heavy precipitation because less precipitation is formed in the lower parts of convective clouds so that more latent heat is released when the cloud glaciates. Globally the dominant effect of aerosols on precipitation is that aerosols cool the surface due to the increased aerosol and cloud optical depth, which then reduces evaporation and, hence, precipitation.

1 Introduction

The burning of fossil fuels and biofuels due to human activities has greatly increased the amount of particulate matter in the atmosphere. The major aerosol components are mineral dust, sea salt, sulfates, nitrates, black carbon (also termed soot) and particulate organic matter. The natural aerosol species, mineral dust and sea salt dominate the mass concentration in the atmosphere. Aerosols also act as condensation nuclei and ice nuclei, thereby changing cloud microphysical properties. If more aerosol particles compete for the uptake of water vapor, the resulting cloud droplets do not grow as large. More numerous but smaller cloud droplets have a larger surface area than fewer larger cloud droplets for the same amount of cloud water. Thus, a polluted cloud reflects more solar radiation back to space, resulting in a negative radiative forcing at the top-of-the-atmosphere (cloud albedo effect). The cloud albedo effect is purely a radiative forcing and does not cause any change in precipitation (Table 1).

Institute of Atmospheric and Climate Science, Universitätstrasse
16, 8092 Zurich, Switzerland
ulrike.lohmann@env.ethz.ch

S. Brönnimann et al. (eds.), *Climate Variability and Extremes during the Past 100 Years.*

Table 1 Overview of the different aerosol indirect effects and their sign of the change in surface precipitation

Effect	Cloud types affected	Process	Sign of change in surface precipitation	Potential magnitude
Cloud albedo effect	All clouds	For the same cloud water or ice content more but smaller cloud particles reflect more solar radiation	n/a	n/a
Cloud lifetime effect	All clouds	Smaller cloud particles decrease the precipitation efficiency thereby presumably prolonging cloud lifetime	Negative	Small
Glaciation indirect effect	Mixed-phase clouds	An increase in ice nuclei increases the precipitation efficiency	Positive	Medium
Thermodynamic effect	Mixed-phase clouds	Smaller cloud droplets delay freezing causing supercooled clouds to extend to colder temperatures	Positive or negative	Medium
Semidirect effect	All clouds	Absorption of solar radiation by absorbing aerosols affects static stability, the surface energy budget and may lead to an evaporation of cloud particles	Negative	Large

In addition, these more numerous but smaller cloud droplets collide less efficiently with each other, which reduces the precipitation efficiency of polluted clouds (Fig. 1) and presumably prolongs their lifetime (cloud lifetime effect). The absence of drizzle size drops decreases the amount of precipitation at the surface in all climate model simulations (Table 1 and Sect. 4) but the scientific understanding of this process is still very low. The cloud lifetime effect also implies more scattering of solar radiation back to space, thus reinforcing the cloud albedo effect. Whether the cloud lifetime or the cloud albedo effect is more important, is still an open question. While some models predict that the cloud albedo effect is four times as important as the cloud lifetime effect, other models predict that the cloud lifetime effect dominates over the cloud albedo effect (Lohmann and Feichter 2005).

As shown in Fig. 1, a proposed counteracting effect could include the ice phase (glaciation effect). Here increases in ice nuclei since preindustrial times can result in more frequent freezing of supercooled clouds. As the precipitation formation in ice clouds is faster than in water clouds, this would increase the overall amount of precipitation at the surface (Table 1). General circulation model (GCM) studies suggest that if, in addition to mineral dust, hydrophilic black carbon aerosols are assumed to act as ice nuclei at temperatures between 0°C and −35°C, they can

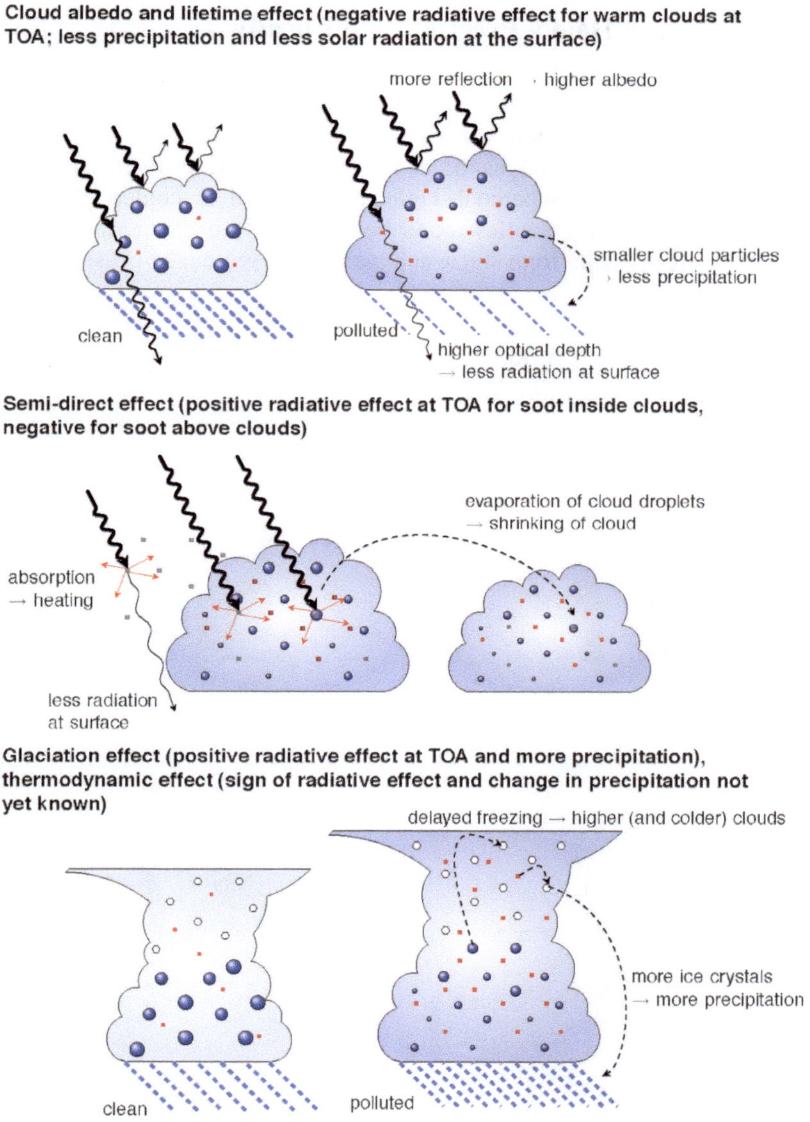

Fig. 1 Schematic description of the aerosol effects on the energy and water balance. TOA denotes top-of-the-atmosphere.

cause a glaciation indirect effect and increase the amount of precipitation via the ice phase (Lohmann 2002). Whether the glaciation effect or cloud lifetime effect is larger depends on the chemical nature of the dust aerosols (Lohmann and Diehl 2006).

2 Aerosols and Heavy Precipitation Locally

Possibilities regarding how aerosols could increase heavy precipitation are related to aerosol effects on convective clouds (e.g., Rosenfeld 1999; Rosenfeld and Woodley 2000). The thermodynamic effect refers to a delay in freezing by the smaller droplets causing supercooled clouds to extend to colder temperatures (Table 1). Here modeling studies found both increases or decreases in surface precipitation. Precipitation from single cell warm and even mixed-phase (water and

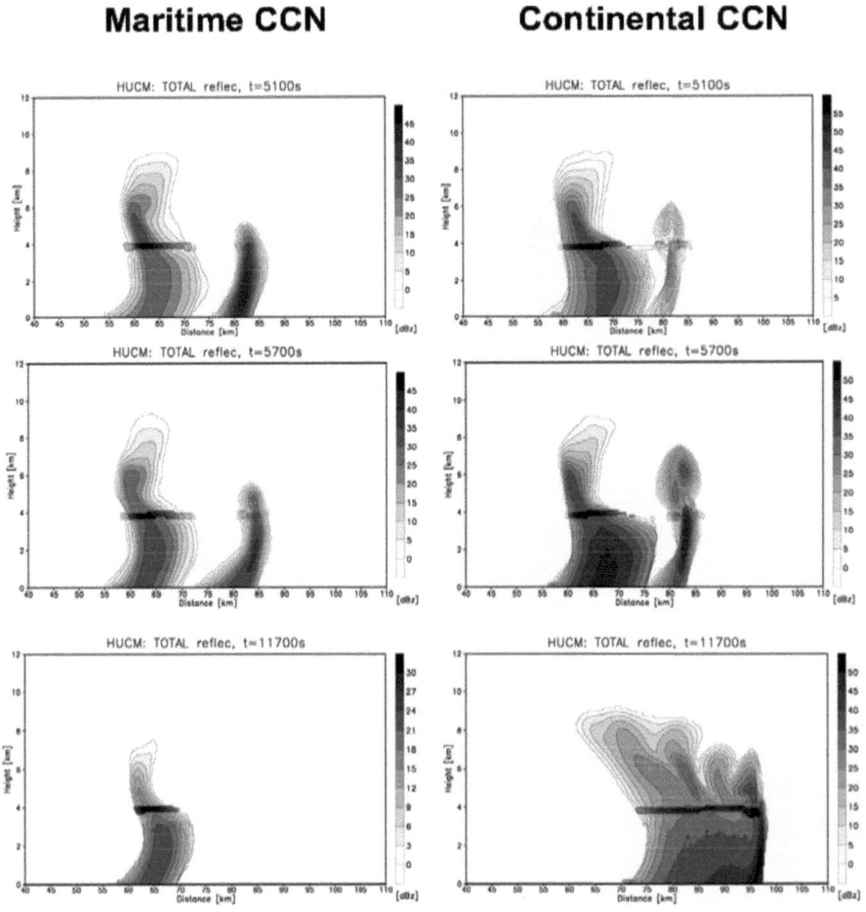

Fig. 2 Time evolution of radar reflectivity in the microphysically maritime conditions (left panels) and in the microphysically continental conditions (right panels). The formation of a squall line in the microphysically continental conditions is seen (reproduced from Khain et al. 2005, with permission).

ice) convective clouds is reduced under continental and maritime conditions due to the cloud lifetime effect as discussed above when aerosol concentrations are increased (Khain et al. 2004; Seifert and Beheng 2006). In the modeling study by Cui et al. (2006) this is caused by drops evaporating more rapidly in the high aerosol case, which eventually reduces ice mass and hence precipitation.

On the other hand, Khain et al. (2005) postulated that smaller cloud droplets, such as those originating from human activity, would change the thermodynamics of deeper convective clouds. More numerous smaller droplets would reduce the production of rain in the lower parts of deep convective clouds so that more liquid water is available for freezing. When these droplets freeze, the increase in the associated latent heat release would then result in more vigorous convection. In a clean cloud, on the other hand, rain would have left the cloud so that less latent heat is released when the cloud glaciates resulting in less vigorous convection. Thus, as shown in Fig. 2, they found that a squall line did not form under clean conditions, whereas a squall line developed under continental aerosol conditions and produced more precipitation after 2 h (not shown). Similar results were obtained by Koren et al. (2005) and Zhang et al. (2005) and for the multicell cloud systems studied by Seifert and Beheng (2006). For a thunderstorm in Florida in the presence of Saharan dust, the simulated precipitation enhancement only lasted 2 h after which precipitation decreased as compared with clean conditions (Van den Heever et al. 2006). This highlights the complexity of the system and indicates that the sign of the local change in precipitation due to anthropogenic aerosols is not yet known.

3 Aerosol Effects on Temperature and Atmospheric Circulation Associated with Changes in Precipitation

Absorption of solar radiation by aerosols can change the cloud amount (semi-direct effect; Grassl 1975; Hansen et al. 1997; Fig. 1). The semi-direct effect has been simulated with GCMs and high-resolution cloud-resolving models, since it is implicitly accounted for whenever absorbing aerosols coupled to the radiation scheme are included (Hansen et al. 1997; Lohmann and Feichter 2001; Hansen et al. 2005). Aerosol heating within cloud layers reduces cloud fractions, whereas aerosol-heating above the cloud layer tends to increase cloud fractions. When diagnosed within a GCM framework, the semi-direct effect can also include cloud changes due to circulation effects and/or surface albedo effects. Moreover, the semidirect effect is not exclusive to absorbing aerosol, as potentially any radiative heating of the mid-troposphere can produce a similar response in a GCM (Hansen et al. 2005). Using a large eddy simulation, Feingold et al. (2005) show that the reduction in net surface radiation and in surface latent and sensible heat fluxes explains most simply the reduction in cloudiness associated with absorbing aerosols. This reduction in the surface latent heat flux will then decrease the global amount of precipitation (Table 1). This could potentially be the largest effect of aerosols on precipitation, but the scientific understanding of all these processes is still very low. Moreover, also increases in scattering aerosols can cause a reduction in net surface radiation and in surface latent and sensible heat fluxes (Liepert et al. 2004). This mechanism has been suggested as having contributed to the Sahel drought as discussed by Paeth (2007) in this book.

Menon et al. (2002), Wang (2004) and Cheng et al. (2005) found that circulation changes could be caused by aerosols in Southeast China. Cheng et al. (2005) observed a strengthening of the West Pacific Subtropical High, with the high pressure system extending farther westward over the continent in southern China, over the last 40 years in a drought region in southern China. Because the early summer average temperature contrast between the land and ocean decreased, the southwesterly monsoon from the ocean onto mainland China weakened. This led to a surface horizontal wind divergence anomaly over southern China, which stabilized the boundary layer. Thus, less moisture was transported to southern China, causing a drying trend. Despite this trend, surface observations show that the aerosol optical depth and low level cloud cover have increased, while visibility has decreased. Precipitation has decreased in this region in the early summer, consistent with both the aerosol lifetime effect and dynamically induced changes from convective to more stratiform clouds. The GCM results (Cheng et al. 2005) suggest that both effects contribute to the changes in low-level cloud cover and precipitation in the drought region in Southern China. The flooding trend in eastern China, however, is more likely caused by strengthened convective precipitation associated with increases in sea surface temperature and greenhouse gases.

In a GCM study Menon et al. (2002) added absorbing aerosols over India and China to understand their effect on the circulation. They obtained increased rising motions where absorbing aerosols were added as well as increased subsidence to the south and north. Figure 3 shows simulated summer (JJA) precipitation changes. Simulation A with absorbing aerosols (representative for having a large amount of black carbon) yields increased precipitation in southern China and over India and Myanmar where the increase in anthropogenic aerosols was largest. There is a broad band of decreased precipitation to the south of the region with increased precipitation, with a lower decrease to the north. The magnitude of the precipitation changes, 0.5 mm/day or 5 cm for the season, is comparable to changes based on observed trends over several decades (Xu 2001). In comparison, precipitation changes are small in experiment B with only scattering aerosols such as sulfates and, contrary to observations, no increase in precipitation is found over south China.

In contrast, Ramanathan et al. (2005) found that convection was suppressed due to increased stability resulting from black carbon heating. Drier conditions resulting from suppressed rainfall can induce more dust and smoke due to the burning of drier vegetation (Ramanathan et al. 2001), thus affecting both regional and global hydrological cycles (Wang 2004).

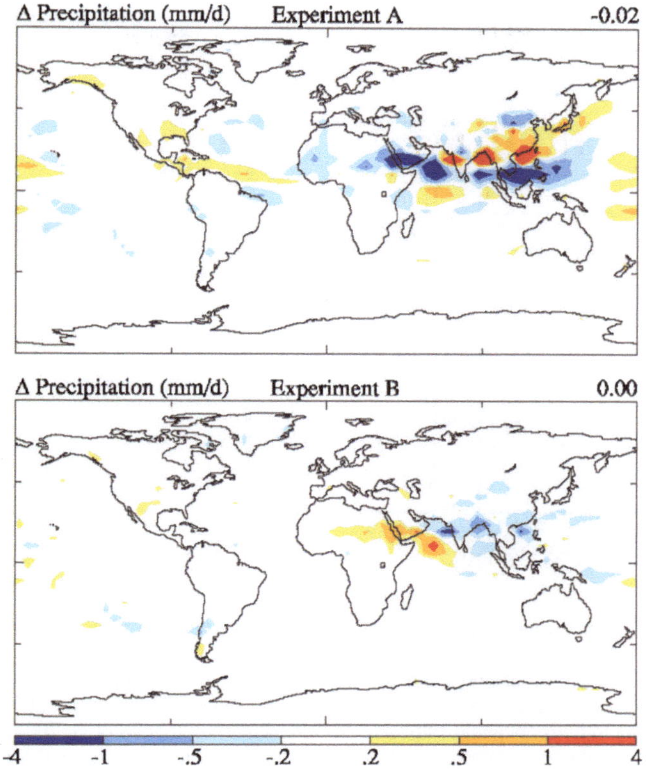

Fig. 3 Simulated June–August surface precipitation change for experiments with increased black carbon aerosols (experiment A) and with increased scattering aerosols (experiment B). Global mean changes are in the upper right corner (reproduced from Menon et al. 2002, with permission from AAAS).

4 Global Effects of Anthropogenic Aerosols on Precipitation

Global climate model estimates of the change in global mean precipitation between pre-industrial times and present day due to the total anthropogenic aerosol effect (sum of the direct, semidirect and indirect aerosol effects) are summarized in Fig. 4. In these simulations, the greenhouse gas concentrations are held constant and only the aerosol emissions change between pre-industrial and present-day times. The climate models include different changes in anthropogenic aerosol compounds (sulfate, organic and black carbon aerosols). While some change all anthropogenic aerosol compounds, others use the increase in sulfate aerosols as surrogates for all anthropogenic aerosols. Some simulations include aerosol effects via the ice phase. While most simulations stem from pure atmospheric GCMs, the green bars refer to estimates from coupled atmosphere-mixed layer ocean GCMs. These coupled

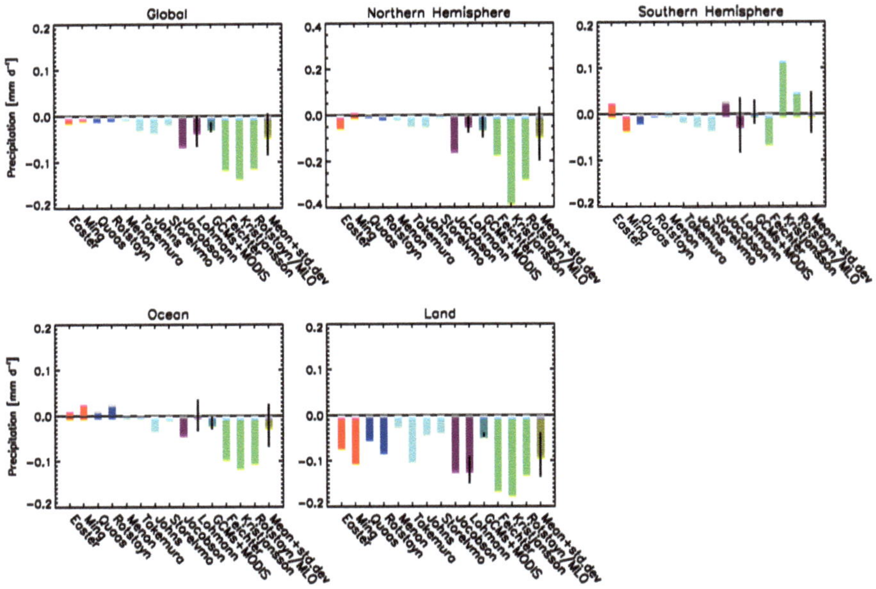

Fig. 4 Global mean change in precipitation due to the total anthropogenic aerosol effect (direct, semidirect and indirect cloud albedo and lifetime effects) from preindustrial times to present day and its contribution over the northern hemisphere (NH), southern hemisphere (SH) and over oceans and over land. Red bars refer to anthropogenic sulfate (Easter et al. 2004; Ming et al. 2005), blue bars to anthropogenic sulphate and organic carbon (Quaas et al. 2004; Rotstayn and Liu 2005), turquoise bars to anthropogenic sulphate, black, and organic carbon (Menon and Genio 2007; Takemura et al. 2005; Johns et al. 2006; Storelvmo et al. 2006), dark purple bars to the mean and standard deviations of anthropogenic sulphate, black, and organic carbon effects on water and ice clouds (Jacobson 2006; Lohmann and Diehl 2006), teal bars refer to a combination of GCM and satellite results (LMDZ/ECHAM+MODIS, Quaas and Boucher 2005), green bars refer to results from coupled atmosphere/mixed-layer ocean (MLO) experiments (Feichter et al. (2004) – sulphate, black, and organic carbon; Kristjansson et al. (2005) – sulphate and black carbon; Rotstayn and Lohmann (2002)[+] – sulphate only) and olive bars to the mean plus standard deviation from all simulations. Vertical black lines refer to ± one standard deviation; [+]Refers to estimates solely from the indirect effects.

simulations are equilibrium simulations separately for pre-industrial and present-day times, in which the last decades of the simulations are used in the analysis.

The change in global mean precipitation between the different GCMs varies between 0 mm and –0.13 mm day^{-1}. The decrease in precipitation in GCMs is due to the cloud lifetime effect because the parameterization of the autoconversion process (the precipitation formation process in warm clouds that contain no ice) depends inversely on the cloud droplet number concentration. Thus, an increase in the cloud droplet number concentration prolongs the autoconversion process, increases the amount of cloud liquid water in the atmosphere and decreases the amount of precipitation reaching the surface.

Differences in the change in precipitation are amplified over the southern hemisphere, ranging from –0.06 mm day^{-1} to 0.12 mm day^{-1}. The decreases in precipitation are larger, when the atmospheric GCMs are coupled to mixed-layer ocean models, where the sea surface temperature and, hence, evaporation is allowed to vary. Here, decreases in the surface temperature cause larger decreases in evaporation and, hence, also in precipitation than when the sea surface temperature is held constant. The decrease in precipitation is most pronounced over land because the land surface temperature is allowed to decrease and where the increase in the cloud droplet number concentration and, hence, the decrease in the autoconversion rate is most pronounced.

When aerosol effects on warm convective clouds are included in addition to their effect on warm stratiform clouds, the overall indirect aerosol effect and the change in surface precipitation can be larger or smaller than if just the aerosol effect on stratiform clouds is considered (Nober et al. 2003; Menon and Rotstayn 2006). Besides changes in the distribution of precipitation the frequency of extreme events may also be reduced by the presence of aerosols (Paeth and Feichter 2006).

Greenhouse warming, primarily as a result of enhanced CO_2 concentrations, is amplified when the global hydrological cycle is enhanced and greater amounts of water vapor are evaporated into the air principally over the oceans but also over land. Because water vapor is a much more powerful greenhouse gas than CO_2, the increased amount of water vapor in the air, in turn, results in a strong positive feedback to CO_2 warming. Recent GCM simulations of greenhouse warming and direct and indirect aerosol effects (Liepert et al. 2004) show that the indirect and direct cooling effects of aerosols reduce surface latent and sensible heat transfer and, as a consequence, act to spin-down the hydrological cycle and thereby substantially weaken greenhouse gas warming. This is important since most investigators compare top of the atmosphere radiative differences for greenhouse gas warming and aerosol direct and indirect effects separately. Since greenhouse warming causes a spin-up of the hydrological cycle, and aerosol direct and indirect cooling counteracts this spin-up, the potential influence of aerosols on global climate could be far more significant than previously thought. The simulated decrease in global mean precipitation from pre-industrial times to the present in GCMs is in contrast to the observed precipitation evolution in the last century and points to an overestimation of aerosol influences on precipitation in these simulations. In a GCM simulation by Roeckner et al. (2006), the decrease in precipitation changes sign to an increase of about 1% in simulations for 2031–2050 as compared to 1981–2000, because the increased warming due to black carbon and greenhouse gases then dominates over the sulfate cooling.

5 Summary and Conclusions

In summary, the cloud-scale effect of aerosols on convective precipitation can go in both directions. Some studies found an enhancement in precipitation while others found a decrease.

Globally the most important aerosol effect on precipitation results from the increase in cloud and aerosol optical depth due to anthropogenic activities. This slows down the precipitation formation in warm clouds and reduces the global mean precipitation in all climate models. However, one problem is that most climate models suggest an increase in liquid water when adding anthropogenic aerosols, whereas newer ship track studies show that polluted marine water clouds can have less liquid water than clean clouds (Platnick et al. 2000; Coakley Jr. and Walsh 2002). Ackerman et al. (2004) attribute this effect to enhanced entrainment of dry air in polluted clouds in these instances with subsequent evaporation of cloud droplets. Also, when cloud lifetime is analyzed on the cloud scale, an increase in aerosol concentration from very clean to very polluted does not increase cloud lifetime, even though precipitation is suppressed (Jiang et al. 2006). This effect is due to the competition between precipitation suppression and enhanced evaporation of the more numerous smaller cloud droplets in polluted clouds.

In addition, the increase in the cloud and aerosol optical depth decreases the amount of solar radiation reaching the surface. In GCMs that are coupled to mixed-layer ocean models, this negative radiative forcing competes with the greenhouse gas warming for determining the change in surface evaporation, atmospheric stability, large-scale circulation changes and hence, changes in surface precipitation.

At present GCMs cannot account for all microphysical processes that are relevant for aerosol–cloud interactions because the cloud microphysical processes are highly parameterized. Thus, the net global aerosol effect on clouds and precipitation as deduced from GCMs is not yet conclusive.

References

Ackerman, A. S., M. P. Kirkpatrick, D. E. Stevens, and O. B. Toon, 2004: The impact of humidity above stratiform clouds on indirect climate forcing. *Nature*, **432**, 1014–1017.

Cheng, Y. J., U. Lohmann, and J. H. Zhang, 2005: Contribution of changes in sea surface temperature and aerosol loading to the decreasing precipitation trend in Southern China. *J. Climate*, **18**, 1381–1390.

Coakley Jr. J. A. and C. D. Walsh, 2002: Limits to the aerosol indirect radiative forcing derived from observations of ship tracks. *J. Atmos. Sci.*, **59**, 668–680.

Cui, Z. Q., K. S. Carslaw, Y. Yin, and S. Davies, 2006: A numerical study of aerosol effects on the dynamics and microphysics of a deep convective cloud in a continental environment. *J. Geophys. Res.-Atmos.*, **111**.

Easter, R. C., S. J. Ghan, Y. Zhang, R. D. Saylor, E. G. Chapman, N. S. Laulainen, H. Abdul-Razzak, L. R. Leung, X. Bian, and R. A. Zaveri, 2004: MIRAGE: model description and evaluation of aerosols and trace gases. *J. Geophys. Res.*, 109:doi:10.1029/2004JD004571.

Feichter, J., E. Roeckner, U. Lohmann, and B. Liepert, 2004: Nonlinear aspects of the climate response to greenhouse gas and aerosol forcing. *J. Climate*, **17**, 2384–2398.

Feingold, G., H. L. Jiang, and J. Y. Harrington, 2005: On smoke suppression of clouds in Amazonia. *Geophys. Res. Lett.*, **32**.

Grassl, H. 1975: Albedo reduction and radiative heating of clouds by absorbing aerosol particles. *Contrib. Atmos. Phys.*, **48**, 199–210.

Hansen, J., M. Sato, and R. Ruedy, 1997: Radiative forcing and climate response. *J. Geophys. Res.*, **102**, 6831–6864.

Hansen, J., M. Sato, R. Ruedy, L. Nazarenko, A. Lacis, G. A. Schmidt, G. Russell, I. Aleinov, M. Bauer, S. Bauer, N. Bell, B. Cairns, V. Canuto, M. Chandler, Y. Cheng, A. Del Genio, G. Faluvegi, E. Fleming, A. Friend, T. Hall, C. Jackman, M. Kelley, N. Kiang, D. Koch, J. Lean, J. Lerner, K. Lo, S. Menon, R. Miller, P. Minnis, T. Novakov, V. Oinas, J. Perlwitz, J. Perlwitz, D. Rind, A. Romanou, D. Shindell, P. Stone, S. Sun, N. Tausnev, D. Thresher, B. Wielicki, T. Wong, M. Yao, and S. Zhang, 2005: Efficacy of climate forcings. *J. Geophys. Res.-Atmos.*, **110**.

Jacobson, M. Z., 2006: Effects of externally-through-internally-mixed soot inclusions within clouds and precipitation on global climate. *J. Phys. Chem.*, **110**, 6860–6873.

Jiang, H. L., H. W. Xue, A. Teller, G. Feingold, and Z. Levin, 2006: Aerosol effects on the lifetime of shallow cumulus. *Geophys. Res. Lett.*, **33**.

Johns, T. C., C. F. Durman, H. T. Banks, M. J. Roberts, A. J. McLaren, J. K. Ridley, C. A. Senior, K. D. Williams, A. Jones, G. J. Rickard, S. Cusack, W. J. Ingram, M. Crucifix, D. M. H. Sexton, M. M. Joshi, B. W. Dong, H. Spencer, R. S. R. Hill, J. M. Gregory, A. B. Keen, A. K. Pardaens, J. A. Lowe, A. Bodas-Salcedo, S. Stark, and Y. Searl, 2006: The new Hadley Centre Climate Model (HadGEM1): evaluation of coupled simulations. *J. Climate*, **19**, 1327–1353.

Khain, A., A. Pokrovsky, M. Pinsky, A. Seifert, and V. Phillips, 2004: Simulation of effects of atmospheric aerosols on deep turbulent convective using a spectral microphysics mixed-phase cumulus cloud model. Part I: model description and possible applications. *J. Aerosol Sci.*, **61**, 2963–2982.

Khain, A., D. Rosenfeld, and A. Pokrovsky, 2005: Aerosol impact on the dynamics and microphysics of deep convective clouds. *Q. J. Roy. Meteorol. Soc.*, **131**, 2639–2663.

Koren, I., Y. J. Kaufman, D. Rosenfeld, L. A. Remer, and Y. Rudich, 2005: Aerosol invigoration and restructuring of Atlantic convective clouds. *Geophys. Res. Lett.*, **32**.

Kristjansson, J. E., T. Iversen, A. Kirkevag, O. Seland, and J. Debernard, 2005: Response of the climate system to aerosol direct and indirect forcing: role of cloud feedbacks. *J. Geophys. Res.-Atmos.*, **110**.

Liepert, B. G., J. Feichter, U. Lohmann, and E. Roeckner, 2004: Can aerosols spin down the water cycle in a warmer and moister world. *Geophys. Res. Lett.*, **31**, doi:10.1029/2003GL019060.

Lohmann, U., 2002: Possible aerosol effects on ice clouds via contact nucleation. *J. Atmos. Sci.*, **59**, 647–656.

Lohmann, U. and K. Diehl, 2006: Sensitivity studies of the importance of dust ice nuclei for the indirect aerosol effect on stratiform mixed-phase clouds. *J. Atmos. Sci.*, **63**, 968–982.

Lohmann, U. and J. Feichter, 2001: Can the direct and semi-direct aerosol effect compete with the indirect effect on a global scale? *Geophys. Res. Lett.*, **28**, 159–161.

Lohmann, U. and J. Feichter, 2005: Global indirect aerosol effects: a review. *Atmos. Chem. Phys.*, **5**, 715–737.

Menon, S. and A. D Genio, 2007: Evaluating the impacts of carbonaceous aerosols on clouds and climate. In: Human induced climate change: an interdisciplinary assessment. Ed: M. F. Schlesinger, Cambridge University Press.

Menon, S., J. Hansen, L. Nazarenko, and Y. Luo, 2002: Climate effects of black carbon aerosols in China and India. *Science*, **297**, 2250–2252.

Menon, S. and L. Rotstayn, 2006: The radiative influence of aerosol effects on liquid-phase cumulus and stratiform clouds based on sensitivity studies with two climate models. *Clim. Dyn.*, **27**, 345–356.

Ming, Y., V. Ramaswamy, P. A. Ginoux, L. W. Horowitz, and L. M. Russell, 2005: Geophysical fluid dynamics laboratory general circulation model investigation of the indirect radiative effects of anthropogenic sulfate aerosol. *J. Geophys. Res.-Atmos.*, **110**.

Nober, F. J., H.-F. Graf, and D. Rosenfeld, 2003: Sensitivity of the global circulation to the suppression of precipitation by anthropogenic aerosols. *Global and Planetary Change*, **37**, 57–80.

Paeth, H., 2007: Human activity and climate change in Africa, this volume.

Paeth, H. and J. Feichter, 2006: Greenhouse-gas versus aerosol forcing and African climate response. *Clim. Dyn.*, **26**, 35–54.

Platnick, S., P. A. Durkee, K. Nielsen, J. P. Taylor, S.-C. Tsay, M. D. King, R. J. Ferek, P. V. Hobbs, and J. W. Rottman, 2000: The role of background cloud microphysics in the radiative formation of ship tracks. *J. Atmos. Sci.*, **57**, 2607–2624.

Quaas, J. and O. Boucher, 2005: Constraining the first aerosol indirect radiative forcing in the LMDZ GCM using POLDER and MODIS satellite data. *Geophys. Res. Lett.*, **32**.

Quaas, J., O. Boucher, and F. M. Breon, 2004: Aerosol indirect effects in POLDER satellite data and the Laboratoire de Meteorologie Dynamique-Zoom (LMDZ) general circulation model. *J. Geophys. Res.-Atmos.*, **109**.

Ramanathan, V., P. J. Crutzen, J. Lelieveld, A. P. Mitra, D. Althausen, J. Anderson, M. O. Andreae, W. Cantrell, G. R. Cass, C. E. Chung, A. D. Clarke, J. A. Coakley, W. D. Collins, W. C. Conant, F. Dulac, J. Heintzenberg, A. J. Heymsfield, B. Holben, S. Howell, J. Hudson, A. Jayaraman, J. T. Kiehl, T. N. Krishnamurti, D. Lubin, G. McFarquhar, T. Novakov, J. A. Ogren, I. A. Podgorny, K. Prather, K. Priestley, J. M. Prospero, P. K. Quinn, K. Rajeev, P. Rasch, S. Rupert, R. Sadourny, S. K. Satheesh, G. E. Shaw, P. Sheridan, and F. P. J. Valero, 2001: Indian Ocean experiment: an integrated analysis of the climate forcing and effects of the great Indo-Asian haze. *J. Geophys. Res.-Atmos.*, **106**, 28371–28398.

Ramanathan, V., C. Chung, D. Kim, T. Bettge, L. Buja, J. T. Kiehl, W. M. Washington, Q. Fu, D. R. Sikka, and M. Wild, 2005: Atmospheric brown clouds: impacts on South Asian climate and hydrological cycle. *Proc. of the National Academy of Sciences of the United States of America*, **102**, 5326–5333.

Roeckner, E., P. Stier, J. Feichter, S. Kloster, M. Esch, and I. Fischer-Bruns, 2006: Impact of carbonaceous aerosol emissions on regional climate change. *Clim. Dyn.*, **27**, 553–571.

Rosenfeld, D., 1999: TRMM observed first direct evidence of smoke from forest fires inhibiting rainfall. *Geophys. Res. Lett.*, **26**, 3105–3108.

Rosenfeld, D. and W. L. Woodley, 2000: Deep convective clouds with sustained supercooled liquid water down to-37.5 degrees C. *Nature*, **405**, 440–442.

Rotstayn, L. D. and Y. Liu, 2005: A smaller global estimate of the second indirect aerosol effect. *Geophys. Res. Lett.*, **32**, doi:10.1029/2004GL021922.

Rotstayn, L. D. and U. Lohmann, 2002: Simulations of the tropospheric sulfur cycle in a global model with a physically based cloud scheme. *J. Geophys. Res.*, **108**, doi:10.1029/2002JD002128.

Seifert, A. and K. D. Beheng, 2006: A two-moment cloud microphysics parameterization for mixed-phase clouds. Part 2: maritime vs. continental deep convective storms. *Meteorol. Atmos. Phys.*, **92**, 67–82.

Storelvmo, T., J. E. Kristjansson, S. J. Ghan, A. Kirkevag, and O. Seland, 2006: Predicting cloud droplet number concentration in CAM-Oslo. *J. Geophys. Res.*, **111**(D24), D24208, doi:10.1029/2005JD006300.

Takemura, T., T. Nozawa, S. Emori, T. Y. Nakajima, and T. Nakajima, 2005: Simulation of climate response to aerosol direct and indirect effects with aerosol transport-radiation model. *J. Geophys. Res.-Atmos.*, **110**.

Van den Heever, S. C., G. G. Carrio, W. R. Cotton, P. J. DeMott, and A. J. Prenni, 2006: Impacts of nucleating aerosol on Florida convection. Part I: mesoscale simulations. *J. Aerosol. Sci.*, **63**, 1752–1775.

Wang, C., 2004: A modelling study on the climate impacts of black carbon aerosols. *J. Geophys. Res.*, **109**, doi:10.1029/2003JD004084.

Xu, Q., 2001: Abrupt change of the mid-summer climate in central east China by the influence of atmospheric pollution. *Atmos. Environ.*, **35**, 5029–5040.

Zhang, J. H., U. Lohmann, and P. Stier, 2005: A microphysical parameterization for convective clouds in the ECHAM5 climate model: single-column model results evaluated at the Oklahoma Atmospheric Radiation Measurement Program site. *J. Geophys. Res.-Atmos.*, **110**.

Section C
Climate Variability and Extremes

Human Activity and Climate Change in Africa

H. Paeth

Abstract The severe drought during the second half of the 20th century has demonstrated that livelihood in Africa is highly sensitive to climatic conditions. From previous studies the complexity of the key factors in African climate variability can be inferred. This chapter discusses the various key factors and highlights the need of more complex scenarios to assess the human influence on African climate. In particular, the effect of ongoing land-use changes in the course of population growth, shifting cultivation and overgrazing has to be taken into account. New ensemble simulations with a regional climate model are presented, which are subject to increasing greenhouse gas concentrations and land cover changes. The model predicts dryer conditions, near-surface warming and an intensification of heat stress. There is some indication that reforestation and regional protection of natural vegetation may be more effective for mitigating climate change in Africa than reduced greenhouse gas emissions.

1 The Special Case of Africa

Changes in climate may have dramatic implications for livelihood and ecosystems, particularly in the low latitudes. This chapter is dedicated to climate change in Africa where many regions are characterized by a limited natural potential (Achenbach 1994). Among the physical causes the crucial limiting factor is deficient freshwater availability due to intense rainfall variability at various time scales (Nicholson et al. 2000; Herbers 1999). The most prominent feature in recent African climate variability and probably the most apparent anomaly at the climatological timescale observed during the 20th century was the severe drought in the 1970s and 1980s (Fig. 1), which caused enormous economic loss, large-scale migration processes, and irreversible land degradation (Findley 1994; Benson and Clay 1998; Hammer 2000, Richter 2000; Nicholson 2001).

Geographical Institute, University of Würzburg, Am Hubland, 97074 Würzburg, Germany
Heiko.Paeth@uni-wuerzburg.de

S. Brönnimann et al. (eds.), *Climate Variability and Extremes during the Past 100 Years.*

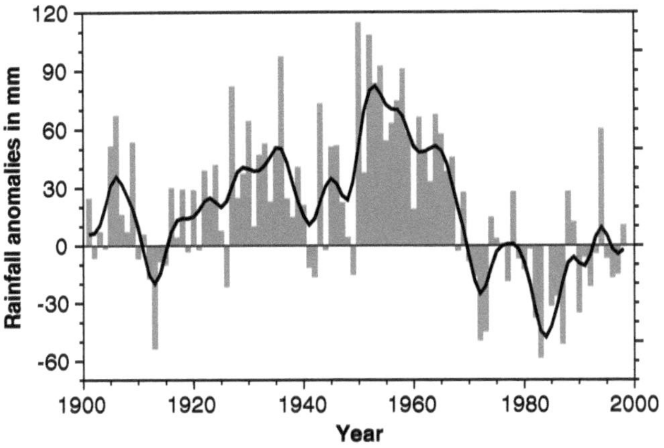

Fig. 1 Time series of observed summer monsoon rainfall anomalies in the Sahel Zone (12°–21°N, 15°W–20°E) with respect to the 1961–1990 reference period. The thick line is smoothed with a 9-year binomial filter.

This disastrous development is tied to a complex interaction between climatic, social and economic factors (Achenbach 1994; Braun and Scholz 1997; Sturm 1999; Hammer 2000). Beside the drought years, severe floods after heavy-rain events represent a further problem in tropical and subtropical Africa, causing soil erosion, damage and even loss of life (Richter 2000; Balzarek et al. 2003). From a demographic point of view livelihood in tropical Africa is supposed to deteriorate progressively into the 21st century, given the strongest population growth on earth and accelerated urbanization (Gaebe 1994; Schulz 2001). Thus, large importance is assigned to the future availability of freshwater in Africa.

2 Key Factors of Climate Variability and Change in Africa

After the prominent drought period during the second half of the 20th century, rainfall is partly recovering since the 1990s (Nicholson et al. 2000), leading to the question of the future development of African precipitation. Thus, improved knowledge of the natural and anthropogenic key factors in African climate variability and change is required in order to provide a scientific basis for political measures, for instance in the form of reforestation and emission protection (Azar and Rodhe 1997). It is obvious that this objective can only be addressed by means of climate model simulations.

There are a number of studies dealing with climate and, especially, rainfall anomalies in Africa. The northern and north-western parts of Africa are closely tied to extratropical ocean dynamics (Rodriguez-Fonseca and de Castro 2002; Rotstayn and Lohmann 2002; Knippertz et al. 2003) and atmospheric phenomena like the

North Atlantic Oscillation (Cullen and de Menocal 2000). Sub-saharan West Africa is mainly affected by tropical sea surface temperatures (SSTs) (Chang et al. 2000; Camberlin et al. 2001). This region is characterized by a prominent monsoon system with a dry north-westerly flow in boreal winter, named Harmattan, and southwesterlies with advection of humid air masses, atmospheric instability and deep convection in summer (Saha and Saha 2001). In general, oceanic heating in the tropical Atlantic is associated with more abundant precipitation during the summer rainy season in the southernmost part of West Africa (Paeth and Hense 2004). Simultaneously, the Sahel Zone is subject to more arid conditions (Mo et al. 2001; Paeth and Stuck 2004). Ward (1998) described the relationship between West African rainfall anomalies and an interhemispheric SST dipole in the Atlantic Ocean: A relative cooling (warming) of the North (South) Atlantic usually triggers dryer conditions in sub-saharan Africa by favouring a more southward position of the rain-bearing Atlantic intertropical convergence zone (ITCZ). Furthermore, there is a teleconnection to the El Niño-Southern Oscillation phenomenon in the eastern tropical Pacific (Nicholson et al. 2000; Janicot et al. 2001). Usually, warm events (El Niño) are associated with less monsoonal rainfall over South-West Africa. This remote response is governed by the Walker circulation and a time-delayed anomaly in the tropical Atlantic (Latif and Grötzner 2000). The Sahel Zone is again reacting in the opposite way, emphasizing the existence of a dipole feature in African rainfall fluctuations (Saravanan and Chang 2000). In addition, large parts of tropical East Africa are affected by SSTs in the Indian Ocean (Bader and Latif 2003; Black et al. 2003). Thus, African rainfall variability is largely embedded in the SST anomalies of the tropical ocean basins (Paeth and Friederichs 2004).

There is also some indication from global climate model simulations that increasing greenhouse gas (GHG) concentrations may play a role in recent and future climate changes in tropical Africa (Houghton et al. 2001; Hulme et al. 2001; Paeth and Hense 2004). Radiative heating is supposed to warm up the tropical Atlantic, which in turn intensifies the African summer monsoon circulation via enhanced latent heat fluxes into the atmosphere, surface wind convergence and deep convection over the southern part of West Africa, whereas the subtropical dry belt may experience more arid conditions in the context of global warming. In North-West Africa, a general strengthening and stronger eastward extent of the Azores High (AH) is expected under the greenhouse, inducing negative rainfall anomalies over Morocco and large parts of the Mediterranean Basin (Cullen and de Menocal 2000; Knippertz et al. 2003).

In addition, human activity leads to enhanced aerosol emissions. In tropical Africa, biomass burning in the context of shifting cultivation and domestic fuel is responsible for high atmospheric aerosol burden, particularly during the dry season in boreal winter and spring (Stier et al. 2005). In summer, the peaks are observed over southern Africa. In addition, mineral dust is originating from the Saharan dry belt and mostly transported towards the tropical North Atlantic and partly to the Carribean (Jones et al. 2003). Aerosols cause a net cooling at the surface, which partly counterbalances the GHG-induced warming (Feichter et al. 2004). Moreover, the indirect aerosol impacts on microphysical cloud properties modify the formation of rainfall. As a consequence, precipitation tends to decrease in African regions with high aerosol burden (Paeth and Feichter 2006). This process is further intensified by a reduction in sensible and latent heat fluxes from the surface. Rotstayn and Lohmann (2002) postulated an indirect mechanism with enhanced aerosol emissions in the northern hemisphere altering the interhemispheric SST gradient in the Atlantic Ocean towards a relatively cool (warm) North (South) Atlantic. According to Ward (1998), this leads to a weakening of the West African summer monsoon flow and negative preciptiation anomalies in sub-saharan Africa.

Drought conditions during the second half of the 20th century were probably related as well to land cover changes in tropical and subtropical Africa (Semazzi and Song 2001; Zeng et al. 2002). Indeed, Texier et al. (2000) demonstrated that the African monsoon system is much more sensitive to low-frequency changes in vegetation cover than for instance the South Asian monsoon. Especially, Sahelian rainfall appears to be closely tied to vegetation cover (Wang and Eltahir 2000), whereas the Guinean Coast region is less sensitive due to the compensating effect of stronger near-surface wind convergence and deep convection (Clark et al. 2001). Satellite data reveal a distinct relationship between the occurrence of drought years in the Sahel region and vegetation (Eklundh and Olsson 2003). Changes in vegetation cover affect the energy balance at the land surface, including albedo and radiation budget as well as latent and sensible heat fluxes into the atmosphere (Bounoua et al. 2000). In addition, soil moisture, surface run-off, and the turbulent mixing in the atmospheric boundary layer are modified. The relationship between rainfall and vegetation cover is fully interactive, a feature, which is so far barely represented in state-of-the-art climate models (Schnitzler et al. 2001).

Soil moisture is a further key factor in African climate (GLACE Team 2004). In general, increasing soil moisture favours the formation of rainfall at the local to regional scale through intensified latent heat fluxes. The African monsoon is much more sensitive to soil moisture than for instance the Indian monsoon (Douville et al. 2001). Fontaine et al. (2002) have revealed a lead-lag relationship between the West African rainy season and preceding soil moisture anomalies. An indirect impact via the African Easterly Jet (AEJ) has been described by Cook (1999).

Although several studies tend to blame human activity for the severe drought period since the 1960s (e.g., Hammer 2000), proxy-data from paleo-studies indicate that comparable dry conditions also occurred in ancient times, for instance in the early 19th century (Nicholson 2001). The strong response of African rainfall to solar variability during the Holocene has also been reported by Texier et al. (2000). At millennial timescales, orbital forcing (Kutzbach and Liu 1997) and slow shifts in the vegetation zones (Texier et al. 1997) represent crucial driving forces for African climate. The role of aerosols from volcanic eruptions is less documented for African climate.

In summary, the detection of individual influences on African climate variations is quite complex. Nevertheless, most authors agree that rainfall fluctuations are primarily imposed by changes in the tropical SSTs and secondarily enhanced by numerous feedbacks with the land cover, involving vegetation, soil moisture and surface albedo (Long et al. 2000; Giannini et al. 2003; Wang et al. 2004). This

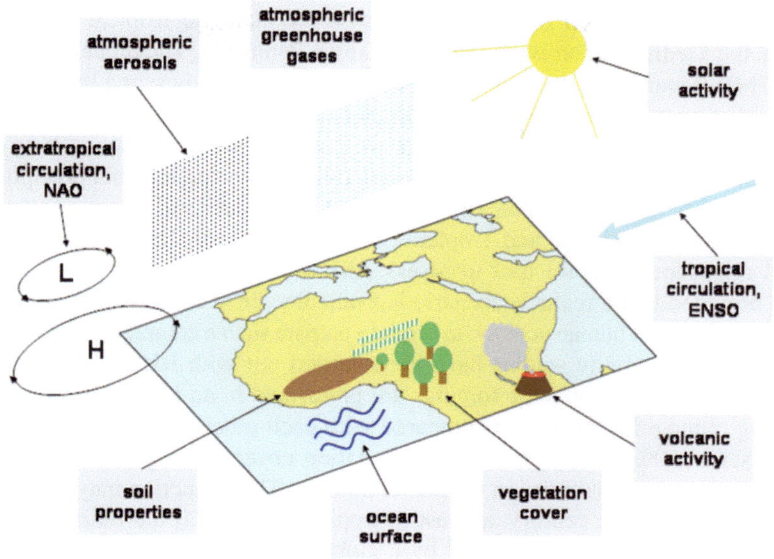

Fig. 2 Presumed natural and anthropogenic factors in African climate.

results in an intensification of rainfall variability in terms of the amplitude and duration of the anomalies (Nicholson 2001). In fact, the Sahel region is characterized by the most pronounced persistence of climate anomalies in Africa (Long et al. 2000), as well as compared to other parts of the earth.

Figure 2 summarizes the presumed players in African climate and highlights the complexity of the problem when detecting the human influence. An objective analysis of African climate change must allow for all the impact factors mentioned above. Particularly, the relative effects of land degradation and greenhouse warming need to be evaluated and compared with each other.

3 Sensitivity Studies with a Regional Climate Model

Most of the studies cited above refer to coarse-grid global climate model experiments which usually do not meet the needs of decision-makers at the national and regional scale (Jenkins et al. 2002). A useful climatological basis for political measures and planning has to fulfil several basic requirements: (1) spatial information with high resolution over Africa, (2) fairly realistic scenarios of future climate change, (3) a long-term simulation period beyond the presumed timescales of natural variability, and (4) a proper treatment of uncertainty. For this purpose, the hydrostatic regional climate model REMO has been implemented over Africa (Jacob 2001). The model is run on a 0.5° grid with 20 atmospheric levels and five soil layers. The model domain extends from 30°W to 60°E and from 15°S to 45°N, including tropical and northern Africa, the Mediterranean Basin and the Arabian Peninsula. The model is nested in the ECMWF reanalysis data for hindcast and sensitivity studies, and in the coupled global climate model ECHAM5/MPI-OM for climate-change simulations.

REMO is able to reproduce all main features of African climate (Paeth et al. 2005), including the typical patterns and variability of temperature and precipitation, the seasonality of the monsoon circulation and the middle and upper tropospheric jetstreams like the African and Tropical Easterly Jets (Hastenrath 2000). Thus, this model represents a useful tool to assess African climate change at the regional scale, provided that a realistic scenario is available which accounts for all relevant processes related to human activity. In order to prepare such a comprehensive scenario, a number of sensitivity studies have been carried out with REMO (Paeth 2004). Three different anthropogenic forcings are prescribed in an idealized way and the resulting climate responses are compared with each other (Fig. 3): (1) vegetation loss is set to 50% of the present-day vegetation cover, homogeneously modified over all land masses included in the model domain. This reduction may be achieved at the end of the 21st century, at least in tropical Africa; (2) the radiative heating refers to the heating rate simulated by a global climate model simulation for the decade 2085–2095 under the IS92a business-as-usual emission scenario (Paeth 2004); (3) soil degradation is parameterized by a shift in the ratio between infiltration (decreased) and surface run-off (increased). The amplitude of this shift is assumed to be associated with a 50% reduction of vegetation cover, although no observational data are available for a proper quantification throughout Africa. Therefore the results in Fig. 3 can only be interpreted in a qualitative way.

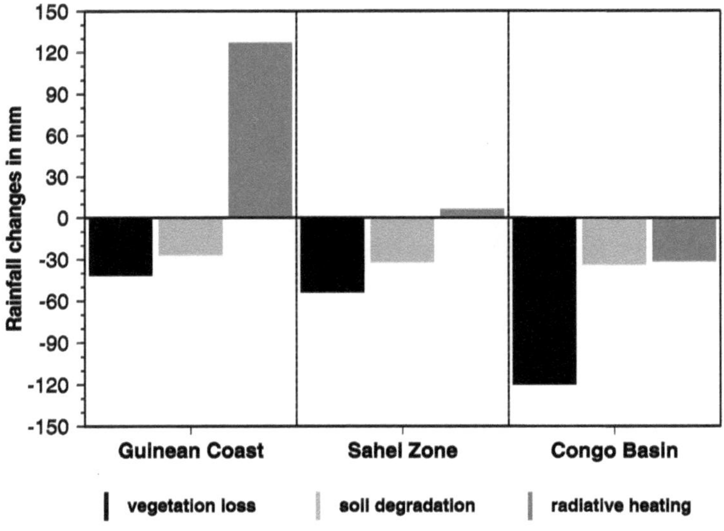

Fig. 3 Simulated changes in regional-mean summer monsoon rainfall in response to three forcings arising from human activity.

Figure 3 illustrates the changes in precipitation during the West African summer monsoon season (June to September) in three main regions of sub-Saharan Africa. Each bar refers to one of the three anthropogenic impacts. Enhanced greenhouse conditions lead to more abundant rainfall along the Guinean Coast region, whereas a negative anomaly is simulated over the Congo Basin. The Sahel region is barely affected by the greenhouse forcing. Soil degradation and, particularly, reduced vegetation cover cause dryer conditions all over tropical Africa. In coastal West Africa the process of land degradation may compensate the effect of the greenhouse forcing. In the other subregions vegetation loss clearly dominates.

It is obvious that a realistic scenario of future African climate change can not only rely on increasing GHG concentrations. The effects of reduced vegetation cover and soil degradation have to be taken into account as well. Both processes are related to land-use changes in Africa which arise from increasing population density, urbanisation, intensive agriculture including cash crops and excessive pasture (Achenbach 1994). As a consequence, deforestation due to shifting cultivation and domestic fuel as well as desertification due to overgrazing is steadily increasing (Hammer 2000).

4 Realistic Scenarios of Future Climate Change

Estimates of future land-use changes in Africa are provided by the United Nations (UNEP 2003). These estimates are derived from the present-day relationship between global vegetation and population, extrapolated into the 21st century under the assumption of future population growth rates by CIESIN (1996). The land-use change data are only given for large regions. Based on these regional-mean estimates a stochastic model has been developed in order to construct a spatial pattern of land degradation at the scale of REMO. The stochastic model accounts for all processes of human activity mentioned above and produces a random pattern of reduced forest fraction and vegetation cover, enhanced surface albedo, reduced leaf area index and decreasing roughness length. Thus, the complexity of land–atmosphere interactions is well represented. The resulting pattern of future land-use changes is heterogeneous in space, according to the real process of shifting cultivation and agricultural extent, and consists of a main belt of desertification in the southern Sahel region as well as a deforestation area along the borders of the dense woodlands. Based on these land-use forcing patterns several ensembles of REMO simulations have been carried out in order to predict the future behaviour of African climate. At the same time, increasing GHG concentrations are prescribed according to the IPCC SRES scenarios A1b and B1 (Houghton et al. 2001). Three ensemble members are forced with the observed GHG increase between 1960 and 2000, building a validation period for the REMO runs. Between 2001 and 2050 the A1b and B1 emission scenarios are combined with the process of land-use changes, each ensemble consisting of three members. Finally, three runs are realized under the A1b scenario 2001–2050 but without land-use changes. This experimental setup allows for the treatment of several important issues: (1) How is African climate responding to a more realistic anthropogenic scenario of enhanced greenhouse conditions plus land-use changes? (2) What is the uncertainty arising from the initial conditions in the climate-change signals? (3) What is the reduced effect between the A1b and B1 scenarios? (4) How much of the climate-change signals can be assigned to the land-use changes alone? (5) Which climate protection strategy is most effective for Africa? Some results of these experiments are summarized in Fig. 4, referring to the ensemble-mean under the combined A1b scenario between 2001 and 2050. The combined scenario has been chosen here in order to highlight the total climate changes in Africa arising from various human activities. The left panels display the linear changes of seasonal-mean temperature and total seasonal precipitation during the West African summer monsoon season. Grid boxes are only plotted if the linear regression coefficient is statistically significant at the 5% level. The regional model simulates a strong near-surface warming with highest amplitude of more than + 3°C in sub-Saharan West Africa. The warming is less pronounced in tropical East Africa and the Arabian Peninsula (around + 1°C). Simultaneously, the total amount of summer monsoon rainfall is decreasing. The pattern of significant changes is fairly coherent in space only in western and central tropical Africa. The rainfall reduction amounts to more than 150 mm, being equivalent to 25–40% of the present-day total amount.

Except for the western Mediterranean Basin, all other regions do not exhibit significant changes in summer precipitation, although the same radiative forcing is

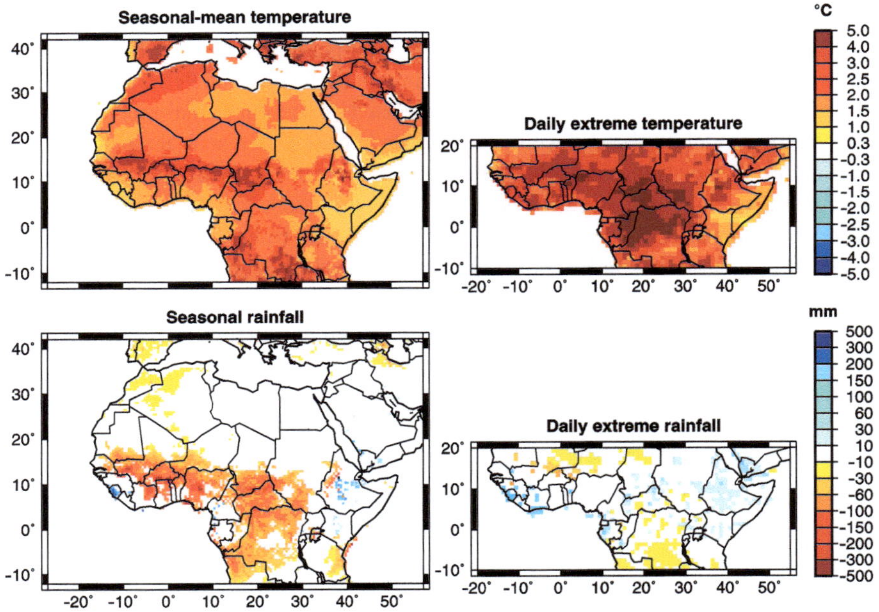

Fig. 4 Simulated changes in mean and extreme temperature and rainfall during the summer monsoon season under enhanced greenhouse conditions according to IPCC scenario A1b and ongoing land-use changes according to U.N. figures. Only values significant at the 5% level are plotted.

implemented throughout. This finding reflects the importance of land-use changes because the major precipitation signal occurs exactly where the land-use changes are prescribed: in sub-Saharan Africa. More evidence of the crucial relevance of land-use changes in tropical Africa is given by the direct comparison between the A1 ensembles with greenhouse forcing alone and greenhouse forcing plus land-use changes. The difference patterns between the respective climate change signals reveal that land degradation accounts for around half of the near-surface warming and most of the drying trend (not shown). In terms of temperature, the same difference is found between the A1 and B1 scenarios, including both anthropogenic forcing (not shown). In contrast, rainfall reduction is hardly mitigated under the combined B1 scenario compared with the combined A1 scenario. Thus, omitting land degradation appears to be more appropriate for climate protection in Africa than reducing global greenhouse gas emissions as given by the transition from A1 to B1 scenario.

The right panels illustrate the corresponding changes in extremes of daily warm temperature and heavy rainfall events. The extreme value estimate is based on the Generalized Pareto distribution using daily values beyond the 90% quantile. The maps show statistically significant changes in the 1-year return values between the last and first decade of the simulation period 1960–2050. Statistical significance at the 5% level is inferred from a parametric Monte Carlo approach with 1,000 random samples (Paeth and Hense 2005). REMO predicts a prominent intensification of extreme heat anomalies in tropical Africa. The increase amounts to up to 5°C and is almost twice as much as the warming in the seasonal mean. In terms of precipitation, the pattern is neither coherent in space nor qualitatively explicit. The estimate of rainfall extremes is very uncertain and, hence, the confidence intervals of the Monte Carlo samples do overlap. In some regions, there is some indication that a reduction of total precipitation amount may come along with an intensification of individual rain events (cf. Richter 2000; Balzarek et al. 2003).

In summary, strong near-surface warming, enhanced heat stress, and reduced precipitation point to a deterioration of livelihood in tropical Africa. Under the weaker B1 emission scenario and slightly weaker land-use changes the same picture can be drawn qualitatively. When the land-use changes are switched off in the 21st century simulations, the warming is reduced by around 50% and the rainfall reduction can be prevented almost entirely. Thus, the local protection of vegetation and reforestation measures may be more effective for the mitigation of African climate change than the global reduction in GHG emissions.

References

Achenbach, H., 1994: Die agraren Produktionszonen der Erde und ihre natürlichen Risikofaktoren. *Geogr. Rundschau*, **46**, 58–64.

Azar, C. and H. Rodhe, 1997: Targets for stabilization of atmospheric CO_2. *Science*, **276**, 1818–1819.

Bader, J. and M. Latif, 2003: The impact of decadal-scale Indian Ocean sea surface temeprature anomalies on Sahelian rainfall and the North Atlantic Oscillation. *Geophys. Res. Lett.*, **30**, 10.1029/2003GL018426.

Balzarek, H., W. Fricke, J. Heinrich et al., 2003: Man-made flood disaster in the Savanna town of Gombe/NE Nigeria. The natural hazard of gully erosion caused by urbanization dynamics and their peri-urban footprints. *Erdkunde*, **57**, 94–109.

Benson, C. and E. J. Clay, 1998: The impact of drought on sub-Saharan economies. World Bank Tech Paper 401, Wold Bank, Washingtion DC.

Black, E., J. Slingo, and K. R. Sperber, 2003: An observational study of the relationship between excessively strong short rains in coastal East Africa and Indian Ocean SST. *Mon Wea Rev.*, **131**, 74–94.

Bounoua, L., G. J. Collatz, S. O. Los et al., 2000: Sensitivity of climate to changes in NDVI. *J. Climate*, **13**, 2277–2292.

Braun, T. and U. Scholz, 1997: Shifting Cultivation – Krebsschaden aller Tropenländer? *Geogr. Rundschau*, **49**, 4–10.

Camberlin, P., S. Janicot, and I. Poccard, 2001: Seasonality and atmospheric dynamics of the teleconnection between African rainfall and tropical sea-surface temperature: Atlantic vs. ENSO. *Int. J. Climateatol*, **21**, 973–1004.

Chang, P., R. Saravanan, Ji L, and G. C. Hegerl, 2000: The effect of local sea surface temperature on atmopsheric circulation over the tropical Atlantic sector. *J. Climate*, **13**, 2195–2216.

CIESIN, 1996: *Gridded Population Density of the World*. Consortium for International Earth Science Information Network. Michigan.

Clark, D. B., Y. Xue, R. J. Harding, and P. J. Valdes, 2001: Modeling the impact of land surface degradation on the climate of tropical North Africa. *J. Climate*, **14**, 1809–1822.

Cook, K. H., 1999: Generation of the African easterly jet and its role in determining West African precipitation. *J. Climate*, **12**, 1165–1184.

Cullen, H. M. and P. B. de Menocal, 2000: North Atlantic influence on Tigris-Euphrates streamflow. *Int. J. Climateatol*, **20**, 853–863.

Douville, H., F. Chauvin, and H. Broqua, 2001: Influence of soil moisture on the Asian and African monsoons. Part I: Mean monsoon and daily precipitation. *J. Climate*, **14**, 2381–2402.

Eklundh, L. and L. Olsson, 2003: Vegetation index trends for the African Sahel 1982–1999. *Geophys. Res. Lett.*, **30**, 10.1029/2002GL016772.

Feichter, J., E. Roeckner, U. Lohmann, and B. G. Liepert, 2004: Nonlinear aspects of the climate response to greenhouse gas and aerosol forcing. *J. Climate*, **17**, 2384–2398.

Findley, S. E., 1994: Does drought increase migration? A study of migration from rural Mali during the 1983–1985 drought. *Int Migr Rev*, **28**, 539–553.

Fontaine, B., N. Philippon, S. Trzaska, and R. Roucou, 2002: Spring to summer changes in the West African monsoon through NCEP/NCAR reanalyses (1968–1998). *J. Geophys. Res.*, **107**, 10.1029/2001JD000834.

Gaebe, W., 1994: Urbanisierung in Afrika. *Geogr Rundschau*, **46**, 570–576.

Giannini, A., R. Saravanan, and P. Chang, 2003: Oceanic forcing of Sahel rainfall on interannual to interdecadal time scales. *Science*, **302**, 1027–1030.

GLACE Team, 2004: Regions of strong coupling between soil moisture and precipitation. *Science*, **305**, 1138–1140.

Hammer, T., 2000: Desertifikation im Sahel. Geogr. Rundschau, **52**, 4–10.

Hastenrath, S., 2000: Interannual and longer term variability of upper-air circulation over the tropical Atlantic and West Africa in boreal summer. *Int. J. Climateatol*, **20**, 1415–1430.

Herbers, H., 1999: Wasser als kritischer Faktor der Ernährungssicherheit. Über die Bedeutung eines essentiellen Lebensmittels. *Erdkunde*, **53**, 35–47.

Houghton, J. T., Y. Ding, D. J. Griggs et al., eds., 2001: Climate Change 2001: *The Scientific Basis*. Cambridge.

Hulme, M., R. Doherty, T. Ngara et al., 2001: African climate change: 1900–2100. *Climate Res.*, **17**, 145–168.

Jacob, D., 2001: A note to the simulation of the annual and interannual variability of the water budget over the Baltic Sea drainage basin. *Meteorol Atmos. Phys.*, **77**, 61–74.

Janicot, S., S. Trzaska, and I. Poccard, 2001: Summer Sahel-ENSO teleconnection and decadal time scale SST variations. *Clim. Dynam.*, **18**, 303–320.

Jenkins, G. S., A. Kamga, A. Garba et al., 2002: Investigating the West African climate system using global/regional climate models. *Bull. Am. Meteorol Soc.*, **83**, 583–595.

Jones, C., N. Mahowald, and C. Luo, 2003: The role of Easterly Waves on African desert dust transport. *J. Climate*, **16**, 3617–3628.

Knippertz, P., M. Christoph, and P. Speth, 2003: Long-term precipitation variability in Morocco and the link to the large-scale circulation in recent and future climate. *Meteor Atmos. Phys.*, **83**, 67–88.

Kutzbach, J. E. and Z. Liu, 1997: Response of the African monsoon to orbital forcing and ocean feedbacks in the middle Holocene. *Science*, **278**, 440–443.

Latif, M. and A. Grötzner, 2000: The equatorial Atlantic oscillation and its response to ENSO. *Clim. Dynam.*, **16**, 213–218.

Long, M., D. Entekhabi, and S. E. Nicholson, 2000: Interannual variability in rainfall, water vapor flux, and vertical motion over West Africa. *J. Climate*, **13**, 3827–3841.

Mo, K., G. D. Bell, and W. M. Thiaw, 2001: Impact of sea surface temperature anomalies on the Atlantic tropical storm activity and West African rainfall. *J. Atmos. Sci.*, **58**, 3477–3496.

Nicholson, S. E., 2001: Climatic and enviromental change in Africa during the last two centuries. *Climate Res.*, **17**, 123–144.

Nicholson, S. E., B. Some, and B. Kone, 2000: An analysis of recent rainfall conditions in West Africa, including the rainy season of the 1997 El Niño and the 1998 La Ni/ na years. *J. Climate*, **13**, 2628–2640.

Paeth, H., 2004: Key factors in African climate change evaluated by a regional climate model. *Erdkunde*, **58**, 290–315.

Paeth, H. and J. Feichter, 2006: Greenhouse-gas versus aerosol forcing and African climate response. *Clim. Dynam.*, **26**, 35–54.

Paeth, H. and P. Friederichs, 2004: Seasonality and timescales in the relationship between global SST and African rainfall. *Clim. Dynam.*, **23**, 815–837.

Paeth, H. and A. Hense, 2004: SST versus climate change signals in West African rainfall: 20th century variations and future projections. *Clim Change*, **65**, 179–208.

Paeth, H. and A. Hense, 2005: Mean versus extreme climate in the Mediterranean region and its sensitivity to future global warming conditions. *Meteorol Z*, **14**, 329–347.

Paeth, H. and J. Stuck, 2004: The West African dipole in rainfall and its forcing mechanisms in global and regional climate models. *Mausam*, **55**, 561–582.

Paeth, H., K. Born, R. Podzun, and D. Jacob, 2005: Regional dynamical downscaling over West Africa: model evaluation and comparison of wet and dry years. *Meteorol Z*, **14**, 349–367.

Richter, R. E., 2000: Umweltflüchtlinge in Westafrika. *Geogr. Rundschau*, **52**, 12–17.

Rodriguez-Fonseca, B. and M. de Castro, 2002: On the connection between winter anomalous precipitation in the Iberian Peninsula and North West Africa and the summer subtropical Atlantic sea surface temperature. *Geophys. Res. Lett.*, **29**, 1863, doi:10.1029/2001GL014421.

Rotstayn, L. D. and U. Lohmann, 2002: Tropical rainfall trends and the indirect aerosol effect. *J. Climate*, **15**, 2103–2116.

Saha, K. and S. Saha, 2001: African monsoons. Part I: Climatological structure and circulation. *Mausam*, **52**, 479–510.

Saravanan, R. and P. Chang, 2000: Interaction between tropical Atlantic variability and El Niño-Southern Oscillation. *J. Climate*, **13**, 2177–2194.

Schnitzler, K. G., W. Knorr, M. Latif et al., 2001: Vegetation feedback on Sahelian rainfall variability in a coupled climate land-vegetation model. Max-Planck-Inst. f. Meteor., Report No. 329, Hamburg.

Schulz, R., 2001: Neuere Trends der Weltbevölkerungsentwicklung. *Geogr. Rundschau*, **53**, 4–9.

Semazzi, F. H. M. and Y. Song, 2001: A GCM study of climate change induced by deforestation in Africa. *Climate Res.*, **17**, 169–182.

Stier, P., J. Feichter, S. Kinne et al., 2005: The aerosol-climate model ECHAM5-HAM. *Atmos. Chem. Phys.*, **5**, 1125–1156.

Sturm, H. J., 1999: Weidewirtschaft in Westafrika. *Geogr. Rundschau*, **51**, 269–274.

Texier, D., N. de Noblet, S. P. Harrison et al., 1997: Quantifying the role of biosphere-atmosphere feedbacks in climate change: coupled model simulations for 6000 years BP and comparison with palaeodata for northern Eurasia and northern Africa. *Clim. Dynam.*, **13**, 865–882.

Texier, D., N. de Noblet, and P. Braconnot, 2000: Sensitivity of the African and Asian monsoons to mid-holocene insolation and data-inferred surface changes. *J. Cim.*, **13**, 164–181.

UNEP, 2003: Global environmental outlook 3. http://www.grida.no/geo/geo3/english/

Wang, G. and A. B. Eltahir, 2000: Ecosystem dynamics and the Sahel drought. *Geophys. Res. Lett.*, **27**, 795–798.

Wang, G., E. A. B. Eltahir, J. A. Foley et al., 2004: Decadal variability of rainfall in the Sahel: results from the coupled GENESIS-IBIS atmosphere-biosphere model. *Clim. Dynam.*, **22**, 625–637.

Ward, M. N., 1998: Diagnosis and short-lead time prediction of summer rainfall in tropical North Africa at interannual and multi-decadal time scales. *J. Climate*, **11**, 3167–3191.

Zeng, N., K. Hales, and J. D. Neelin, 2002: Nonlinear dynamics in a coupled vegetation-atmosphere system and implications for desert-forest gradient. *J. Climate*, **15**, 3474–3487.

Impacts of Climate Variability, Trends and NAO on 20th Century European Plant Phenology

A. Menzel, N. Estrella, and C. Schleip

Abstract We provide here a brief overview of the impacts of climate variability and recent climate change on the European plant phenology across the 20th century. Facing recent climate changes, phenology has two major functions. Firstly, it reveals measurable impacts of climate change on nature, which at the same time clearly demonstrate global climate change in people's backyards. Secondly, long-term phenological data allow the reconstruction of temperature and its variability in the last centuries. The most prominent temperature driven changes in plant phenology are an earlier start of spring in the last three to five decades of, on average, 2.5 days/decade, mainly observed in midlatitudes and higher latitudes of the northern hemisphere. More heterogeneous changes in autumn are not as pronounced as in spring and cannot be linked to climate factors. A marked spatial and temporal variability of spring and summer onset dates and their changes can be mainly attributed to regional and local temperature. In this context, we discuss the temperature responses of the growing season and other phenological phases and their relation to the North Atlantic Oscillation. These results illustrate main feedbacks in biogeochemical cycles and land–surface interactions of the climate system.

1 Introduction to European Plant Phenology

Europe has a substantial and long history of phenological observation. Today, this relatively old science has gained new importance as phenological records are valuable climate proxies which allow temperature over the last centuries to be reconstructed. In addition, current phenological observations most easily reveal impacts of recent climate change on nature.

Chair of Ecoclimatology, Technical University Munich, Germany
menzel@forst.tu-muenchen.de
estrella@met.forst.tu-muenchen.de
Christoph.Schleip@web.de

S. Brönnimann et al. (eds.), *Climate Variability and Extremes during the Past 100 Years.*

Although the term phenology – the timing of seasonal activities in plants and animals – was not defined until 1853 by the Belgian botanist Morren, its history is considerably longer. In Japan and China the flowering of cherry and peach trees is associated with ancient festivals and some of the data can be traced to the 8th century (Menzel 2002; Chen 2003). Nowadays, phenology can be defined as "the study of the timing of recurring biological phases, the causes of their timing with regard to biotic and abiotic forces, and the interrelation among phases of the same or different species" (Global Phenological Monitoring www.dow.wau.nl/msa/gpm/). In phenology, annual recurring events, such as leaf unfolding, flowering, fruit ripening, drilling, harvest, leaf colouring and leaf fall, are often observed by volunteers interested in nature. Figure 1 shows two of these events, start of flowering of snowdrop (*Galanthus nivalis* L.) and of lilac (*Syringa vulgaris* L.).

In Europe, many phenological networks are run by national weather services and a few new Internet-based schemes exist (for a recent survey see www.dow.wau.nl/msa/epn/). Because network manuals contain detailed descriptions and photos of the phases to be observed, phenological data, although based on no more than visual observations, provide valuable climate related information, equivalent to physical measurements. Statistical analyses, manipulation experiments as well as semi-physiological models demonstrate that the onset of spring and summer events in plants is mainly triggered by the temperature of the preceding months (Sparks and Menzel 2002). Thus, long phenological records now form an important source for (1) temperature proxies of the past climate and (2) observed impacts of recent climate warming, as phenology is perhaps the simplest process in which to track changes in the ecology of species in response to climate change.

The most prominent changes in plant phenology during the last 100 years are an earlier start of spring in recent decades of, on average, 2.5 days/decade, as observed in midlatitudes and higher latitudes of the northern hemisphere (Menzel et al. 2006a; Parmesan and Yohe 2003). Changes in autumn are not as pronounced as in spring and are, in general, more heterogeneous. In total, a lengthening of the growing season by up to 2 weeks in the last three decades is evident in ground phenological observations. These changes are confirmed by remotely sensed vegetation indices, climatological data and atmospheric CO_2-concentrations (Keeling et al. 1996; Myneni et al. 1997; Menzel and Fabian 1999; Lucht et al. 2002).

In this paper we give an overview of European phenological data in the 20th century. We start with a brief overview of observed changes derived from phenological data. Then we give an overview of spatial and temporal variability of these changes. Finally, we assess the temperature response of these data for climate research and their relation to the North Atlantic Oscillation. For more detailed information on observed changes in natural and managed terrestrial ecosystems, the reader is referred to reviews by Hughes (2000), Menzel and Estrella (2001), Walther et al. (2002), Sparks and Menzel (2002), Linderholm (2006) and Parmesan (2006).

Fig. 1 Start of flowering of snowdrop (*Galanthus nivalis* L.) and of lilac (*Syringa vulgaris* L.).

2 Observed Growing Season Changes in Recent Decades

2.1 Ground-based Phenological Observations

Analyses of long-term observational series, such as the cherry flowering record at Kyoto in Japan (Fig. 2) or the grape harvest dates in Burgundy (France), suggest major multi-decadal variability in onset dates linked to climate. From the middle of the 20th century (Kyoto) or even from the 1980s (France) onwards, a marked advance in these records is clearly evident. This altered timing of spring and summer events has been reported for a wide range of species and locations in Europe (see reviews). Traditionally, changes in onset dates have been determined by linear regression and are then reported in days/year or days/decade. Network studies show that leaf unfolding and flowering in spring and summer have, on average, advanced by 1–3 days per decade in Europe, North America, and Japan over the last 30–50 years.

For Europe, within the COST725 action "Establishing a European Phenological Data Platform for Climatological Applications" (www.cost725.org), all available phenological records for Europe have been analysed for the period 1971–2000. In total, the linear trend of 103199 series of 15 years and more were analysed, covering 542 plant species in 21 countries (Menzel et al. 2006a). Spring leaf unfolding and flowering of wild plants changed on average by −2.5 days/decade, ranging from −8.3 to +4.1 days/decade (5th and 95th percentiles). Seventy-eight per cent of all

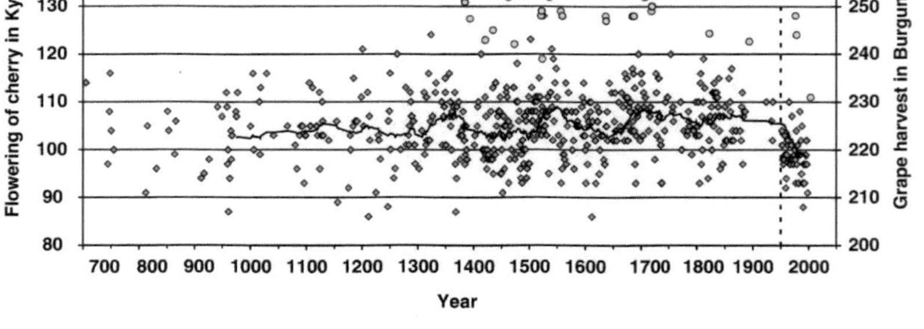

Fig. 2 Examples of long-term phenological records. Onset of flowering of cherry in Kyoto (left y-axis, lower plot with diamonds) and of grape harvest in Burgundy, France (right y-axis, upper plot with circles) with 30-year moving averages (Menzel and Dose 2005; Chuine et al. 2004). The dotted vertical line indicates the beginning of obvious advances in onset dates mid of the 20th century.

records displayed advancing onset dates (see Fig. 3 for further percentages). This earlier flowering also implies an earlier start of the pollen season, often related to a later end of the pollen season and higher pollen sums, which are monitored by pollen counting stations (Beggs 2004, European Aeroallergen Network).

There are also indications that the onset of fruit ripening in early autumn has advanced in many cases (Jones and Davis 2000; Peñuelas et al. 2002; Menzel 2003). The comprehensive COST725 study revealed that fruit ripening exhibited very similar changes with a mean change of −2.4 days/decade, ranging from −9.1 to +4.3 days/decade (5th and 95th percentiles, 75% advancing, see Fig. 3). Due to the enormous variety of plant species observed, detailed differences among species and groups of species can be observed. Concerning life forms, it has been found that annual plants respond more strongly than congeneric perennials, insect-pollinated plants more than wind-pollinated plants, herbaceous plants more than woody plants (Fitter and Fitter 2002), but there are no significant differences between Raunkiaer life forms or different origins (Peñuelas et al. 2002).There are indications that autumn leaf colouring and leaf fall have been delayed in recent decades, however the signal is not as clear as in spring and summer (e.g., Menzel and Fabian 1999; Matsumoto et al. 2003). The results for Europe (COST725) suggest that autumn is delayed by 1.3 days/decade (country averages, see also Fig. 3).

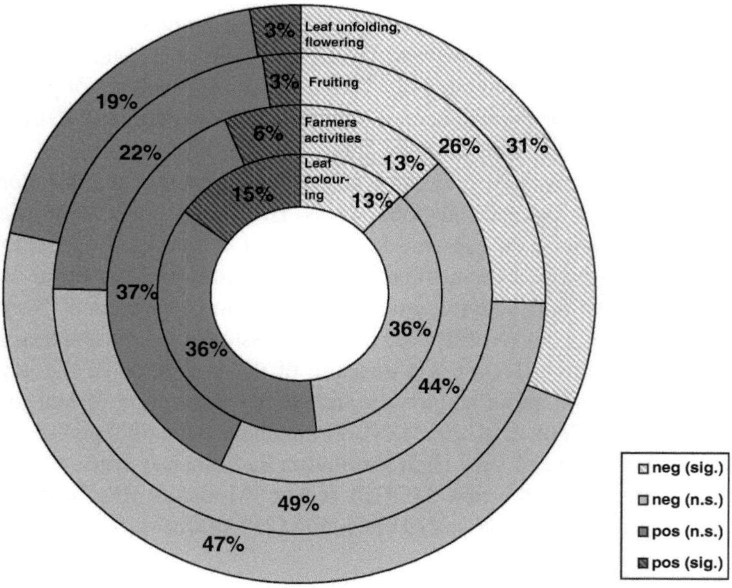

Fig. 3 Summary of phenological trends in Europe (COST725, Menzel et al. 2006a). All temporal trends (1971–2000, time series of 15+ years, linear regression) which have been systematically reported to the COST725 meta-analysis for four different groups (leaf unfolding and flowering, fruiting, farmers activities, leaf colouring) with the proportions of negative (light grey) and positive (dark grey) trends and significantly negative and positive trends (hatched, $P < 0.05$). Negative trends over time indicate advancing onset dates, positve trends delayed onset dates.

Phenological ground observations in networks allow a more detailed analysis of events observed in agricultural crops. Here, some phases, such as drilling and harvest, are called "false" phases since their onset is mainly triggered by farmers' decisions (Menzel et al. 2006c; Estrella et al. 2007). In the COST725 study, farmers' activities have only slightly changed by –0.41 days/decade (–7.4 to +7.1 days/decade, 5th and 95th percentiles, 57% advancing, see Fig. 3), which is considerably less than the reaction of wild plants. This is the other major advantage of phenological ground observations, that site and especially species-specific information is given.

2.2 *Satellite-based and Measurement-derived Changes*

Besides ground-based observations, at least three other techniques can be used to derive measures of the length and the intensity of the growing season. In general, all these methods provide comparable results: A lengthening of the growing season in midlatitudes and higher latitudes of the northern hemisphere and higher photosynthetic activity in the last two decades (Penuelas and Filella 2001; Lucht et al. 2002). The methods are (1) analyses of remotely sensed vegetation indices (e.g. Myneni et al. 1997; Zhou et al. 2001; Tucker et al. 2001; Badeck et al. 2004), (2) analysis of the atmospheric CO_2 signal (Keeling et al. 1996) (both providing spatially and species-averaged information), (3) analyses of meteorological and climatological data, such as frost days, frost season or the Bowen ratio (e.g., Menzel et al. 2003; Schwartz and Chen 2002; Schwartz et al. 2006), and (4) the analyses of phenological ground observations described in Sect. 2.1 (reviewed by Menzel and Estrella 2001; Sparks and Menzel 2002; Walther et al. 2002; Walther 2004).

Remotely sensed data of the growing season offer a great potential for monitoring vegetation dynamics at regional and global scales. The spring "green wave" (Schwartz 1998) can be especially well captured in time series of satellite images. Different vegetation indices allow a quantification of the spatial and temporal variation in vegetation growth and activity, among them the most prominent Normalized Difference Vegetation Index (NDVI). Due to the contrast between red and near-infrared reflectance of vegetation, NDVI is a measure of the greenness of the vegetation comprising information about abundance and energy absorption of leaf pigments, such as chlorophyll (Zhou et al. 2001). NDVI data have been obtained mainly from NOAA-AVHRR (Advanced Very High Resolution Radiometer) instruments, starting in 1981. In comparison, the new MODIS sensor (Moderate Resolution Imaging Spectroradiometer) offers higher spatial resolution (250–500 m vs. 1.1–8 km), but over a shorter time period.

Major regional and global studies on changes in the start and length of the growing season have reported a longer growing season, for example, by 19 days for Europe (1982–2000, Stöckli and Vidale 2004), by 18 days for Eurasia and by 12 days for North America (1981–1999, Zhou et al. 2001), by 12 days for the Globe (1981–1991, Myneni et al. 1997). At least 50% of this lengthening can be attributed to an earlier start of the "green season".

3 Temporal and Spatial Variability

Shifts in spring activities, mostly earlier starting of spring events, differ by species and by time of season with early-season plant species exhibiting the stronger reactions (Abu-Asab et al. 2001; Menzel et al. 2001; Fitter and Fitter 2002; Sparks and Menzel 2002; Menzel 2003). These early-season plant species also exhibit a higher year-to-year variation in onset dates. But even within groups of species reacting at the same time of the year, there are species-specific differences (Menzel 2002).

In Fig. 4 various measures of spatial variability of onset dates, temporal variability, temporal trends and spatial variability of those trends are described. Network studies, comprising numerous sites, suggest that there is considerable variation in observed trends, although the overall signal is quite obvious (e.g., Defila and Clot 2001). Two methods of deriving regional trends, condensing/averaging single site onset dates and subsequent trend analysis or averaging single-site trends, lead to comparable results (Menzel 2003). A recently recognised trait is that warmer springs with an earlier start of leaf unfolding and flowering exhibit a higher spatial intra-annual (geographical) variability (Menzel et al. 2006b).

When comparing changes derived from time series of varying length (varying starting and ending years) in Europe, a marked differentiation is apparent: Earlier

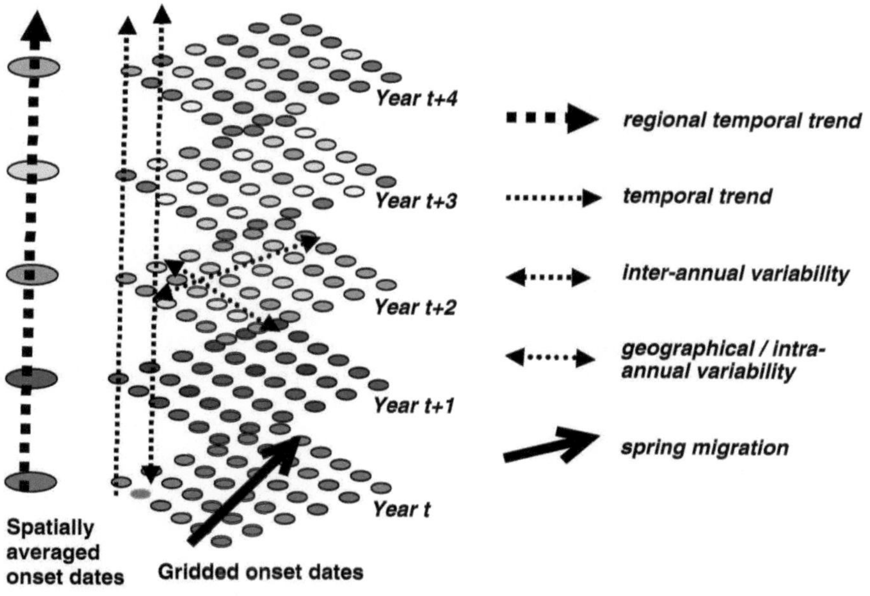

Fig. 4 Scheme of temporal and spatial variability of onset dates and trends. The scheme displays hypothetical onset dates at grid points in several years. Measures of inter-annual variability of onset dates as well as direction and speed of spring migration are restricted to single years or mean dates. The calculation of temporal trends requires longer observational time series at specific sites. Regional means of trends can be derived from spatially averaged onset dates.

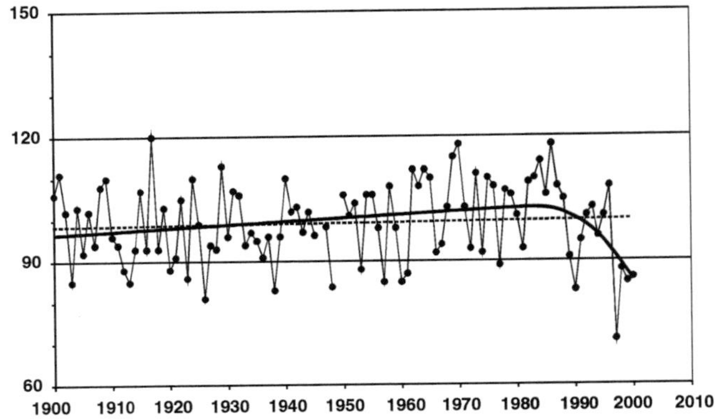

Fig. 5 Onset dates (day of the year, y-axis) of flowering of redcurrant (*Ribes rubrum* L.) at Geisenheim, near Frankfurt am Main, in the period 1900–2000. The constant model (not shown) has a probability of ~5%, the linear model of only ~2% (dashed line), and the one-change point model of ~93% (solid line). The linear model even indicates a delay of onset dates, whereas the one-change point model emphasises the strong advancing onset dates from the mid-1980s onwards.

spring events and a longer growing season in Europe are most evident for time-series ending in the mid-1980s or later (Schaber 2002; Scheifinger et al. 2002; Badeck et al. 2004; Dose and Menzel 2004; Menzel and Dose 2005; Schleip et al. 2006), which matches the turning points in the respective spring temperature series (Dose and Menzel 2006). Figure 5 displays the 1900–2000 time series of observed flowering dates of redcurrant (*Ribes rubrum* L.) at Geisenheim, near Frankfurt am Main. Considerable varying onset dates over the last century result in a linear trend indicating a slightly delayed onset in time (regression coefficient 0.152 days/decade). Employing a Bayesian approach of describing the functional behaviour (see Dose and Menzel 2004 for further details), three different model types were compared, (1) constant, (2) linear, and (3) one turning point/change point model, possessing (1) 5%, (2) 2%, and (3) 93% probability, respectively. This means that the one-change point model is by far the best in explaining the functional behaviour of the century-long record. The change point probability distribution exhibits a marked peak in the last two decades (1980–2000) with a maximum of ~11% in 1986. This feature is quite commonly observed in phenological records (see Schleip et al. 2006).

4 Temperature and NAO Response

Spring and summer phenology is very sensitive to climate and local weather; within photoperiodically defined constraints the trigger is nearly exclusively air temperature (Sparks et al. 2000; Lucht et al. 2002; Menzel 2003). In contrast to autumn phenology (Estrella and Menzel 2006), the spring climate signal is fairly well understood: nearly all spring and summer changes in plants, including agricultural crops (Estrella et al. 2007) correlate highly with spring temperatures of the preceding months. This simplified approach assumes that the winter chilling requirement of lower temperatures to break dormancy is met in every case when (warmer) forcing temperatures accelerate the plant development in spring. Over a broad band of temperature conditions, the relationship between spring onset dates and spring temperature is almost linear (see Fig. 6). In this example, mean onset dates of horse chestnut (*Aesculus hippocastanum* L.) and silver birch (*Betula pendula* Roth) leaf

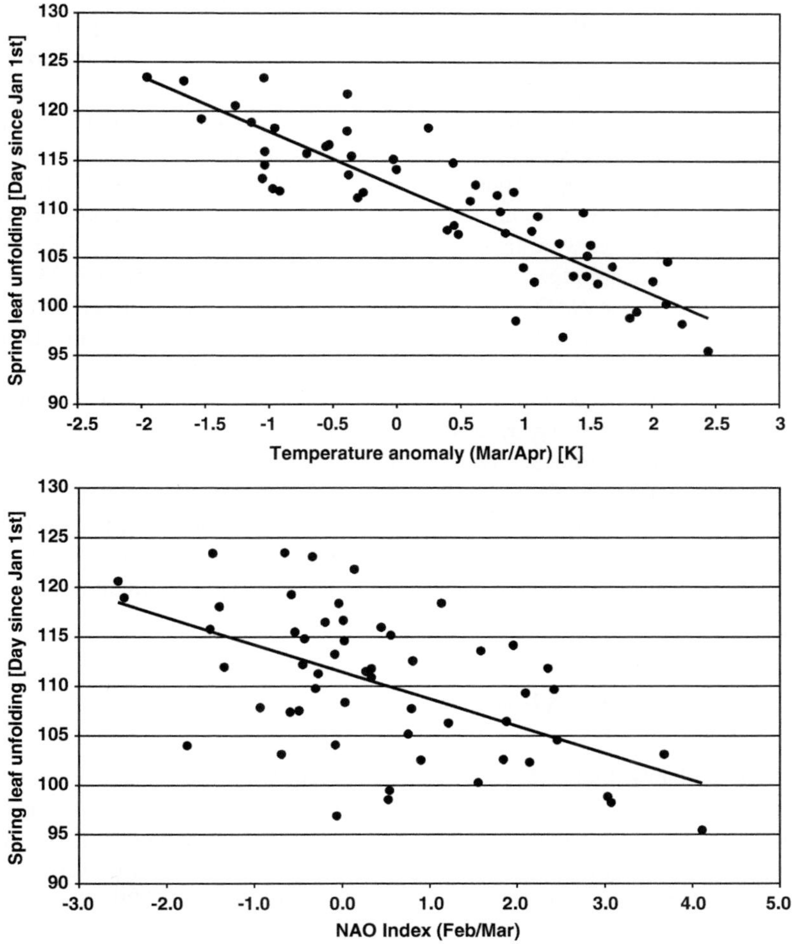

Fig. 6 Onset dates of spring leaf unfolding in Germany: national mean onset of leaf unfolding of horse chestnut (*Aesculus hippocastanum* L.) and silver birch (*Betula pendula* Roth) and anomalies of mean spring air temperature (HadCRUT3v, Brohan et al. 2006, upper panel) and North Atlantic Oscillation index NAO (http://www.cru.uea.ac.uk/cru/data/, lower panel) for 1951–2005 period (updated after Walther et al. 2002).

unfolding in Germany are plotted against anomalies of mean spring air temperature (HadCRUT3v) for Germany ($R^2 = 0.777$).

At the extreme temperatures, one would expect asymptotic functional behaviour, meaning that the onset dates cannot go beyond certain minimum and maximum onset limits. However, this has not been found in long-term records in Europe so far. Even for century-long records, no changes in this temperature response over time could be detected (Menzel et al. 2005a).

The advancement is estimated as 1–12 days for every 1°C increase in spring temperature, with average values ranging between 2.5 and 6 days/°C (e.g., Sparks et al. 2000; Chmielewski and Rötzer 2001; Menzel 2003; Menzel et al. 2006a). There are discernible differences in the temperature response among species, which might lead to a temporal decoupling of formerly synchronised events in ecosystem processes, e.g., food webs.

The North Atlantic Oscillation (NAO) Index in late winter and spring over the last three decades has represented a good proxy measure for spring temperatures in western and central Europe. Thus, a widespread influence of the NAO on diverse ecological processes was found (e.g., Blenckner and Hillebrand 2002). Equally, phenological onset dates were also found to correlate well with the NAO (Chmielewski and Rötzer 2001; Scheifinger et al. 2002; Walther et al. 2002; Menzel 2003). In the example of Fig. 6, the NAO index of February–March is less well correlated with spring leaf unfolding ($R^2 = 0.298$) than spring temperature. Even the direction and velocity of the spring green wave through Europe varies with NAO (Menzel et al. 2005b). As NAO only accounts for a small part of the inter-annual phenological variability, it is necessary to provide some ideas about other possible triggers. Besides this large scale NAO influence on climate, regional climate contributions, such as regional temperatures, and local factors, such as microclimate at the site, slope, exposition, soil factors, may also be of influence. In addition, there is considerable phenological variation with genetics (Menzel 2002). However, the exact contribution of local and species specific effects can only be assessed by experiments.

5 Conclusions and Outlook

Phenology probably holds the longest historical records of weather dependent natural phenomena. It provides valuable information about long-term changes in the timing of life-cycle events, whose recent changes can be linked to recent temperature increases.

Phenological data are an excellent index of spring temperatures with warm periods closely corresponding to early flowering and vice versa (Bradley 1999). Phenological records can often be calibrated and thus constitute an excellent paleoclimatic temperature proxy (e.g., Chuine et al. 2004; Menzel 2005).

Phenology has also gained new importance in the last decade as it is a brilliant bio-indicator for recent temperature increases, which allow an easy illustration of man's anthropogenic influence on the climate systems. The second major function of phenology is related to global change impact research. In the newest, fourth

IPCC report, Working Group II found broead observational evidence that recent regional climate changes, particularly temperature increases, have already affected many physical and biological systems (Rosenzweig et al. 2007). Among the systems and sectors affected is phenology where the changes are quite obvious, strongly temperature related and almost no non-climate driving factors exist. The consequences of altered phenology may include a longer growing season with possible effects on growth and carbon sequestration, disentangled synchrony in ecosystem processes, altered competition among species, and an earlier, longer and more intense pollen season (for further information see Sparks and Menzel 2002; Walther et al. 2002; Parmesan and Yohe 2003; Root et al. 2003; Parmesan 2006). In addition, many of the land–atmosphere interactions, which can modulate the dynamics of the climate system (e.g., Seneviratne and Stöckli 2007), are based on energy and water balances, which themselves vary with the leaf stages throughout the year.

The phenological data situation is sound for the period since 1981, when remote sensing data allowed a global perspective of those changes, as in general, historical phenological observation involve a marked paucity of studies in developing countries. Currently, the challenges include maintaining the phenological networks, enlarging them to new areas and regions, integrating other sources, such as eddy flux network data, and incorporating the research on applications and consequences of changes in the growing season, including vulnerability and adaptation studies.

References

Abu-Asab, M. S., P. M. Peterson, S. G. Shetler, and S. S. Orli, 2001: Earlier plant flowering in spring as a response to global warming in the Washington, DC, area. *Biodivers. Conserv.*, **10**, 597–612.

Badeck, F. W., A. Bondeau, K. Bottcher, D. Doktor, W. Lucht, J. Schaber, and S. Sitch, 2004: Responses of spring phenology to climate change. *New Phytol.*, **162**, 295–309.

Beggs, P. J., 2004: Impacts of climate change on aeroallergens: past and future. *Clin. Exp. Allergy*, **34**, 1507–1513.

Blenckner, T. and H. Hillebrand, 2002: North Atlantic Oscillation signatures in aquatic and terrestrial ecosystems – a meta-analysis. *Glob. Change Biol.*, **8**, 203–212.

Bradley, R. S., 1999: *Paleoclimatology*. Harcourt Academic Press, San Diego, CA, London, Boston, MA.

Brohan, P., J. J. Kennedy, I. Haris, S. F. B. Tett, and P. D. Jones, 2006: Uncertainty estimates in regional and global observed temperature changes: a new dataset from 1850. *J. Geophys. Res.*, **111**, D12106.

Chen, X., 2003: Phenological observation and research in China. In: *Phenology: An integrative Environmental Science*, M. D. Schwartz, ed., Kluwer, Dordrecht/Boston/London, 11–25.

Chmielewski, F. M. and T. Rötzer, 2001: Response of tree phenology to climate change across Europe. *Agr. Forest Meteorol*, **108**, 101–112.

Chuine, I., P. Yiou, N. Viovy, B. Seguin, V. Daux, E. Le Roy Ladurie, 2004: Grape ripening as a past climate indicator. *Nature*, **432**, 289–290.

Defila, C. and B. Clot, 2001: Phytophenological trends in Switzerland. *Int. J. Biometeorol*, **45**, 203–207.

Dose, V. and A. Menzel, 2004: Bayesian analysis of climate change impacts in phenology. *Glob. Change Biol.*, **10**, 259–272.

Dose, V. and A. Menzel, 2006: Bayesian correlation between temperature and blossom onset data. *Glob. Change Biol.*, **12**, 1451–1459.
Estrella, N. and A. Menzel, 2006: Responses of leaf colouring of four deciduous tree species to climate and weather in Germany. *Climate Res.*, **32**, 253–267.
Estrella, N., T. Sparks, and A. Menzel, 2007: Trends and temperature response in the phenology of crops in Germany. *Glob. Change Biol*, **13**, 1737–1747. (accepted).
Hughes, L., 2000: Biological consequences of global warming: is the signal already apparent? *Trends Ecol. Evol.*, **15**, 56–61.
Fitter, A. H. and R. S. R. Fitter, 2002: Rapid changes in flowering time in British plants. *Science*, **296**, 1689–1691.
Jones, G. V. and R. E. Davis, 2000: Climate Influences on grapevine phenology, grape composition, and wine production and quality for Bordeaux, France. *Am. J. Enol. Viticult.*, **51**, 249–261.
Keeling, C. D., J. F. S. Chin, and T. P. Whorf, 1996: Increased activity of northern vegetation inferred from atmospheric CO_2 measurements. *Nature*, **382**(6587), 146–149.
Linderholm, H. W., 2006: Growing season changes in the last century. *Agr. Forest Meteorol*, **137**, 1–14.
Lucht, W., I. C. Prentice, R. B. Myneni, S. Sitch, P. Friedlingstein, W. Cramer, P. Bousquet, W. Buermann, and B. Smith, 2002: Climatic control of the high-latitude vegetation greening trend and Pinatubo effect. *Science*, **296**, 1687–1689.
Matsumoto, K., T. Ohta, M. Irasawa, and T. Nakamura, 2003: Climate change and extension of the Ginkgo biloba L. growing season in Japan. *Glob. Change Biol.*, **9**, 1634–1642.
Menzel, A. and N. Estrella, 2001: Plant phenological changes. In: *"Fingerprints" of Climate Change – Adapted behaviour and shifting species ranges*, G. R. Walther, C. A. Burga, P. J. Edwards, eds., Kluwer/Plenum, New York/London, 123–137.
Menzel, A. and P. Fabian, 1999: Growing season extended in Europe. *Nature*, **397**, 659.
Menzel, A., N. Estrella, and P. Fabian, 2001: Spatial and temporal variability of the phenological seasons in Germany from 1951–1996. *Glob. Change Biol.*, **7**, 657–666.
Menzel, A., 2002: Phenology, its importance to the Global Change Community. Editorial Comment. *Clim Change*, **54**, 379–385.
Menzel, A., 2003: Phenological anomalies in Germany and their relation to air temperature and NAO. *Clim Change*, **57**, 243–263.
Menzel, A., 2005: A 500 year pheno-climatological view on the 2003 heatwave in Europe assessed by grape harvest dates. *Meteorol Z*, **14**, 75–77.
Menzel, A. and V. Dose, 2005: Analysis of long-term time-series of beginning of flowering by Bayesian function estimation. *Meteorol Z*, **14**(3), 429–434.
Menzel, A., G. Jakobi, R. Ahas, H. Scheifinger, and N. Estrella, 2003: Variations of the climatological growing season (1951–2000) in Germany compared with other countries. *Int. J. Climateatol*, **23**, 793–812.
Menzel, A., N. Estrella, and A. Testka, 2005a: Temperature response rates from long-term phenological records. *Climate Res.*, **30**, 21–28.
Menzel, A., T. H. Sparks, N. Estrella, and S. Eckhardt, 2005b: 'SSW to NNE' – NAO affects the progress of seasons across Europe. *Glob. Change Biol.*, **11**, 909–918.
Menzel, A., T. Sparks, N. Estrella, E. Koch, A. Aasa, R. Ahas, K. Alm-Kübler, P. Bissolli, O. Braslavská, A. Briede, F. M Chmielewski, Z. Crepinsek, Y. Curnel, Å. Dahl, C. Defila, A. Donnelly, Y. Filella, K. Jatczak, F. Måge, A. Mestre, Ø Nordli, J. Peñuelas, P. Pirinen, V. Remišová, H. Scheifinger, M. Striz, A. Susnik, A. J. H. van Vliet, F. E. Wielgolaski, S. Zach, and A. Zust, 2006a: European phenological response to climate change matches the warming pattern. *Glob. Change Biol.*, **12**, 1969–1976.
Menzel, A., T. H. Sparks, N. Estrella, and D. B. Roy, 2006b: Geographic and temporal variability in phenology. *Glob. Ecol. Biogeogr.*, **15**, 498–504.
Menzel, A., J. von Vopelius, N. Estrella, C. Schleip, and V. Dose, 2006c: Farmers' annual activities are not tracking speed of climate change. *Climate Res.*, **32**:201–207.
Myneni, R. B., C. D. Keeling, G. Asrar, R. R. Nemani, 1997: Increased plant growth in the northern high latitudes from 1981 to 1991. *Nature*, **386**, 698–702.

Parmesan, C., 2006: Ecological and evolutionary responses to recent climate change. *Annu. Rev. Ecol. Evol. S*, **37**, 637–669.

Parmesan, C. and G. Yohe, 2003: A globally coherent fingerprint of climate change impacts across natural systems. *Nature*, **421**, 37–42.

Peñuelas, J. and I. Filella, 2001: Phenology – responses to a warming world. *Science*, **294**, 793.

Peñuelas, J., I. Filella, and P. Comas, 2002: Changed plant and animal life cycles from 1952 to 2000 in the Mediterranean region. *Glob. Change Biol.*, **8**, 531–544.

Root, T. L., J. T. Price, K. R. Hall, S. H. Schneider, C. Rosenzweig, J. A. Pounds, 2003: Fingerprints of global warming on wild animals and plants. *Nature*, **421**, 57–60.

Rosenzweig, C., G. Casassa, D.J. Karoly, A. Imeson, C. Liu, A. Menzel, S. Rawlins, T.L. Root, B. Seguin, P. Tryjanowski, 2007: Assessment of observed changes and responses in natural and managed systems. *Climate Change 2007: Impacts, Adaptation and Vulnerability. Contribution of Working Group II to the Fourth Assessment Report of the Intergovernmental Panel on Climate Change*, M.L. Parry, O.F. Canziani, J.P. Palutikof, P.J. van der Linden and C.E. Hanson, Eds., Cambridge University Press, Cambridge, UK, 79–131.

Schaber, J., 2002: Phenology in Germany in the 20th century: methods, analyses and models. Dissertation, University of Potsdam.

Scheifinger, H., A. Menzel, E. Koch, C. Peter, and R. Ahas, 2002: Atmospheric mechanisms governing the spatial and temporal variability of phenological phases in Central Europe. *Int. J. Climateatol*, **22**, 1739–1755.

Schleip, C., A. Menzel, N. Estrella, and V. Dose, 2006: The use of Bayesian analysis to detect recent changes in phenological events throughout the year. *Agr. Forest Meteorol*, **141**, 179–191.

Schwartz, M. D., 1998: Green-wave phenology. *Nature*, **394**, 839–840.

Schwartz, M. D. and X. Chen, 2002: Examining the onset of spring in China. *Climate Res.*, **21**, 157–164.

Schwartz, M. D., R. Ahas, and A. Aasa, 2006: Onset of spring starting earlier across the northern hemisphere. *Glob. Change Biol.*, **12**, 343–351.

Sparks, T. and A. Menzel, 2002: Observed changes in seasons: an overview. *Int. J. Climateatol*, **22**, 1715–1725.

Sparks, T. H., E. P. Jeffree, and C. E. Jeffree, 2000: An examination of the relationship between flowering times and temperature at the national scale using long-term phenological records from the UK. *Int. J. Biometeorol*, **44**, 82–87.

Stöckli, R. and P. L.Vidale, 2004: European plant phenology and climate as seen in a 20-year AVHRR land-surface parameter dataset. *Int. J. Remote Sens*, **25**, 3303–3330.

Tucker, C. J., D. A. Slayback, J. E. Pinzon, S. O. Los, R. B. Myneni, and M. G. Taylor 2001: Higher northern latitude normalized difference vegetation index and growing season trends from 1982 to 1999. *Int. J. Biometeorol*, **45**, 184–190.

Walther, G. R., 2004 Plants in a warmer world. *Perspec. Plant Ecol.*, **6**, 169–185.

Walther, G. R., E. Post, P. Convey, A. Menzel, C. Parmesan, T. J. C. Beebee, J. M. Fromentin, O. Hoegh-Guldberg, and F. Bairlein 2002: Ecological responses to recent climate change. *Nature*, **416**, 389–395.

Watson, R. T. and the Core Writing Team, eds., 2001: Climate Change 2001: Synthesis Report, SPM. IPCC, Geneva, Switzerland, 184.

Zhou, L. M., C. J. Tucker, R. K. Kaufmann, D. Slayback, N. V. Shabanov, and R. B. Myneni, 2001: Variations in northern vegetation activity inferred from satellite data of vegetation index during 1981 to 1999. *J. Geophys. Res.*, **106**, 20069–20083.

Summer Heat Waves in Western Europe, Their Past Change and Future Projections

P. M. Della-Marta[1] and M. Beniston[2]

Abstract Summer heat waves and extremely hot temperatures are a serious threat to society, the environment and the economy of Europe. In this chapter we present an overview of selected recent literature which looks specifically at European heat waves and extreme temperature events, their past change and expected future change from 1880 to 2100. Della-Marta et al. (2007b) show that over the period 1880–2005 the length of summer heat waves over western Europe has doubled and the frequency of hot days has almost tripled. These changes are seen in the probability density function (PDF) of western European daily summer maximum temperature (DSMT) as a significant change in the mean (+1.6 ± 0.4°C) and variance (+6 ± 2%). The relatively small change in variance over the last 126 years can explain approximately 40% of the change in hot days. We see a continuation of the observed trends in the future regional projections. Beniston et al. (2007) show that regional surface warming causes the frequency, intensity and duration of heat waves to increase over Europe. By the end of the 21st century, countries in central Europe will experience the same number of hot days as are currently experienced in southern Europe. The intensity of extreme temperatures increases more rapidly than the intensity of more moderate temperatures over the continental interior due to increases in temperature variability.

1 Introduction

Extreme weather and climate events have a large impact on our society and environment (IPCC 2001). Heat waves are especially devastating for societies that are not used to coping with such extremes. The 1995 Chicago heat wave was

[1] Climate Services, Federal Office of Meteorology and Climatology MeteoSwiss, Krähbühlstrasse 58, Postfach 514, CH-8044 Zürich
paul.della-marta@meteoswiss.ch

[2] Chair for Climate Research, University of Geneva, Site de Battelle/D, 7 Route de Drize, CH-1227 Carouge/Geneva, Switzerland
Martin.Beniston@unige.ch

such an event (Karl and Knight, 1997) where over 500 people died from heat-related illnesses. The 2003 heat wave in Europe is unprecedented in terms of loss of life, with over 30,000 deaths in Europe (Poumadère et al. 2005) attributable to the excessive and persistent heat (Milligan 2004). This extreme event also affected other parts of our society and environment, such as the destruction of large areas of forest by fire, the drying out of rivers causing damage to water ecosystems and the excessive melting of glaciers (e.g., Gruber et al. 2004; Koppe et al. 2004; Schär and Jendritzky 2004; Kovats and Koppe 2005). The sobering results of Stott et al. (2004) show that humans have contributed to at least a doubling of risk of such an extreme event in Europe over the last 150 years. Reducing the impacts of future heat waves requires addressing fundamental questions, such as whether heat waves can be predicted and whether their impacts can be mitigated. A fundamental basis for answering these questions is the analysis of the observed climate record. Comparing observations and model results allow us to test hypotheses on the physical mechanisms that lead to such extreme events (IPCC 2001). Of particular interest in the debate on future extreme temperature events in western Europe is the change in the probability density function (PDF). For example, an increase in both the mean and the variance of a temperature PDF has greater consequences for the duration and frequency of temperature extremes than if the PDF were to only increase in mean. These questions are also explored in two other chapters of this volume, Scherrer et al., who look at changes in the seasonal temperature PDF during the last 45 years compared with projections of future PDF changes, and Seneviratne et al., who detail the underlying physical mechanisms responsible for expected future changes in extreme summer heat events in Europe.

In this chapter we consider a longer-term perspective than the other chapters mentioned above. We present detailed evidence of the increase in the frequency of extremely hot temperatures and an increase in the length of heat wave events in western Europe which have been associated with significant changes in the PDF of the daily maximum temperature from 1880 up to now (Della-Marta et al. 2007b). We also present results of the latest extreme temperature and heat wave regional climate projections for the end of the 21st century (Beniston et al. 2007). Furthermore, this chapter serves as an introduction to the latest literature on the topic of heat waves in Europe. We present results from our collaborations and place them in the context of recent literature. The chapter is separated into three sections. Firstly, we give special emphasis on the data used to assess the changes in summer extreme temperatures in western European over the last 126 years. In the following chapters we show the results of the latest studies concerning observed past and expected future changes in summer hot extreme temperature and heat waves over western Europe. Whilst the extreme temperature indices discussed in these sections are not directly comparable (due to technical differences), they all characterize the extreme hot tail of the daily temperature probability distribution and are aimed at determining the change in frequency and/or persistence of extreme hot temperatures.

2 The Quality of Western European Observed Daily Temperature Data

In order to obtain an accurate estimate of possible climate change all climate data must be rigorously checked for poor quality data and inhomogeneities caused by changes in instruments, observation site, or observation practices. The IPCC Second Assessment Report (Nicholls et al. 1996) raised the awareness of the important consequence of climate change on extreme events and helped to focus greater attention of climate research on this field. Extreme events are often short-lived (minutes to days) and usually require higher temporal and spatial resolution data to quantify their change, compared to quantifying the change in the climatological mean. The analysis of extreme events highlighted the problems with existing data sets and the need to maintain long-term climate stations as part of a consistent high quality network (Karl et al. 1995). In order to detect and attribute changes in the frequency of extreme events it is crucial that, (a) reliable and continuous long-term observations have to be performed with a regular control of the quality standards of the measurements (e.g., by National Weather Services), (b) the quality-controlled data has to be stored for long-term use in a data format which will still be easily accessible in the future. Frich et al. (2002) provided a comprehensive look at changes in extreme temperature events for many parts of the globe for the IPCC Third Assessment Report, which has been updated by Alexander et al. (2006) to cover many more regions of the world, including Africa, the Asia-Pacific region and South America to be included in the IPCC Fourth Assessment Report. A major contribution to the global extremes papers of Frich et al. (2002) and Alexander et al. (2006) in Europe has been the European Climate Assessment (Klein Tank et al. 2002) project which is detailed below as it is the basis of more recent studies which have been made on the analysis of extreme temperature events in Europe (e.g., Della-Marta et al. 2007a, b).

The work of Klein Tank et al. (2002) was the result of extensive collaboration between European national meteorological services which involved the exchange of daily climate data for use in the assessment of climate extremes. The European Climate Assessment data set (Klein Tank et al. 2002) is a collection of many countries' highest quality and longest daily temperature observations. However, in such a large data set, the task of ensuring high data quality and homogeneity has been a huge undertaking for which several strategies have been used. Some of the individual national meteorological services or data providers have performed detailed data quality/homogeneity testing and homogeneity correction their monthly or daily data (e.g., Böhm et al. 2001; Auer et al. 2001; Brunet et al. 2006; Parker and Horton 2005; Begert et al. 2005; Demarée et al. 2002; Moberg et al. 2002; Bergström and Moberg 2002; Herzog and Müller-Westermeier 1998). Other studies have assessed the homogeneity of time series within the data set using the results as a basis for including or excluding particular time series from the data set (Wijngaard et al. 2003; Begert et al. 2007). A further number of studies have improved the quality of the data by screening the data for inconsistencies and outliers (Klein Tank et al. 2002; Moberg et al. 2006; Begert et al. 2007).

These activities have, amongst others, ensured a reliable assessment of changes in extremes over the last 50 years (e.g., Klein Tank and Können 2003). However, in order to assess changes in daily climate extremes over longer periods for which daily data exist there remain considerable challenges in the homogenisation of these data. At the daily timescale there are inhomogeneities in temperature time series which are not adequately accounted for by traditional correction methods (Della-Marta and Wanner 2006). For example, experiments comparing modern instrument shelters with older ones usually show that the earlier shelters allowed the thermometers to be exposed to greater extremes of short- and long-wave radiation (e.g., Parker 1994; Nordli et al. 1997) with nonlinear effects on temperature records throughout the seasonal cycle. In a statistical context, these non-linearities are expressed as a change in the PDF, often with changes in higher-order moments of the PDF. Therefore, there is a need to make adjustments to the entire probability distribution when homogenising daily data (Della-Marta and Wanner 2006). Presently there are only a few methods which have been used to homogenize daily temperature. Current methods either consist of interpolating monthly mean adjustment values to the daily timescale using simple statistical interpolation methods (e.g., Vincent et al. 2002) or more sophisticated methods which can adjust the entire PDF (Trewin and Trevitt 1996; Della-Marta and Wanner 2006). These methods adjust the PDF non-parametrically. Della-Marta et al. (2007a) have applied the method of Della-Marta and Wanner (2006) to 25 long-term daily maximum temperature records contained in the data set described in Moberg et al. (2006). This study is the first to apply a PDF homogenisation method to daily European maximum temperature. Della-Marta et al. (2007a) find that many of the 25 homogenized time series require negative adjustments to the mean and variance of their measured values in order to be homogeneous with current observed values from the same long-term time series. Figure 1 shows an example of a daily adjustment made to the August time series measured at Basel, Switzerland from 1895–1929.

The fitted curve in Fig. 1 shows a negative slope from decile 1 to decile 10. This indicates that the variance of daily maximum temperature in August is too high in comparison with modern measurements (Della-Marta and Wanner 2006). Also notice that the mean adjustment over all deciles is negative. This indicates that the mean daily temperature is also too high in comparison with modern measurements. An analysis by Begert et al. (2005) suggests that the correction needed to August mean temperature in Basel during this period was approximate –0.3°C. The average adjustment in Fig. 1 is −1.6°C showing that the sign of the adjustment is in agreement but the magnitude is much greater using the method of Della-Marta and Wanner (2006). There are many possible reasons for these types of discrepancies which are described in Della-Marta and Wanner (2006), however, the adjustments are intuitively correct since from metadata we know that the thermometer shelter was changed from a more open wall-mounted screen to a free-standing screen. This would result in the need to decrease the variance of daily August maximum temperature to be homogeneous with today's temperatures (see Della-Marta et al. 2007a;

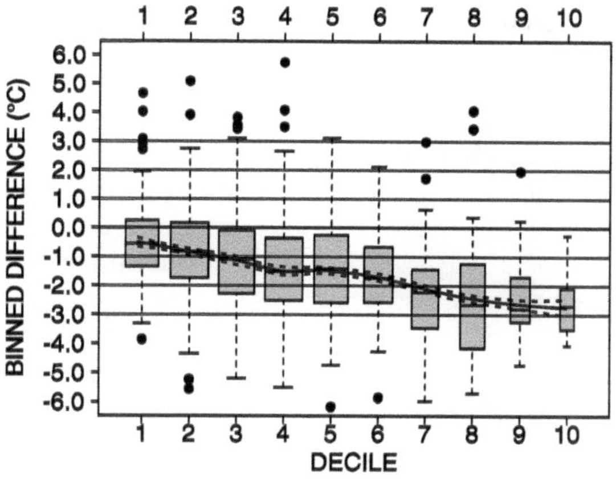

Fig. 1 The adjustment in degree Celsius needed to make August Basel (Switzerland) daily maximum temperature (1895–1929) to be homogeneous with current measurements as a function of the decile of the Basel PDF before the inhomogeneity. The smoothed adjustments (°C) for each percentile shown as a solid black curve fitted using a LOESS function (Cleveland and Devlin 1988). The box plots indicate the mean of the binned differences between modeled and observed daily temperature after the inhomogeneity (black line), the interquartile range (shaded area), 1.5 times the interquartile range (dashed black line), and outliers (dots). The width of the box indicates the relative number of observations in each. The dashed black curved lines show the 95% confidence interval of the fitted curve. See Della-Marta and Wanner (2006) and Della-Marta et al. (2007a) for details of the method and its application respectively.

Begert et al. 2005). Della-Marta et al. (2007a) show that many of the 25 stations needed adjustments not only to their mean but also to their variance and skewness characteristics.

The task of homogenising the European Climate Assessment and related European daily data sets (e.g., Moberg et al. 2006) are far from complete, especially for earlier data, however, given a sufficiently dense network of time series (such as the ENSEMBLES daily data set, Begert et al. 2007, http://www.ensembles-eu.org/), automated inhomogeneity detection and correction techniques (e.g., Caussinus and Mestre 2004) may be coupled with daily temperature correction methods (e.g., Della-Marta and Wanner 2006) to provide daily data sets in which the largest and/or most easily detected inhomogenieties known to affect daily temperature data are removed.

3 Heat Wave and Extremely Hot Temperature Changes over the Past 126 Years

In this section we provide an overview of recent studies which have used long-term observational data of daily temperature to document changes in the frequencies of extremely hot temperatures and heat waves over western Europe. All studies agree that mean summer European temperature has increased over the last 126 years. Some studies show that Europe has experienced an unprecedented rate of summer warming in recent decades (Klein Tank and Können 2003; Luterbacher et al. 2004; Klein Tank et al. 2005). The selected studies mentioned below usually used daily temperature data which have different temporal and spatial extent and different levels of data quality. This makes a quantitative intercomparison difficult.

Kyselý (2002) presents a regionally focused paper on the changes in the frequency of heat waves in Prague over the 20th century using homogenized data. The summer daily time series has been shown to be free of urban influences. Using a heat wave index based on fixed thresholds (30°C and 25°C) they conclude that the severity of heat waves in Prague has increased, although they did not fit a trend. At the time of writing the most severe heat wave year was 1994 (during the 1901–1997 period).

Hundecha and Bárdossy (2005) present an analysis of changes in extreme temperatures in the Rhine basin. They conclude that the 90th percentile of summer daily maximum temperature has increased by 1.4°C from 1958–2001. No homogeneity tests were performed since the daily data were assumed to be homogeneous.

Domonkos et al. (2003) document 20th-century trends in extreme warm events in time series from south-central Europe. They use a relatively low fixed threshold of 23°C for all 11 stations and count the number of daily exceedances of this value. They conclude that the frequency of extreme warm events contains large inter-decadal variability with only a few series exhibiting statistically significant trends. They also note that some of the individual trends could be contaminated with inhomogeneities caused by station moves.

Yan et al. (2002) study the trends in summer temperature extremes based on various percentile extremes from eight long-term daily temperature stations. All of the records were homogenized and generally show a decrease in the frequency of extreme warm daily summer temperatures in the 19th and early 20th century followed by an increase in the latter half of the 20th century. However, a more recent study by Moberg et al. (2003) indicates that daily summer temperatures in the Stockholm and Uppsala series (the same used in Yan et al. 2002) may be warm biased before the 1860s due to the thermometers being exposed to higher amounts of short-wave radiation, a finding also supported by Della-Marta et al. (2007a).

Klein Tank and Können (2003) analysed a European climate dataset for changes in the frequency of extreme temperature events. They demonstrated that the annual change in the frequency of hot daily temperature extremes increased twice as fast as corresponding cold daily temperature extremes over the period 1976–1999.

Moberg and Jones (2005) also analysed the same European climate data set but concentrated on analysing the seasonal changes of the sparser but longer time series (some from 1901) in the data set. They found that extremely high summer daily minimum temperature (and to a smaller extent daily maximum temperature) has increased significantly during the past century. There is more evidence for summer warming in the first half of the century compared with the second half. Importantly they note that the European Climate Assessment data set has many daily temperature series whose homogeneity is questionable. They removed those series which had been identified as being the most inhomogeneous by Wijngaard et al. (2003) and they described results which were the most spatially consistent, as they believe they are the most trustworthy.

Moberg et al. (2006) presents an update to Moberg and Jones (2005) by including a greater number of long-term time series starting as early as 1850. These longer time series were collected as part of the EMULATE project (see http://www.cru.uea.ac.uk/cru/projects/emulate/). The authors conclude that regional differences in the magnitude of summer warming can be high and that the strongest increases in the extreme warm summer maximum temperature occurred in central Europe.

Della-Marta et al. (2007b) use a subset of the longest and most complete daily maximum temperature time series from Moberg et al. (2006) (54 time series) to look closer at the summer daily maximum temperature changes over the last 126 years. The time series were homogenized at the daily timescale (Della-Marta et al. 2007a) using the method of Della-Marta and Wanner (2006). The homogeneity analysis shows that many summer daily maximum temperature series in western Europe are biased warm in the earlier parts of their time series (Della-Marta et al. 2007a) believed to be caused mainly by early thermometers being exposed in open type, metallic or wall-mounted radiation shields. Compared to modern radiation shields (e.g., the Stevenson screen) the older types generally allowed the thermometers to be exposed to more short- and long-wave radiation.

Della-Marta et al. (2007b) placed the observations of increased daily summer maximum temperature (DSMT) variance in recent years (Klein Tank et al. 2005; Moberg et al. 2006) into a longer-term perspective. They show that the recently observed increase in summer daily and summer seasonal (~45 years, see Scherrer et al., this volume) temperature variability in western Europe is indeed unique over the last 126 years. Over western Europe the variance of DSMT across western Europe has increased significantly by approximately 6 ± 2% and by 11 ± 2% for central-western Europe (see Table 1 and Della-Marta et al. 2007b).

These changes can be seen in the PDF of western European DSMT which has become flatter and wider (see Fig. 4a in Della-Marta et al. 2007b). In order to accurately

Table 1 Daily Summer Maximum Temperature mean, variance and skewness trends from 1880 to 2005, for each of four regions in western Europe (reproduced from Della-Marta et al. 2007b).

Region (Number of stations)	Mean (°*C*)	Variance (%)	Skewness (%)
Western Europe (54)	**1.6 ± 0.4**	**+6 ± 2**	+0 ± 7
Central Western Europe (36)	**1.3 ± 0.5**	**+11 ± 2**	+0 ± 6
Iberian Peninsula (12)	**2.6 ± 0.6**	**−7 ± 3**	−1 ± 12
Scandinavia (6)	**1.7 ± 0.7**	+4 ± 6	**+9 ± 6**

Trends have been calculated over nine independent periods. The mean, variance and skewness trends are expressed in the units of °C, % and % with respect to 1906–1990 period respectively. The error estimates are 95% confidence intervals based on the standard error of the robust linear fit. Figures quoted in bold face are significantly (5% significance) different from zero (see Della-Marta et al. (2007b) for more details).

assess the change in DSMT variance, data are split into independent periods of 14 years and have had the linear trend removed using a technique called piece-wise detrending (Scherrer et al., this volume; Scherrer et al. 2005; Della-Marta et al. 2007b). Della-Marta et al. (2007b) show that western European DSMT has increased by 1.6 ± 0.4°C from 1880 to 2005 consistent with all previous studies, however, they find that as a result of the homogenisation applied to the daily data (Della-Marta et al. 2007a) the estimate of climate change in western Europe is more extreme than any published before. Figure 2 is reproduced from Della-Marta et al. (2007b) and shows the temporal and spatial distribution of changes in hot days (HD) and maximum length heat wave (HW) over the last 126 years. A HD is defined as a day where the daily maximum temperature exceeds the long-term daily 95th percentile of daily maximum temperature. The hot day index (HD) is the number of such days within a June–August season expressed as a percentage of time. A heat wave (HW) is defined as the maximum number of consecutive days where the DSMT exceeds the long-term daily 95th percentile of DSMT within a June–August season. The daily 95th percentile is calculated using a centred 15-day average using the daily data from the normal period of 1906–1990. They find that

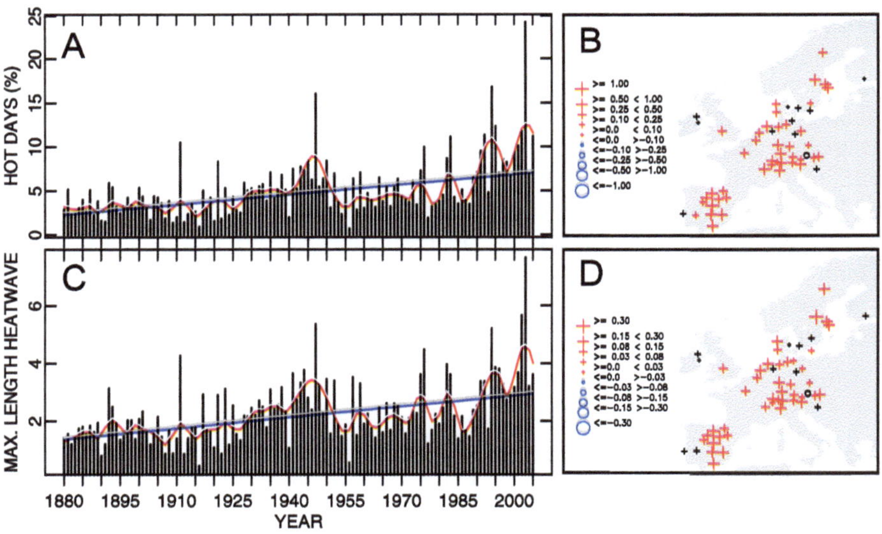

Fig. 2 Reproduced from Della-Marta et al. (2007b): June–August average number of western European hot days (HD) (a) and maximum length heat wave (HW) (c) from 1880 to 2005. The long-term decadal variability (thick red line) and the overall robust linear trend (thick blue line) are shown. The units of HDs are percent of June–August days and the units of HWs are days. The spatial distribution of decadal HD (b) and HW d) trends at each station. The size of the "+" and "o" symbols is proportional to the magnitude of the positive and negative trends respectively according to the legend left of the figure. Red (black) coloured crosses indicate significant (not significant) positive trends at the 5% significance level (similarly for negative trends) (see Della-Marta et al. (2007b) for more details).

almost all station records used in their analysis have a positive trend in the occurrence of summer HDs and HWs from 1880 to 2005 (Fig. 2b, d). Around 80% of the trends are significant (Della-Marta et al. 2007b). The largest trends have been found over the Iberian Peninsula and in central-western Europe. The time evolution of European average HDs (Fig. 2a) shows an overall positive trend of 0.38 ± 0.05% of summer days per decade. This trend is equivalent to approximately 188 ± 54% increase, with an average of 7.3% of summer days classified as HDs at the end of the period, compared with 2.5% in 1880. The 6 ± 2% and the 11 ± 2% increase in DSMT variance over the whole of western Europe and central-western Europe is responsible for approximately 25% and 40% of the increase in HDs in these regions respectively. This confirms that small changes in the intrinsic variance of DSMT have lead to greatly amplified changes in the frequency of extremes. Similarly, the length of HWs has increased by approximately 111 ± 36%, from an average of 1.4–3.0 days per HW (Fig. 2c). As a result, HWs have doubled in length over the period 1880–2005. The changes in the HW index are especially important since this index combines a measure of the extreme temperature magnitude as well as a measure of their persistence (Della-Marta et al. 2007b).

4 Projections of Heat Wave and Extremely Hot Temperatures

In this section we outline some of the recent literature concerning future changes in the frequency of extremely hot temperatures and heat waves in Europe. Some studies are based on the output of general circulation models (GCM) forced by expected changes in greenhouse gases, aerosols and other known forcings. Other studies nest regional climate models (RCM) within GCMs to try and get more realistic estimates of temperature change at higher resolutions. Projections of temperature in Europe all agree that mean temperature and the frequency of extremes will increase during the next 100 years (Wagner 1996; Huth et al. 2000; Beniston 2004; Meehl and Tebaldi 2004; Schär et al. 2004; Weisheimer and Palmer 2005; Clark et al. 2006; Beniston et al. 2007); however, the precise magnitude of this change is uncertain due to uncertainties in the estimates of global greenhouse gas emissions and biases in GCMs/RCMs due to inadequate parameterisations of unresolved physical processes. Many of the simulations performed by RCMs are based either on the IPCC SRES A2 high greenhouse gas emissions scenarios or the lower B2 scenario (Nakicenovic et al. 2000), that both have a large number of underlying assumptions on the future course of global population, economics, technological adjustments and political decision-making. In the following, the A2 scenario results are highlighted as they represent the upper range of greenhouse gas forcing and thus potentially lead to some of the strongest impacts.

RCM projections of summer extreme temperature change in Europe suggest that there will be substantial regional variation linked with complex feedback mechanisms between soil moisture, precipitation and circulation (Schär et al. 2004; Brabson et al. 2005; Ferranti and Viterbo 2006; Seneviratne et al., this volume; Seneviratne et al.

2006; Beniston et al. 2007). Decreased levels of soil moisture in central and eastern future European climate have been shown to be responsible for the projected increases in summer temperature variability (Schär et al. 2004; Seneviratne et al., this volume; Seneviratne et al. 2006), a trend which is seen in many other projections of European climate change (Scherrer et al., this volume; Scherrer et al. 2005; Weisheimer and Palmer 2005; Beniston et al. 2007) and in some cases is expected to be as large as 100% in 2071 compared to the climate of 1961–1990 (Schär et al. 2004).

Beniston et al. (2007) present a summary of future projections of European extreme events (including heat waves) which have been shown to have a high impact on European society and the environment. They summarize the findings of 55, 30-year integrations from a variety of RCMs created as part of the PRUDENCE project (http://prudence.dmi.dk/index.html). The experiments compare present control climate simulations over the period 1961–1990 with 2071–2100 simulations. Beniston et al. (2007) use a number of indices to investigate changes in extreme temperatures and heat waves: the frequency with which daily maximum temperature exceeds 30°C, high temperature percentiles, and four heat wave indices. A heat wave is defined to be a spell of at least six consecutive days with maximum temperature exceeding the 1961–90 calendar day 90th percentile, calculated for each day over a centred 5-day window at each grid point. The four indices, calculated for each year are: Heat Wave Number (HWN); the number of heat waves, Heat Wave Frequency (HWF); the total length (days) of all heat waves, Heat Wave Duration (HWD); the maximum length (days) of all heat waves (very similar to the definition of HW), and, Heat Wave Intensity (HWI); the maximum threshold excess (degree days) of all heat waves.

High percentiles of daily maximum temperature across Europe generally increase more than lower percentiles, implying that changes in the PDFs are more than shifts in location. Changes in the variance (as well as the mean) of daily maximum temperatures have a substantial impact on future extreme daily temperatures. Largest changes in variance were found over the continental interior (a latitude band encompassing France and Hungary; see also Schär et al. 2004) and are most likely caused by a drying out of the land surface in warmer and drier future summer conditions.

Figure 3 shows the changes in the four heat wave indices simulated by HIRHAM-H, expressed as ratios. The mean duration (HWD, Fig. 3a) increases by a factor of between one and eight over most of Europe. Much higher increases of at least a factor of seven are predicted for the mean intensity (HWI, Fig. 3b), the mean number of heat waves (HWN, Fig. 3c) and the frequency of heat wave days (HWF, Fig. 3d), with greatest changes (more than tenfold increases) in the south of France and Spain.

Figure 4 shows the shift in JJA maximum temperature between the 1961–1990 reference period and 2071–2100 for the RCM grid-point closest to Basel (northwestern Switzerland), for the IPCC A2 and B2 scenarios, for both the mean and the 90% quantile. In the reference climate, mean temperatures in Basel are close to 23°C and heat waves can be considered to occur when temperatures exceed 30°C (which is the average level of the 90% quantile for this period). For the scenario

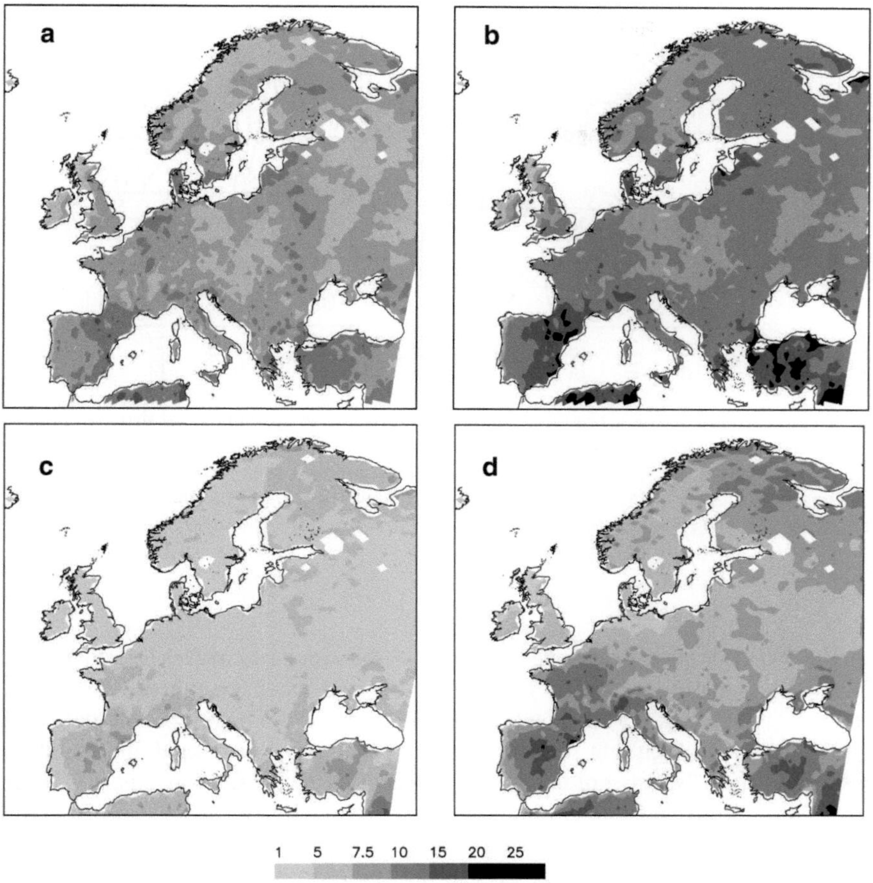

Fig. 3 Changes (expressed as a ratio) in the heat wave indices (a) HWN (b) HWF (c) HWD d) HWI indices between the 1961–1990 and 2071–2100 periods, based on HIRHAM4 simulations. A heat-wave is defined to be a spell of at least six consecutive days with maximum temperature exceeding the 1961–1990 calendar day 90th percentile, calculated for each day over a centred 5-day window at each grid point. The four indices, calculated for each year, are Heat Wave Number (HWN) – the number of heat waves, Heat Wave Frequency (HWF) – the total length (days) of all heat waves, Heat Wave Duration (HWD) – the maximum length (days) of all heat waves, Heat Wave Intensity (HWI) – the maximum threshold excess (degree days) of all heat waves (see Beniston et al. 2007 for more details) (reproduced from Beniston et al. 2007).

climates, whatever the scenario chosen, temperatures are seen to rise on average between 5°C and 7°C over current values; the difference in temperature between the A2 and B2 emissions scenarios is less than between the low emissions scenario and current climate. This implies that even with rather stringent policies to abate greenhouse gas emissions, the increase in temperatures as seen for the B2 scenario may ultimately result in summer heat waves that are as intense, or even stronger,

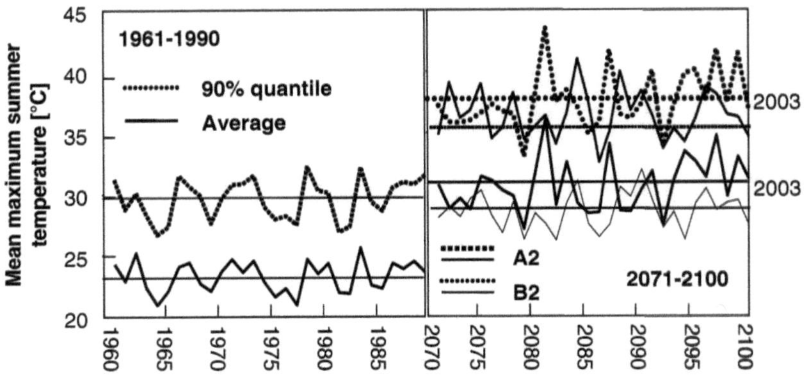

Fig. 4 Comparisons between summer (JJA) mean maximum temperatures and their 90% quantiles for Basel, for each year of the reference 1961–1990 climate and the A2 and B2 scenario climates. The horizontal lines depict the 30-year means for each time series. 2003 refers to the mean and 90% quantile values recorded during the summer of 2003 in Europe, in order to highlight the exceptional nature of that heat wave (reproduced from Beniston and Diaz 2004).

than the 2003 European heat wave; the potential for strong heat waves is even greater for the A2 scenario, as can be intuitively expected when greenhouse forcing is stronger. The mean and 90% quantile statistics of the 2003 heat wave are provided in this diagram to highlight the fact that this event was exceptional and could be considered to be a "summer of the future". Indeed, statistically speaking, the 2003 heat wave could occur one summer out of two in a future climate (Schär et al. 2004).

The 2003 heat wave, by mimicking quite closely the possible course of summers in the latter part of the 21st century, can thus be used within certain limits as an analogue to what may occur with more regularity in the future. The physical processes that characterised the event, such as soil moisture depletion and the positive feedback on summer temperatures as well as the lack of convective rainfall in many parts of the continent that generally occur from June–September, are predicted to take place with greater frequency in the future. In view of the severity of the impacts related to the heat wave, it should help scientists in assessing the course of future climatic impacts, and decision-makers in formulating appropriate response strategies. It is of interest to note, however, that according to the baseline used, the very definition of a heat wave could change in a future, systematically warmer climate. The climate of southern Spain, for example, that is currently characterized by temperatures exceeding 30°C for about 60 days per year on average may in the future experience over 150 days or more (Beniston and Diaz 2004). Under such circumstances, the notion of heat wave loses some of its value when a rare or exceptional feature of today's climate becomes commonplace in tomorrow's climate. This aspect of changing extreme events in a warmer world also needs to be taken into account in forward planning to adapt to the impacts of such events.

5 Conclusions

We presented a brief overview of the latest literature concerning observed and future estimates of the change in the frequency of extreme temperature events and the duration of heat waves in Europe spanning the period between 1880 and 2100. We have given special attention to summarising the current efforts to improve the quality of observed daily temperature series in Europe. A new method (Della-Marta and Wanner 2006) has been used to correct previously uncorrected inhomgeneities in some early European daily temperature measurements (Della-Marta et al. 2007a). We focused the presentation of results from two recent publications on observed changes in heat waves from 1880 to 2005 (Della-Marta et al. 2007b) and future expected changes of heat waves (Beniston et al. 2007). Below we combine and summarize the important findings from the literature.

- New analyses reveal that the increases in western European extreme temperatures from 1880 to 2005 is greater than previously thought (Della-Marta et al. 2007b).
- Observations of regional changes in mean and variance of daily summer maximum temperature (Della-Marta et al. 2007b) fall in line with expected future summer temperatures in Central- western Europe (Schär et al. 2004; Seneviratne et al. 2006; Beniston et al. 2007). This region, where the largest observed increase in daily summer maximum temperature of 11 ± 2% (see Table 1) from 1880 onwards occurred, is also the region with the greatest expected change in summer temperature variability, with changes up to 100% by 2071–2100 (Schär et al. 2004).
- The expected large increase in temperature variability in Central- western Europe is likely to be driven in part by soil moisture, precipitation and atmospheric circulation feedback processes (Schär et al. 2004; Seneviratne et al. 2006; Beniston et al. 2007).
- Approximately 40% of the change in the frequency of hot days over the last 126 years in central-western Europe is due to increases in the daily summer maximum temperature variance. In the future climate up to 60% of change in the frequency of hot days could be due to changes in summer daily temperature variability.
- The duration of heat waves (based on the daily 95th percentile) in each summer season since 1880 have doubled in length (Della-Marta et al. 2007b). The projected future increase in the duration of slightly less extreme heat waves (based on the daily 90th percentile) is estimated to be between a factor of 1 and 8 by 2100 (Beniston et al. 2007).

Acknowledgement This study was supported by the Swiss NSF through the National Centre for Competence in Research Climate (NCCR-Climate).

References

Alexander, L. V. et al., 2006: Global observed changes in daily climate extremes of temperature and precipitation. *J. Geophys. Res.*, **111**, D05109.

Auer, I., R. Böhm, and W. Schöner, 2001: Austrian long-term climate 1767–2000 multiple instrumental climate time series from central Europe, *Techn. Ber., Zentralanstalt für Meteorologie und Geodynamik*, Wien.

Begert, M., T. Schlegel, and W. Kirchhofer, 2005: Homogeneous temperature and precipitation series of Switzerland from 1864 to 2000. *Int. J. Climatol*, **25**, 65–80.

Begert, M., E. Zenklusen, L. Klok, C. Häberli, and C. Appenzeller, 2007: Testing an automated homogenization procedure for a large multivariate dataset (in preparation).

Beniston, M., 2004: The 2003 heat wave in Europe: a shape of things to come? An analysis based on Swiss climatological data and model simulations. *Geophys. Res. Lett.*, **31**, 2202–2202.

Beniston, M. and H. F. Diaz, 2004: The 2003 heat wave as an example of summers in a greenhouse climate? Observations and climate model simulations for Basel, Switzerland. *Glob. Planet Change*, **44**, 73–81.

Beniston, M., D. B. Stephenson, O. B. Christensen, C. A. T. Ferro, C. Frei, S. Goyette, K. Halsnaes, T. Holt, K. Jylhä, B. Koffi, J. Palutikoff, R. Schöll, T. Semmler, and K. Woth, 2007: Future extreme events in European climate; an exploration of Regional Climate Model projections. *Clim. Change*, **81**, 71–95.

Bergström, H. and A. Moberg, 2002: Daily air temperature and pressure series for Uppsala (1722–1998). *Clim. Change*, **53**, 213–252.

Böhm, R., I. Auer, M. Brunetti, M. Maugeri, T. Nanni, and W. Schöner, 2001: Regional temperature variability in the European Alps: 1760–1998 from homogenized instrumental time series. *Int. J. Climatol*, **21**, 1779–1801.

Brabson, B. B., D. H. Lister, P. D. Jones, J. P. Palutikof, 2005: Soil moisture and predicted spells of extreme temperatures in Britain. *J. Geophys. Res.*, **110**, D05104.

Brunet, M., O. Saladié, P. D. Jones, J. Sigró, E. Aguilar, A. Moberg, D. Lister, A. Walther, López, 2006: The development of a new dataset of Spanish daily adjusted temperature series (SDATS) (1850–2003). *Int. J. Climatol*, **26**, 1777–1802, doi:10.1002/joc.1338.

Caussinus, H. and O. Mestre, 2004: Detection and correction of artificial shifts in climate series. *J. Roy. Stat. Soc. C-App*, **53**, 405–425.

Clark, R. T., S. J. Brown, and J. M. Murphy, 2006: Modeling northern hemisphere summer heat extreme changes and their uncertainties using a physics ensemble of climate sensitivity experiments. *J. Climate*, **19**, 4418–4435.

Cleveland, W. S. and S. J. Devlin, 1988: Locally-weighted fitting: an approach to fitting analysis by local fitting. *J. Am. Stat. Assoc.*, **83**, 596–610.

Della-Marta, P. M. and H. Wanner, 2006: A method of homogenising the extremes and mean of daily temperature measurements. *J. Climate*, **19**, 4179–4197.

Della-Marta, P. M., J. Luterbacher, H. von Weissenfluh, E. Xoplaki, M. Brunet, and H. Wanner, 2007a: Summer heat waves over western Europe 1880–2003, their relationship to large scale forcings and predictability. *Clim. Dynam*, **29**, 251–275.

Della-Marta, P. M., M. R. Haylock, J. Luterbacher, and H. Wanner, 2007b: Doubled length of western European summer heat waves since 1880. *J. Geophys. Res.* (**112**, D15103).

Demarée, G., P. Lachaert, T. Verhoeve, and E. Thoen, 2002: The long-term daily Central Belgium Temperature (CBT) series (1767–1998) and early instrumental meteorological observations in Belgium. *Clim. Change*, **53**, 269–293.

Domonkos, P., J. Kysely, K. Piotrowicz, P. Petrovic, and T. Likso 2003) Variability of extreme temperature events in south-central Europe during the 20th century and its relationship with large-scale circulation. *Int. J. Climatol*, **23**, 987–1010.

Ferranti, L. and P. Viterbo, 2006: The European summer of 2003: sensitivity to soil water initial conditions. *J. Climate*, **19**, 3659–3680.

Frich, P., L. Alexander, P. Della-Marta, B. Gleason, M. Haylock, A. Klein Tank, and T. Peterson, 2002: Observed coherent changes in climatic extremes during the second half of the twentieth century. *Climate Res.*, **19**, 193–212.

Gruber, S., M. Hoelzle, and W. Haeberli, 2004: Permafrost thaw and destabilization of Alpine rock walls in the hot summer of 2003. *Geophys. Res. Lett.*, **31**, L13504.

Herzog, J. and G. Müller-Westermeier, 1998: Homogenitätsprüfung und Homogenisierung klimatologischer Messreihen im Deutschen Wetterdienst, Techn. Ber. 202, Deutscher Wetterdienst.

Hundecha, Y., A. Bàrdossy, 2005. Trends in daily precipitation and temperature extremes across western Germany in the second half of the 20th century. *Int. J. Climatol.*, **25**, 1189–1202.

Huth, R., J. Kysely, and L. Pokorna, 2000: A GCM simulation of heat waves, dry spells, and their relationships to circulation. *Clim. Change*, **46**, 29–60.

IPCC, 2001: *Climate Change 2001: The Scientific Basis*. Contribution of Working Group I to the Third Assessment Report of the Intergovernmental Panel on Climate Change (IPCC), Cambridge University Press, Cambridge, UK.

Karl, T. R. and R. W. Knight, 1997: The 1995 Chicago heat wave: how likely is a recurrence? *Bull. Am. Meteorol Soc.*, **78**, 1107–1119.

Karl, T., V. Derr, D. Easterling, C. Folland, D. Hofmann, S. Levitus, N. Nicholls, D. Parker, and G. Withee, 1995:. Critical issues for long-term climate monitoring. *Clim. Change*, **31**, 185–221.

Klein Tank, A. et al., 2002: Daily dataset of 20th-century surface air temperature and precipitation series for the European Climate Assessment. *Int. J. Climatol*, **22**, 1441–1453.

Klein Tank, A. M. G. and G. P. Können, 2003: Trends in indices of daily temperature and precipitation extremes in Europe, 1946–99. *J. Climate*, **16**, 3665–3680.

Klein Tank, A. M. G., G. P. Können, and F. M. Selten, 2005: Signals of anthropogenic influence on European warming as seen in the trend patterns of daily temperature variance. *Int. J. Climatol*, **25**, 1–16.

Koppe, C., R. Kovats, G. Jendritzky, and B. Menne, 2004: Heat-waves: impacts and responses, Techn. Ber., Word Health Organisation Regional Office for Europe., Copenhagen, Sweden., In: *Health and Global Environmental Change Series. No. 2.*

Kovats, R. and C. Koppe, 2005: Integration of public health with adaptation to climate change: lessons learned and new directions. *Kap. Heatwaves: Past and Future Impacts.* Swets & Zeitlinger, Lisse.

Kyselý, J. 2002: Temporal fluctuations in heat waves at Prague-Klementinum, the Czech Replublic, from 1901–97, and their relationships to atmospheric circulation. *Int. J. Climatol*, **22**, 33–50.

Luterbacher, J., D. Dietrich, E. Xoplaki, M. Grosjean, and H. Wanner, 2004: European seasonal and annual temperature variability, trends, and extremes since 1500. *Science*, **303**, 1499–1503.

Meehl, G. and C. Tebaldi, 2004: More intense, more frequent, and longer lasting heat waves in the 21st century. *Science*, **305**, 994–997.

Milligan, J., 2004: Heatwaves: the developed world's hidden disaster, Techn Ber, International Federation of Red Cross and Red Crescent, World Disasters Report.

Moberg, A., H. Bergstrom, J. Krigsman, and O. Svanered, 2002: Daily air temperature and pressure series for Stockholm (1756–1998). *Clim. Change*, **53**, 171–212.

Moberg, A., H. Alexandersson, H. Bergström, and P. D. Jones, 2003: Were southern Swedish summer temperatures before 1860 as warm as measured? *Int. J. Climatol*, **23**, 1495–1521.

Moberg, A. and P. D. Jones, 2005: Trends in indices for extremes in daily temperature and precipitation in central and western Europe, 1901–99. *Int. J. Climatol*, **25**, 1149–1171.

Moberg, A. et al., 2006: Indices for daily temperature and precipitation extremes in Europe analysed for the period 1901–2000. *J. Geophys. Res.*, **111**, D22106.

Nakicenovic, N. et al., 2000: *IPCC Special Report on Emissions Scenarios*, Cambridge University Press, Cambridge, UK/New York, 599 pp.

Nicholls, N. G. Gruza, J. Jouzel, T. Karl, L. Ogallo, D. Parker, 1996: IPCC 1995. The Second IPCC scientific assessment of climate change Observed variability and change. Houghton, J. & Meira Filho, L. (ed.), Cambridge University Press.

Nordli, P. O., H. Alexandersson, P. Frich, E.J. Forland, R. Heino, T. Jonsson, H. Tuomenvirta, O.E. Tveito, 1997: The effect of radiation screens on Nordic time series of mean temperature. *Int. J. Climatol.*, **17**, 1667–1681.

Parker, D. E., 1994: Effects of Changing Exposure of Thermometers at Land Stations. *Int. J. Climatol*, **14**, 1–31.

Parker, D. E. and B. Horton, 2005: Uncertainties in central England temperature 1878–2003 and some improvements to the maximum and minimum series. *Int. J. Climatol*, **25**, 1173–1188.

Poumadère, M., C. Mays, S. Le Mer, and R. Blong, 2005: The 2003 heat wave in France: dangerous climate change here and now. *Risk Analysis*, **25**, 1483–1494.

Schär, C. and G. Jendritzky, 2004: Climate change: hot news from summer 2003. *Nature*, **432**, 559–560.

Schär, C., P. Vidale, D. Luthi, C. Frei, C. Haberli, M. Liniger, and C. Appenzeller, 2004: The role of increasing temperature variability in European summer heatwaves. *Nature*, **427**, 332–336.

Scherrer, S. C., C. Appenzeller, M.A. Liniger, C. Schär, 2005: European temperature distribution changes in observation and climate change scenarios. *Geophys. Res. Lett.*, **32**, L19705.

Scherrer, S. C., M. A. Liniger, and C. Appenzeller, 2007: Distribution Changes of Seasonal Mean Temperature in Observations and Climate Change Scenarios. (this volume)

Seneviratne, S. I., D. Lüthi, M. Litschi, and C. Schär, 2006: Land-atmosphere coupling and climate change in Europe. *Nature*, **443**, 203–206.

Seneviratne, S. I. and R. Stöckli, 2007: The Role of Land-Atmosphere Interactions for Climate Variability in Europe. (this volume).

Stott, P., D. Stone, and M. Allen, 2004: Human contribution to the European heatwave of 2003. *Nature*, **432**, 610–614.

Trewin, B. and A. Trevitt, 1996: The development of composite temperature records. *Int. J. Climatol.*, **16**, 1227–1242.

Wagner, D., 1996: Scenarios of extreme temperature events. *Clim. Change*, **33**, 385–407.

Wijngaard, J., A. M. G. Klein Tank, and G. Können, 2003: Homogeneity of 20th century European daily temperature and precipitation series. *Int. J. Climatol*, **23**, 679–692.

Weisheimer, A. and T. N. Palmer, 2005: Changing frequency of occurrence of extreme seasonal temperatures under global warming. *Geophys. Res. Lett.*, **32**, L20721.

Yan, Z., P. Jones, T. Davies, A. Moberg, H. Bergstrom, D. Camuffo, C. Cocheo, M. Maugeri, G. Demaree, T. Verhoeve, E. Thoen, M. Barriendos, R. Rodriguez, J. Martin-Vide, and C. Yang, 2002: Trends of extreme temperatures in Europe and China based on daily observations. *Clim. Change*, **53**, 355–392.

Distribution Changes of Seasonal Mean Temperature in Observations and Climate Change Scenarios

S. C. Scherrer[1,2], M. A. Liniger[1], and C. Appenzeller[1]

Abstract Distribution changes in seasonal mean 2 m temperature are investigated in Central Europe and the extratropical northern hemisphere for the most recent period 1961–2005 and the 21st century. Data from both observations and climate model runs are used. The latter are taken from scenarios prepared for the fourth IPCC assessment report (AR4). All data sets and model runs are scaled by their corresponding interannual variability to facilitate the comparison between the data sets. Time series are detrended to investigate changes in internal variability. For the last 45 years (1961–2005) the strongest changes in mean are found for in the summer mean temperature, both in observations and AR4 runs. With the exception of autumn, changes in the mean are captured reasonably well by the models on the regional as well as hemispheric scale. For Central Europe, estimates for variability changes show a weak increase (decrease) in summer (winter) observations. Both are not statistically significant at the 10% level. For the 21st century all climate scenario runs suggest large relative increases in mean temperature for all seasons. The AR4 model-to-model differences in the mean changes are largest in summer and not substantially smaller than those from the third assessment report (TAR). Model uncertainties are in the same order or even larger than the uncertainty introduced by the different scenarios. Compared to TAR, the AR4 runs show a much more consistent tendency for increases in Central Europe summer temperature variability especially towards the end of the 21st century (for 2070–2099: ~22–47% increase). No clear changes in seasonal mean temperature variability are found for the other seasons and the averaged northern hemisphere land time series.

[1]Climate Services, Federal Office of Meteorology and Climatology MeteoSwiss, Krähbühlstrasse 58, Postfach 514, CH-8044 Zürich
mark.liniger@meteoswiss.ch, christof.appenzeller@meteoswiss.ch

[2]Now at Climate and Global Dynamics Division, National Center for Atmospheric Research, P.O. Box 3000, Boulder, CO 80307-3000, USA
scherrer@ucar.edu

S. Brönnimann et al. (eds.), *Climate Variability and Extremes during the Past 100 Years.*

1 Introduction

In terms of global mean surface temperature, the 1990s has been the warmest decade, at least since the beginning of instrumental measurements in the early 19th century (Houghton et al. 2001). In Central Europe nine of the ten warmest years of the 1851–2006 temperature record have been observed since 1989 (updated from Scherrer et al. 2006). Reconstructed temperature series show that the late 20th-century European climate is very likely warmer than that of any time during the past 500 years (Luterbacher et al. 2004, 2007; Xoplaki et al. 2005). For the 21st century the third Intergovernmental Panel on Climate Change (IPCC) assessment report suggests changes in the mean that are several times larger than those caused by natural variability in the past millennium (Houghton et al. 2001). Less clear is, whether there are additionally changes in interannual near-surface temperature variability, although the latter might be of importance especially for climate risk assessments, e.g., in the health, agriculture, energy, traffic or infrastructure sector.

Many of these climate risk-related questions can often be related to changes in temperature extremes on a day-to-day basis. Several studies addressed this issue in observations (e.g., Klein Tank and Können 2003; Moberg and Jones 2005; Moberg et al. 2006) and climate scenario runs (e.g., Ferro et al. 2005; Meehl et al. 2000; Sardeshmukh et al. 2000; Zwiers and Kharin 1998). A subject of particular interest is the observed and expected change in heat wave frequency, cf. Della-Marta et al. (2006) and Della-Marta and Beniston (this volume).

On the interannual timescale knowledge is still sparse but previous studies found indications for decreasing temperature variability in most parts of the world (Karl et al. 1995), whereas others report no changes or small increases (Luterbacher et al. 2006; Parker et al. 1994; Xoplaki et al. 2005). For Central Europe, Scherrer et al. (2005) analysed CRUTEM2v (Jones and Moberg 2003) temperature data and reported weak variability increases (decreases) in summer (winter), but these changes are not statistically significant at the 10% level.

Theoretical studies have shown that in terms of sensitivities, the frequency of extreme events depends more on changes in the variability than in changes in the mean (Katz and Brown 1992). On the other hand, the same authors mention that it is possible that the magnitude of the changes in the mean still are the dominant effect for changes in the frequency of extremes. Analyses of the Climate Model Intercomparison Project 2 (CMIP2) $2CO_2$ projections show, that at least on the monthly to yearly time scale, changes in temperature extremes may be essentially determined by the changes in the mean for most parts of the world (Räisänen 2002). On a Central European scale on the other hand, European summer surface temperature variability might increase (by up to 100%) within the current century (Schär et al. 2004). The effect has qualitatively been confirmed by several other models and studies (Brabson et al. 2005; Giorgi and Bi 2005; Kleinn et al. 2005; Meehl and Tebaldi 2004; Seneviratne et al. 2006). Scherrer et al. (2005) used observational and IPCC Third Assessment Report (TAR) near surface temperature data with seasonal resolution and found that beside large relative increases in mean with maximum amplitude in summer there is a tendency for increasing (decreasing) interannual variability for future summers (winters). They also found substantial differences between the different models.

In this chapter, observed and modelled Central European land temperature distributions are investigated in a standardized Gaussian framework for each season separately. Relative changes in the means and standard deviation are compared separately between models and observations including a quantification of the associated uncertainties. The work is based upon the methodology presented in Scherrer et al. (2005) and extends the previous work to scenario runs produced for the fourth Assessment Report (AR4) of the IPCC.

2 Data

2.1 Seasonal Averages of Observed Series

Two-metre temperature averages from the monthly land-surface temperature data set CRUTEM2v (Jones and Moberg 2003, http://www.cru.uea.ac.uk/cru/data/tem2) with a resolution of $5°\times 5°$ are considered. Additionally, 2 m temperature from the European Centre for Medium-Range Weather Forecasts (ECMWF) reanalysis (ERA-40) (Simmons et al. 2004; Uppala et al. 2005) with a resolution of around 1.125° are used as another "proxy" for observations. The time period considered in this chapter is 1961–2005. Special focus is given to the comparison between the 1961–1990 normal period and the most recent 30-year period 1976–2005. The periods overlap by 15 years and therefore not entirely independent, but the latter is expected to be more influenced by global climate change. The ERA-40 values after August 2002 are operational analyses regridded to the ERA-40 grid resolution using a simple lapse rate correction to account for ERA-40's coarse orography. Seasonal averages of the meteorological seasons (DJF, MAM, JJA and SON) are used.

2.2 AR4 Model Runs

To investigate the near past and future climate, IPCC SRES greenhouse gas scenario runs prepared in the context of the IPCC AR4 assessment are used, see http://www-pcmdi.llnl.gov/ipcc/about_ipcc.php for details. For the period 1961–2000, we use the climate of the 20th century (20C3M) runs which (depending on the model) incorporate various natural and anthropogenic forcings including changes greenhouse gases, ozone and aerosols distribution. For details refer to the references in Table 1 and Cordero and de Forster (2006). For the periods 2001–2005, 2010–2039, 2040–2069 and 2070–2099 the following scenario runs are considered: B1 (a convergent world with the same global population $\Rightarrow$ weak changes), A1B (a future world of very rapid economic growth, CO_2 concentrations increase by 1% per year until

Table 1 Some characteristics of the IPCC AR4 models and scenarios considered in this chapter

Model	Resolution (horizontal/vertical)	Runs and scenarios considered 20C3M 1961–2000	A1B 2001–2099	B1 2001–2099	A2 2001–2099	Reference
HadCM 3	3.75° × 2.5°, L19	1	1	1	1	Gordon et al. (2000)
HadGEM 1	1.875° × 1.25°, L38	1	1	–	1	Johns et al. (2006)
CSIRO Mk 3.0	T63, L18	1	1	1	1	Gordon et al. (2002)
CGCM 3.1	T63, L31	5	5	5	5	Flato (2005)
MPI ECHAM 5	T63, L31	3	3	3	3	Jungclaus et al. (2006)
GFDL CM 2.1	2.5° × 2°, L24	1	1	1	1	Delworth et al. (2006)
NCAR PCM 1.1	T42, L18	4	4	–	4	Washington et al. (2000)
Total		16	16	11	16	

it has doubled at a level of 720 ppm ⇒ moderate changes) and A2 (a very heterogeneous world with continuously increasing global population ⇒ strong changes). The differences between the scenarios grow with time (cf., e.g., IPCC 2001). Seven different models with one to five model runs available for each of the three scenarios are used for this chapter (Table 1). The restriction to this relatively small number of models is due to the fact that we wanted to use models that were either already used in the TAR study by Scherrer et al. (2005) or showing good skill over Europe in terms of dynamics (e.g., van Ulden and van Oldenborgh 2006).

2.3 *Regions of Interest*

The 2003 summer heat wave which caused tenths of thousands of excess deaths (Fouillet et al. 2006; Kovats et al. 2004) might be an indicator that Central Europe is especially susceptible in terms of changes in temperature extremes (Schär et al. 2004; Seneviratne et al. 2006). Therefore it seems straight forward to focus on Central Europe more closely. Central Europe, is defined here as the CRUTEM2v, ERA-40 and climate model individual's land grid points in the area covering 3°W–27°E and 44°N–55°N, see box in Fig. 1 or Scherrer et al. (2005). The analysis is repeated for the land grid points of the extratropical northern hemisphere (latitude > 20°N). This allows to discuss the differences between the regional and hemispheric scale changes. Monthly values are converted to seasonal means before taking area weighted average series (details given below).

3 Methodology

Monthly to seasonal 30-year samples of averaged Central European 2 m temperature observations are in general fairly well normally distributed (cf. Scherrer et al. 2006). It is thus justifiable to characterize the distribution solely by the mean

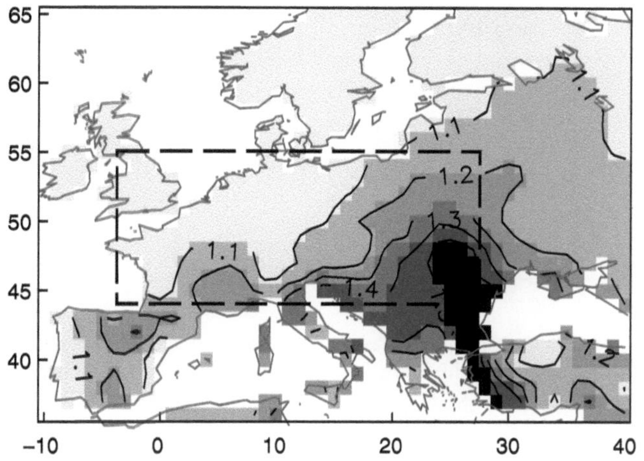

Fig. 1 Map of inflation factor s/s_0 as given by Eq. 1 for ERA-40 JJA 2m temperature data over Europe for the 30-year period 1973–2002. The black dashed box shows the region defined as Central Europe (adapted from Scherrer et al. 2005, supplementary material).

(the normal value) T_0 and standard deviation s_0. The non-stationarity in the series, i.e., trends in time series leads to an inflation of variability as noted, e.g., by Räisänen (2002) and Schär et al. (2004). It can be shown (cf. Scherrer et al. 2005) that the ratio between the standard deviation s of a white noise series of length N with a superimposed linear trend of magnitude α per time step and the standard deviation s_0 of the pure white noise series is

$$\frac{s}{s_0}=\left[1+\gamma\left(\frac{\alpha}{s_0}\right)^2\right]^{1/2} \qquad \text{with} \qquad \gamma=\frac{N(N+1)}{12}. \tag{1}$$

Figure 1 shows s/s_0 values applying Eq. 1 for the 1973–2002 period (N = 30) summer season (JJA) over Europe. For large parts of south-eastern Europe the inflation accounts for up to 40–70% in increase of s and is therefore of concern already in the recent observational period (Scherrer et al. 2005, supplementary material). Further, we are interested in changes in internal interannual variability and not in trend induced variability changes. Therefore estimates of standard deviation are always computed using linearly detrended series.

3.1 Displaying Distribution Changes: The MSC Diagram

For Gaussian distributions, distribution changes are described by a combination of a shift in mean (location) and/or a change in standard deviation (scale), cf. Fig. 2a. In order to properly compare observations against model results and the models

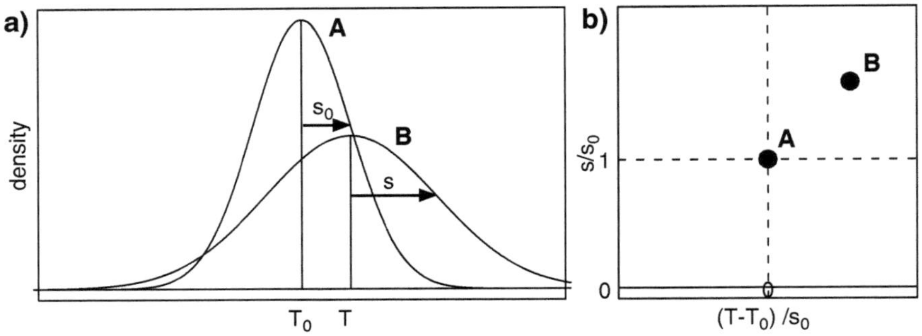

Fig. 2 Illustration of a simultaneous increase in the mean and increase in standard deviation on a Gaussian climate distribution A with mean T_0 and standard deviation s_0 resulting in distribution B with mean T and standard deviation s. (a) Probability density functions, (b) distributions A and B in a standardized MSC plot showing standardized changes in the mean $(T-T_0)/s_0$ against standardized standard deviation s/s_0.

among each other in one graph, standardized changes in mean $(T-T_0)/s_0$ and variability s/s_0 are used henceforth. Here, the subscript 0 indicates the estimates based on the 1961–1990 standard normal reference period. The changes in the mean and standard deviation are computed after computing an area weighted mean time series using all grid points in the domain. This procedure is repeated for all 30-year running windows (N = 30) in the period considered.

Plots with standardized Mean versus Standard deviation Change (referred to as MSC plots hereafter) are used to illustrate the changes in both parameters simultaneously, see Fig. 2b and Sardeshmukh et al. (2000) or Scherrer et al. (2005).

The 5–95% confidence range of the mean and standard deviation change is determined with a bootstrap resampling technique for both estimates together. 5,000 samples in combination with a 2D kernel density estimator (e.g., Venables and Ripley 2002) are used. The confidence range for the 30-year mean and standard deviation value pair estimate is generally an ellipse-like area (grey shading in the figures below) spanning ~±0.2–0.4 s_0 for both the mean and standard deviation.

4 Results

4.1 Distribution Changes in the Observational Period

4.1.1 Central European Land Region (Period 1961–2005)

Figure 3 shows the standardized running 30-year estimates of mean and standard deviation changes for the observed and modelled Central European mean temperature from 1961–1990 onwards (i.e., the 30-year windows 1962–1991, 1963–1992, …, 1976–2005). The observed changes are in general similar for CRUTEM2v (thick

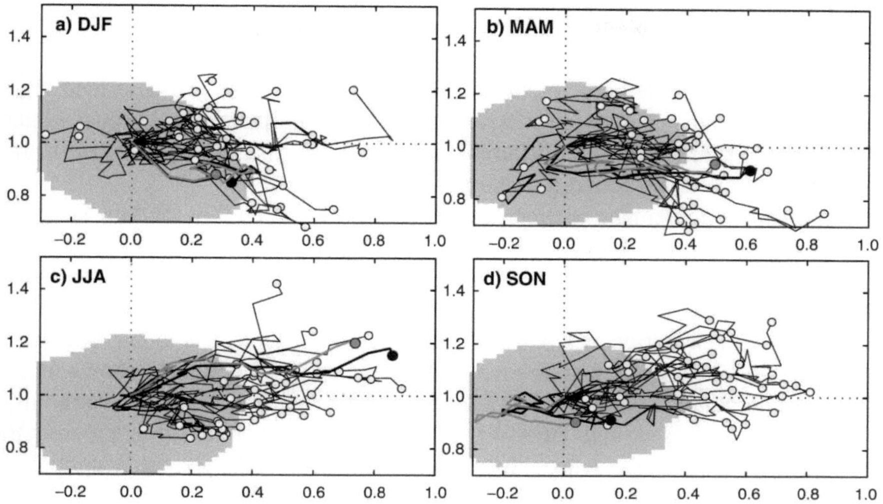

Fig. 3 Evolution of Central European temperature distributions in (a) DJF, (b) MAM, (c) JJA, and (d) SON using standardized MSC plots for the period 1961–2005 for observational temperature data sets (CRUTEM2v: thick black line, extended ERA40: grey line) and AR4 scenario run results for the same period (thin lines). The grey area shows the 5–95% confidence ranges of the 1961–1990 period determined using a bootstrap approach (see text). The circles represent the period 1976–2005. 1961–1990 period is at (0, 1) per definition.

black line) and ERA-40 (thick grey line) in all seasons. The trends are somewhat lower in the ERA-40 data set. The underestimation of the trend, which is probably at least partially caused by inhomogeneous input station data, is in good agreement with results from other studies (Kunz et al. 2007; Scherrer et al. 2006). Significant relative increases in the mean are found for spring (Fig. 3b) and especially for summer (Fig. 3c) where the values are outside the 1961–1990 confidence range. There are only very weak mean changes in autumn (Fig. 3d) confirming results from previous studies (Klein Tank and Können 2003; Scherrer et al. 2006; Scherrer et al. 2005). Increases in variability are particularly found for the JJA season. This increase is still present when excluding the hot summer of 2003. The 10–15% decrease in DJF variability (Fig. 3a) is primarily caused by the extraordinarily cold winter 1962/63 dropping out of the running 30-year period in 1992 (Scherrer et al. 2005).

Most of the 20C3M climate runs of the current climate for the years 1961–2005 show increases in relative mean $(T-T_0)/s_0$. The most consistent increases (up to 0.8 s_0) are found for JJA and SON (Fig. 3c, d). The observed JJA increases are somewhat underestimated by the models, those for SON clearly overestimated. The increases in DJF (Fig. 3a) are mostly comparable with the observed changes, although there are also some runs that show no or even negative trends. The trends in MAM (Fig. 3b) are mostly underestimated. The changes in relative variability (s/s_0) are rather small and stay mostly within the range of uncertainty of the reference data $(0.8 < s/s_0 < 1.2)$.

4.1.2 Extratropical Northern Hemisphere Land Regions

Figure 4 shows the changes for the extratropical northern hemisphere land time series. Note that the changes in relative mean are approximately three times larger than those of Central Europe (different abscissa scaling). Largest changes in the relative mean are again found for summer (Fig. 4c).

The general agreement between the CRUTEM2v and ERA-40 data is again good. The changes in the mean are larger for CRUTEM2v than for ERA-40 in JJA (Fig. 4c), similar for both in MAM (Fig. 4b) and smaller for CRUTEM2v than those of ERA-40 in SON (Fig. 4d) and DJF (Fig. 4a). The relative changes in variability are similar with exception of JJA where for ERA-40 an increase is found. The main reason for this discrepancy is one particular year. In 1992, ERA-40 shows much lower temperatures than CRUTEM2v. As a consequence, ERA-40 variability changes are increased and ERA-40 mean changes are reduced compared to CRUTEM2v. One likely reason for this difference in 1992 is the eruption of Pinatubo in summer 1991 which caused strong tropospheric cooling in 1992 (Hansen et al. 1992; Minnis et al. 1993). Obviously ERA-40 and CRUTEM2v do not agree on the amount of this cooling linked with the eruption. ERA-40 values may be too low. The 20C3M model run changes for the years 1961–2005 are in general consistent with the CRUTEM2v and ERA-40 results with exception of autumn, where most of the models overestimate the changes in the mean (especially wrt CRUTEM2v, cf. Fig. 4d).

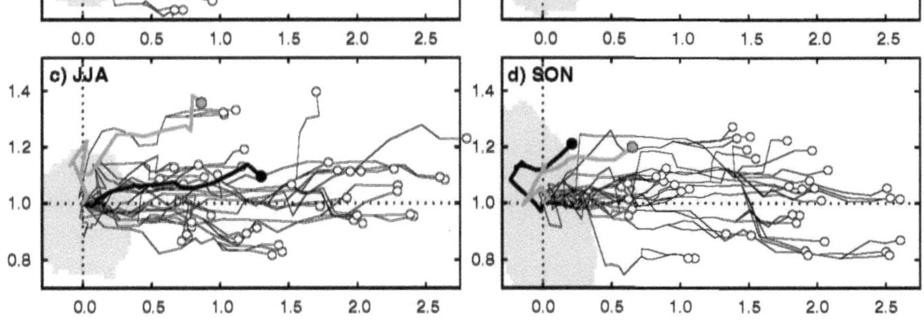

Fig. 4 As Fig. 3, but for the average over the land areas of the extratropical northern hemisphere (latitude > 20°N). Note that the scale of the abscissa is different to the one of Fig. 3.

4.2 21st-century Scenario Distribution Changes

4.2.1 Central European Land Region

The last 30-year period available (2070–2099) from the climate scenario runs prepared for IPCC AR4 are investigated in the same way as the observational period. Results are shown in Fig. 5. The dominant changes in Central European mean temperature are increases in the mean. The largest relative changes in mean are found in summer (Fig. 5c, B1: 1.5–4.5 s_0, A1B: 2–7 s_0, A2: 2–9 s_0) followed by autumn (Fig. 5d, B1: 1.25–4 s_0, A1B: 1.75–5.5 s_0, A2: 2.5–5.75 s_0), spring (Fig. 5b, B1: 1.75–2.75 s_0, A1B: 1.5–3.25 s_0, A2: 1.75–4.25 s_0) and smallest relative changes in winter (Fig. 5a, B1: 1.25–2.25 s_0, A1B: 1.5–3 s_0, A2: 2–3.25 s_0). As expected by construction of the scenarios the changes are in general smallest for the B1 runs, followed by the A1B and A2 scenario runs. However, the distinction between the different scenarios is not always obvious, especially in summer (Fig. 5c). Also the model-to-model spread in the mean change is by far largest in summer, followed by autumn, spring and smallest in the winter season.

Changes in variability are less obvious than the changes in the mean. Roughly half of the values are within the bootstrap range of the reference period variability uncertainties. The most pronounced tendencies are found in summer; with considerable ensemble mean (EM) increases of variability of 22% for B1, 23% for A1B

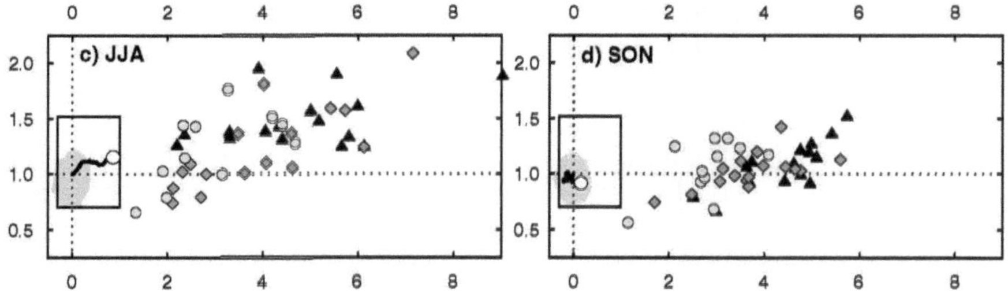

Fig. 5 Standardized MSC plots for Central European AR4 scenario results for the 2070–2099 period (filled symbols). Also shown is the curve for the observational period (1961–2005) in Central Europe (thick black line). The most recent 30-year window value (1976–2005) is shown as white point. The rectangular box shows the boundaries of the area shown in Fig. 3. Grey area: 5–95% confidence ranges of the 1961–1990 period determined using a bootstrap approach. B1 scenario circles are light grey, A1B diamonds are grey and A2 triangles black. (a) DJF, (b) MAM, (c) JJA, and (d) SON.

and 47% for A2. Variability is increasing in most scenarios and models, which is consistent with studies analysing regional climate models (Giorgi et al. 2004; Kleinn et al. 2005; Schär et al. 2004; Seneviratne et al. 2006) and global climate models (Brabson et al. 2005; Meehl and Tebaldi 2004). The results for the other seasons are rather inconclusive. The model-to-model spread for variability changes is large and there are no clear tendencies visible for any of the scenarios.

To get a more clear picture of the expected relative changes in terms of the temperature probability density functions (PDFs) we construct Gaussian PDFs for the B1 (11 runs), A1B (16 runs) and A2 (16 runs) scenarios using EM changes in both relative mean and relative standard deviation. Note that PDFs constructed in this way (PDF_{EM} subsequently) are not a physical solution but rather a simple statistical view of the IPCC ensemble in terms of EM changes. Additionally, PDFs are constructed using the average (PDF_{AVG} subsequently) of the individual model run PDFs, assuming that each model run is equally probable to represent the predicted future climate,

$$PDF_{AVG} = \frac{1}{\sum_{m=1}^{7} R_m} \sum_{m=1}^{7} \sum_{r=1}^{R_m} PDF_{m,r} \quad (2)$$

where m is the index for the seven different models considered, and r the index for the individual model runs. R_m is the number of model runs for model m. No additional smoothing is applied in the PDF_{AVG} construction. In contrast to the Gaussian PDF_{EM} the PDF_{AVG} is in general not Gaussian, but skewed, bumpy or could even be multimodal. It includes a large amount of uncertainties (i.e., model differences, internal variability, model errors). Figure 6 shows examples for JJA A1B and A2 scenario PDF_{EM} (solid line) and PDF_{AVG} (dashed line) for the period 2070–2099. The PDF_{AVG} is unimodal without additional smoothing. The reason for this difference compared with other studies (e.g. Tebaldi et al. 2005) is that we show relative instead of absolute changes in temperature, which makes PDFs smoother by correcting biases and errors in variability. As expected the PDF_{AVG} is broader and flatter than the PDF_{EM} and thus the likelihood for very small or very large changes is considerably increased. Both PDFs show a clear increase in the JJA variability confirming the above results.

Figure 7 shows PDF_{EM}s for all seasons and two different 30-year periods in the observational period (1961–2005, 1976–2005) as well as three future periods (2010–2039, 2040–2069 and 2070–2099). The PDF changes for the most recent 30-year observational period (1976–2005, grey curve) with respect to the 1961–1990 reference period show increases in the mean as well as an increase (small decrease) in variability in summer (winter) as discussed above (cf. Figs. 3a, c and 7a, g).

For the nearer future period 2010–2039 (Fig. 7a, d, g, j), the changes are almost exclusively changes in the mean. These are very similar for all scenarios. There is a weak tendency for smaller relative changes in the mean in winter and spring than in summer and autumn.

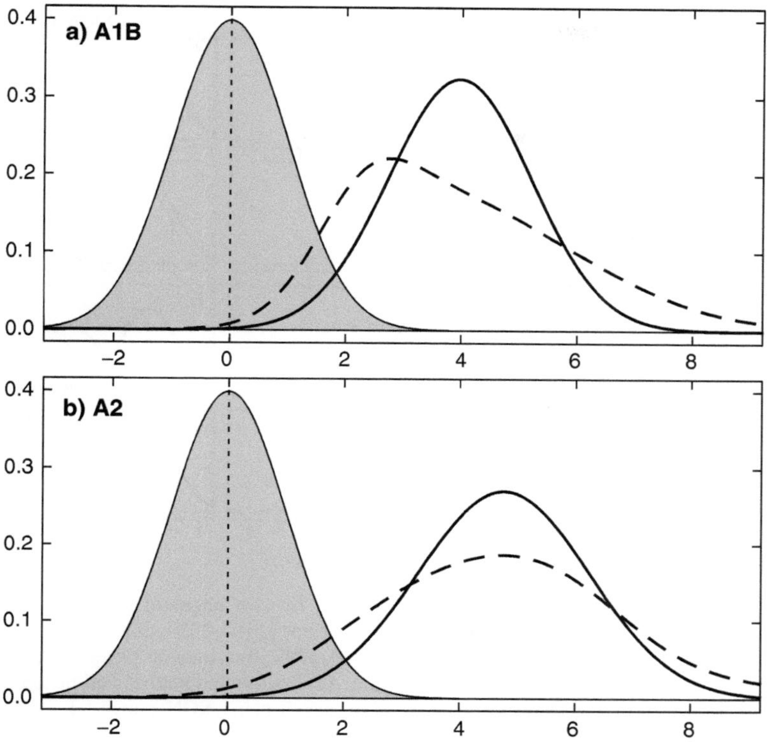

Fig. 6 Central European mean JJA temperature PDFs based on IPCC SRES A1B (panel a) and A2 (panel b) scenarios using (1) a PDF_{EM}, i.e., the ensemble mean of the individual models' relative mean and relative standard deviation changes (solid) and (2) a PDF_{AVG}, i.e., the standardized sum of the individual model PDFs (dashed). No smoothing is applied in the PDF_{AVG} construction. The reference PDF (standard deviation is 1 per definition and representing the 1961–1990 period) is shown in grey.

For the period in the mid-21st century (2040–2069, Fig. 7b, e, h, k) the changes in the mean have increased (still with smaller changes in winter and spring than in summer and autumn). In addition, there is a widening of the distribution in summer and autumn. The B1 changes are now clearly smaller than those of the A1B and A2 scenarios which are still very similar.

For the end of the 21st-century period (2070–2099, Fig. 7c, f, i, l) the changes in the mean have further increased for all scenarios and the differences between the scenarios are now clearly visible also for the A1B and A2 scenario. The increase in the mean is smallest for the B1, followed by the A1B and largest for the A2 scenario. This agrees well with the response on a global scale and the underlying global carbon emissions used to construct these scenarios (IPCC 2001). All scenarios show increased 2070–2099 summer variability, but the largest changes are clearly found for A2 (Fig. 7i). The PDFs in the other seasons are more or less purely

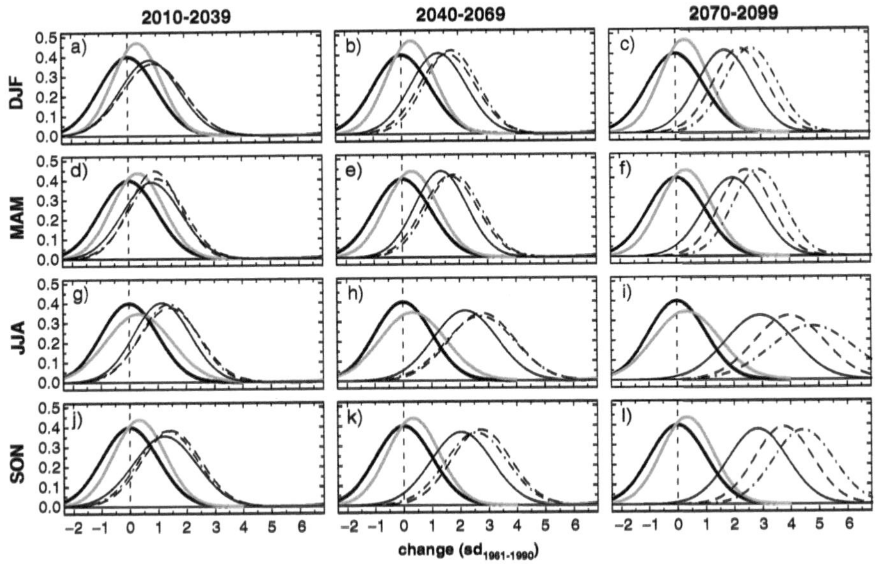

Fig. 7 Central European relative temperature PDF_{EM}s for two observed periods (1961–1990: thick black and 1976–2005: thick grey), three future periods (2010–2039: thin lines in panels a, d, g, j, 2040–2069: thin lines in panels b, e, h, k, 2070–2099: thin lines in panels c, f, i, l, three different scenarios (B1: solid line, A1B: dashed line, A2: dash-dotted line) and all meteorological seasons (DJF: panels a, b, c, MAM: panels d, e, f, JJA: panels g, h, i, SON: panels j, k, l). Abscissa-unit is 1. Refer to text for details on the PDF construction.

shifted (i.e., pure increase in the mean for MAM and SON as well as no change in variability for DJF). This is consistent with the results in terms of variability changes discussed above (cf. Fig. 5). Note that the variability changes discussed are caused by internal variability changes and not by a statistical artefact, since the trend induced changes have been eliminated by the detrending procedure.

4.2.2 Extratropical Northern Hemisphere Land Regions

The relative changes for the extratropical northern hemisphere 2070–2099 period are much larger than those for the Central European region. The scatter of the runs is largest in summer (as for Central Europe, cf. Figs. 5 and 8). The different scenarios (B1, A1B and A2) are still not clearly separated. This is a good indication that interexperiment differences for near-surface temperature are large compared to the differences given by the different scenarios. A similar result is found by Räisänen (2001) for the CMIP2 data set, where most of the interexperiment differences come from model differences and are not caused by internal variability.

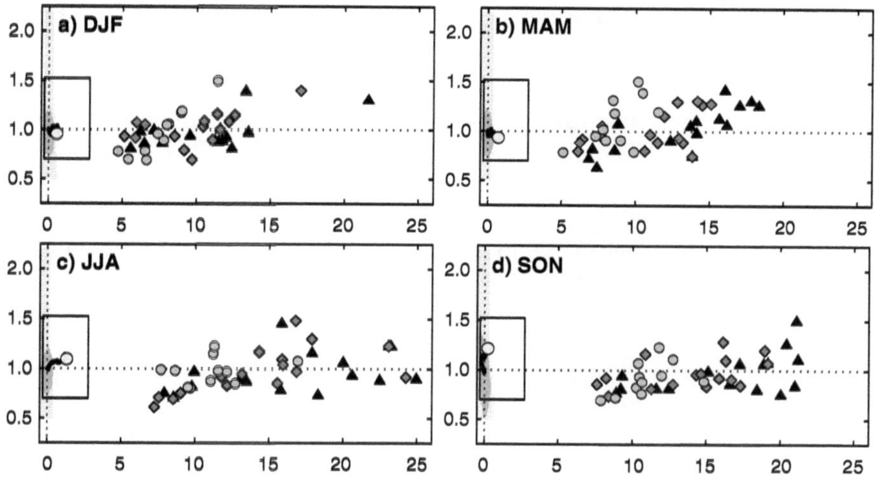

Fig. 8 As Fig. 5, but for land areas of the extratropical northern hemisphere (latitude > 20°N). Note that the scale of the abscissa is different to the one of Fig. 5.

4.3 Differences Between TAR and AR4 Results for Central Europe

Since the publication of the TAR in 2001 the model formulations of coupled climate processes determining local and large-scale variability have been changed in one or other way. Therefore the AR4 results for Central Europe above are compared with those of the TAR presented in Scherrer et al. (2005). The comparison is somewhat difficult and not fully objective since the AR4 results are based on more models, some of them run in ensemble mode. Although numbers might somewhat differ, most TAR results are confirmed by AR4. The general characteristics have been discussed above already. There are three obvious differences:

- The future JJA increase in Central European temperature variability is supported by many more models and model runs in AR4 compared with TAR, cf. Fig. 5c in this chapter with Fig. 4c in Scherrer et al. (2005). The increases are larger and more consistent than in TAR, especially for the A2 scenario.
- The slight future decreases in variability found in the DJF TAR runs on the other hand are not that pronounced in the AR4 results, where both, increases and decreases (no change for the EM, Fig. 7 a, b, c) are found.
- The (insignificant) increase in the 1961–2005 period Central European JJA temperature variability, which was supported by most of the TAR models, is only weakly present in the AR4 runs.

5 Summary and Outlook

In this chapter we examined the relative temporal changes in Central European and extratropical northern hemisphere seasonal land temperature in terms of mean and interannual variability. Observations and state of the art IPCC scenario runs prepared for AR4 are used to investigate (1) the recent past (1961–2005) and (2) 21st-century projections. All data sets and model runs have been scaled by their corresponding interannual variability to facilitate the comparison between the data sets. Time series were detrended to investigate changes in internal variability.

For the 1961–2005 period, the analysis shows that the dominant changes in seasonal temperature distributions are an increase in the mean values in both observations and scenario runs for Central Europe and the extratropical northern hemisphere land regions. This is consistent with earlier studies that investigated changes in mean and variability. The largest observed and modelled relative changes in the mean are found for summer. However, also the largest model-to-model differences are occurring in the summer scenarios. These model-to-model differences are not smaller in the AR4 runs compared with those in the TAR runs. Changes in the mean are captured reasonably well by the models on the regional as well as hemispheric scale although models tend to overestimate changes in the mean in autumn for both spatial scales. The insignificant small decrease in recent observed winter temperature variability is not found in the AR4 climate runs. Similar the insignificant small increase in recent observed summer temperature variability is only weakly present in the AR4 models. As noted above the decrease in winter variability is linked to the extraordinary winter 1962/63, whereas the increase in summer variability lasts even without the extraordinary summer 2003. But it remains unclear whether the observed variability changes are natural decadal variability only or the beginning of a continuing trend.

The projected PDF changes for the 2010–2039 period are changes in the mean. The differences between the different scenarios are negligible. For the mid-century 2040–2069 period, the mean increases further in all seasons, but also a widening of the summer PDF becomes visible. Towards the end of the century (2070–2099), significant increases in Central European summer variability (EM values for B1: +22%, A1B: +23%, and A2: +47%) are found. The largest changes in the mean, but also the largest model-to-model differences are found for the summer season. No clear changes in variability are found on the hemispheric scale and for the other seasons using the AR4 model results available.

Overall the results show that present-day global models are able to reproduce recent Central European and extratropical relative 2 m temperature trends relatively accurately. However, model-to-model differences on a regional to continental scale are in the same order or even larger than the uncertainty introduced by the different scenarios (B1, A1B, A2). This is especially the case for the distant future (2070–2099) summer seasons. The differences could be mainly caused by model errors, but closer investigations are needed to make more general statements. The above point as well as the very small number of ensemble members probably not representing the full uncertainty of the system, makes it very difficult to construct a meaningful probabilistic projection of climate change.

We show that the differences between simple methods to construct estimates of PDFs using the ensembles from different models are large even for a key climate parameter such as mean temperature. More advanced techniques have been applied to produce probabilistic projections of climate change recently. They all determine some sort of weights for the different model runs using among other things the model's capability to model present day or past conditions. Examples for statistical verification approaches can be found, e.g., in Min and Hense (2006) and Tebaldi et al. (2005). Other authors determine the credibility of the models using their thermodynamical or dynamical capabilities (Knutti et al. 2006; Kunkel et al. 2006; van Ulden and van Oldenborgh 2006) or constrain predictions using past climate change (Allen et al. 2000; Lopez et al. 2006).

Besides improving the statistical techniques to construct PDFs which are useful for all kinds of climate risk assessments, impact studies and policy making, we think that it is also crucial to improve the single models participating in the IPCC process with a high priority. A major task should be the identification of the processes which are badly represented in the models and are causing large model errors. In a second step these insights could be used to substantially reduce model errors by implementing better parameterizations and/or more appropriate physics.

Acknowledgements This study was supported by the Swiss NSF through the National Centre for Competence in Research Climate (NCCR-Climate). We acknowledge the modelling groups for providing their data for analysis, the Program for Climate Model Diagnosis and Intercomparison (PCMDI) for collecting and archiving the model output, and the JSC/CLIVAR Working Group on Coupled Modelling (WGCM) for organizing the model data analysis activity. The multi-model data archive is supported by the Office of Science, US Department of Energy.

References

Allen, M. R., P. A. Stott, J. F. B. Mitchell, R. Schnur, and T. L. Delworth, 2000: Quantifying the uncertainty in forecasts of anthropogenic climate change. *Nature*, **407**, 617–620.

Brabson, B. B., D. H. Lister, P. D. Jones, and J. P. Palutikof, 2005: Soil moisture and predicted spells of extreme temperatures in Britain. *J. Geophys. Res.*, **110**, D05104.

Cordero, E. C. and P. M. de F Forster, 2006: Stratospheric variability and trends in IPCC model simulations. *Atmos. Chem. Phys.*, **6**, 5369–5380.

Della-Marta, P. M. and Beniston, M., 2007: Summer Heat Waves in Western Europe, Their Past Change and Future Projection (this volume).

Della-Marta, P. M., M. R. Haylock, J. Luterbacher, and H. Wanner, 2007: The length of western European summer heat waves has doubled since 1880. *J. Geophys. Res*, **112**, DI5103, doi:10.1029/2007JD00810.

Delworth, T. L. et al., 2006: GFDL's CM2 global coupled climate models. Part 1: formulation and simulation characteristics. *J. Climate*, **19**, 643–674.

Ferro, C. A. T., A. Hannachi, and D. B. Stephenson, 2005: Simple non-parametric techniques for exploring changing probability distributions of weather. *J. Climate*, **18**, 4344–4354.

Flato, G., 2005: *The Third Generation Coupled Global Climate Model (CGCM3)*. http://www.cccma.ec.gc.ca/models/cgcm3.shtml.

Fouillet, A. et al., 2006: Excess mortality related to the August 2003 heat wave in France. *Int. Arch. Occ. Env. Hea.*, **80**, 16–24.

Giorgi, F. and X. Q. Bi, 2005: Regional changes in surface climate interannual variability for the 21st century from ensembles of global model simulations. *Geophys. Res. Lett.*, **32**, L13701, doi:10.1029/2005GL023002.

Giorgi, F., X. Q. Bi, and J. S. Pal, 2004: Mean, interannual variability and trends in a regional climate change experiment over Europe. II: climate change scenarios (2071–2100). *Clim. Dyn.*, **23**, 839–858.

Gordon, C. et al., 2000: The simulation of SST, sea ice extents and ocean heat transport in a version of the Hadley Centre coupled model without flux adjustments. *Clim. Dyn.*, **16**, 147–168.

Gordon, H. B. et al., 2002: The CSIRO Mk3 climate system model. CSIRO Atmospheric Research Technical Paper No. 60.

Hansen, J., A. Lacis, R. Ruedy, and M. Sato M, 1992: Potential climate impact of Mount Pinatubo eruption. *Geophys. Res. Lett.*, **19**, 215–218.

Houghton, J. T. et al. (eds.), 2001: *Climate Change 2001: The Scientific Basis, IPCC Third Assessment Report*. Cambridge University Press, 944 pp.

IPCC, 2001: *Emissions Scenarios. 2000: Special Report of the Intergovernmental Panel on Climate Change*. Cambridge University Press, 570 pp.

Johns, T. C. et al., 2006: The new Hadley Centre climate model HadGEM1: evaluation of coupled simulations. *J. Climate*, **19**, 1327–1353.

Jones, P. D. and A. Moberg, 2003: Hemispheric and large-scale surface air temperature variations: an extensive revision and an update to 2001. *J. Climate*, **16**, 206–223.

Jungclaus, J. H. et al., 2006: Ocean circulation and tropical variability in the coupled model ECHAM5/MPI-OM. *J. Climate*, **19**, 3952–3972.

Karl, T. R., R. W. Knight, and N. Plummer, 1995: Trends in high-frequency climate variability in the 20th-century. *Nature*, **377**, 217–220.

Katz, R. W. and B. G. Brown, 1992: Extreme events in a changing climate – variability is more important than averages. *Clim. Change*, **21**, 289–302.

Klein Tank, A. M. G. and G. P. Können, 2003: Trends in indices of daily temperature and precipitation extremes in Europe, 1946–99. *J. Climate*, **16**, 3665–3680.

Kleinn, J., C. Frei, J. Gurtz, D. Lüthi, P. L. Vidale, and C. Schär, 2005: Hydrologic simulations in the Rhine basin driven by a regional climate model. *J. Geophys. Res.*, **110**, D04102, doi:10.1029/2004JD005143.

Knutti, R., G. A. Meehl, M. R. Allen, and D. A. Stainforth, 2006: Constraining climate sensitivity from the seasonal cycle in surface temperature. *J. Climate*, **19**, 4224–4233.

Kovats, S., T. Wolf, and B. Menne, 2004: Heat waves of August 2003 in Europe: provisional estimates of the impact on mortality. *Eurosurveillance Weekly*, **8**(11).

Kunkel, K. E., X. Z. Liang, J. H. Zhu, and Y. R. Lin, 2006: Can CGCMs simulate the twentieth-century "warming hole" in the central United States? *J. Climate*, **19**, 4137–4153.

Kunz, H., S. C. Scherrer, M. A. Liniger, C. Appenzeller, 2007: Changes in surface temperatures and total ozone in the Alpine region: an intercomparison of ERA-40 data and observed Swiss time series. *Meteorol. Z.*, **16**, 171–181.

Lopez, A., C. Tebaldi, M. New, D. A. Stainforth, M. R. Allen, and J. Kettleborough, 2006: Two approaches to quantifying uncertainty in global temperature changes. *J. Climate*, **19**, 4785–4796.

Luterbacher, J., D. Dietrich, E. Xoplaki, M. Grosjean, and H. Wanner, 2004: European seasonal and annual temperature variability, trends, and extremes since 1500. *Science*, **303**, 1499–1503.

Luterbacher, J. et al., 2006: Mediterranean climate variability over the last centuries: a review. In: *The Mediterranean Climate: An Overview of the Main Characteristics and Issues*, P. Lionello, P. Malanotte-Rizzoli, and R. Boscolo, eds., Elsevier, 27–148.

Luterbacher, J., M. A. Liniger, A. Menzel, N. Estrella, P. M. Della-Marta, C. Pfister, T. Rutishauser, and E. Xoplaki, 2007: Exceptional European warmth of autumn 2006 and winter 2007: Historical context, the underlying dynamics, and its phenological impacts. *Geophys. Res. Lett.*, **34**, L12704, doi:10.1029/2007GL029951.

Meehl, G. A. and C. Tebaldi, 2004: More intense, more frequent, and longer lasting heat waves in the 21st century. *Science*, **305**, 994–997.

Meehl, G. A., F. Zwiers, J. Evans, T. Knutson, L. Mearns, and P. Whetton, 2000: Trends in extreme weather and climate events: issues related to modeling extremes in projections of future climate change. *B. Am. Meteorol. Soc.*, **81**, 427–436.

Min, S.-K. and A. Hense, 2006: A Bayesian approach to climate model evaluation and multi-model averaging with an application to global mean surface temperatures from IPCC AR4 coupled climate models. *Geophys. Res. Lett.*, **33**, L08708, doi:10.1029/2006GL025779.

Minnis, P., E. F. Harrison, G. G. Gibson, F. M. Denn, D. R. Doelling, and W. L. Smith Jr, 1993: Radiative forcing by the eruption of Mt. Pinatubo. *Science*, **259**, 1411–1415.

Moberg, A. and P. Jones, 2005: Trends in indices for extremes in daily temperature and precipitation in central and western Europe, 1901–1999. *Int. J. Climatol.*, **25**, 1149–1171.

Moberg, A. et al., 2006: Indices for daily temperature and precipitation extremes in Europe analysed for the period 1901–2000. *J. Geophys. Res.*, **111**, D22106.

Parker, D. E., P. D. Jones, C. K. Folland, and A. Bevan, 1994: Interdecadal changes of surface-temperature since the late-19th-century. *J. Geophys. Res.*, **99**, 14373–14399.

Räisänen, J., 2001: CO2-induced climate change in CMIP2 experiments: quantification of agreement and role of internal variability. *J. Climate*, **14**, 2088–2104.

Räisänen, J., 2002: CO2-induced changes in interannual temperature and precipitation variability in 19 CMIP2 experiments. *J. Climate*, **15**, 2395–2411.

Sardeshmukh, P. D., G. P. Compo, and C. Penland, 2000: Changes of probability associated with El Nino. *J. Climate*, **13**, 4268–4286.

Schär, C., P. L. Vidale, D. Lüthi, C. Frei, C. Häberli, M. A. Liniger, and C. Appenzeller, 2004: The role of increasing temperature variability in European summer heatwaves. *Nature*, **427**, 332–336.

Scherrer, S. C., C. Appenzeller, and M. A. Liniger, 2006: Recent temperature trends in Switzerland and Europe: implications for climate normals. *Int. J. Climatol.*, **26**, 565–580, doi:10.1002/joc.1270.

Scherrer, S. C., C. Appenzeller, M. A. Liniger, and C. Schär, 2005: European temperature distribution changes in observations and climate change scenarios. *Geophys. Res. Lett.*, **32**, L19705, doi:10.1029/2005GL024108.

Seneviratne, S. I., D. Lüthi, M. Litschi, and C. Schär, 2006: Land-atmosphere coupling and climate change in Europe. *Nature*, **443**, 205–209, doi:10.1038/nature05095.

Simmons, A. J. et al., 2004: Comparison of trends and low-frequency variability in CRU, ERA-40, and NCEP/NCAR analyses of surface air temperature. *J. Geophys. Res.*, **109**, D24115, doi:10.1029/2004JD005306.

Tebaldi, C., R. L. Smith, D. Nychka, and L. Mearns, 2005: Quantifying uncertainty in projections of regional climate change: a Bayesian approach to the analysis of multimodel ensembles. *J. Climate*, **18**, 1524–1540.

Uppala, S. M. et al., 2005: The ERA-40 reanalysis. *Q. J. Roy Meteorol. Soc.*, **131**, 2961–3012.

van Ulden, A. P. and G. J. van Oldenborgh, 2006: Large-scale atmospheric circulation biases and changes in global climate model simulations and their importance for climate change in Central Europe. *Atmos. Chem. Phys.*, **6**, 863–881.

Venables, W. N. and B. D. Ripley, 2002: *Modern Applied Statistics with S*. 4th edn., Springer, 495pp.

Washington, W. M. et al., 2000: Parallel climate model (PCM) control and transient simulations. *Clim. Dyn.*, **16**, 755–774.

Xoplaki, E., J. Luterbacher, H. Paeth, D. Dietrich, N. Steiner, M. Grosjean, and H. Wanner, 2005: European spring and autumn temperature variability and change of extremes over the last half millenium. *Geophys. Res. Lett.*, **32**, L15713, doi:10.1029/2005GL023424.

Zwiers, F. W. and V. V. Kharin, 1998: Changes in the extremes of the climate simulated by CCC GCM2 under CO2 doubling. *J. Climate*, **11**, 2200–2222.

Section D
Chemical Changes and the Variability of the Stratosphere

Long-term Tropospheric Ozone Trends: A Critical Review

J. Staehelin and C. Schnadt Poberaj

Abstract In this review we attempt to critically evaluate the availability of reliable tropospheric ozone measurements suitable for long-term trend analysis. The focus is on large-scale changes deduced from measurements, which are used for comparison with numerical simulations of the tropospheric ozone cycle. These are required to quantify the influence of anthropogenic ozone precursor emission changes on climate. Long-term tropospheric ozone measurements show that ozone over Europe has increased by more than a factor of two between World War II and the early 1990s which is consistent with the large increase in anthropogenic ozone precursor emissions in the industrialized world. However, the further increase in background ozone over Europe and North America since the early 1990s cannot be solely explained by regional ozone precursor changes because anthropogenic ozone precursor emissions decreased in the industrialized countries as consequence of air pollution legislation. Measurements also indicate large increases in ozone in the planetary boundary layer over the tropical Atlantic since the late 1970s, which have been attributed to large increases in fossil fuel related emissions. Measurements at southern midlatitudes, which are limited in number, show a moderate increase in tropospheric ozone since the middle of the 1990s.

1 Introduction

Ozone was discovered by C. F. Schönbein in 1842, and he was able to document that ozone is present in ambient air already two years after its discovery (Schönbein 1844). Around the end of World War II it was found that ozone can be produced in the troposphere by photochemical air pollution (Haagen-Smith, 1952). Today it is well known that elevated ozone concentrations occur downwind of most large cities all over the world (e.g., Staehelin 2002). Ozone is an important precursor of the OH-radical which limits the tropospheric lifetimes of many gaseous species including

Institute for Atmospheric and Climate Science, ETH Zentrum, Zürich, Switzerland
Johannes.Staehelin@env.ethz.ch
Christina.Schnadt@env.ethz.ch

S. Brönnimann et al. (eds.), *Climate Variability and Extremes during the Past 100 Years.*

important greenhouse gases such as methane. Ozone is also an important greenhouse gas: its increase due to anthropogenic precursor emissions has significantly contributed to changes in radiative forcing since preindustrial times. It has therefore significantly added to the driving force of anthropogenic changes in climate (e.g., IPCC 2001). The effect of ozone as greenhouse gas maximizes at tropopause altitude. However, determining anthropogenic radiative forcing by tropospheric ozone is a difficult task, since ozone is formed in a series of complex reactions from its precursors, which need to be described by complex numerical simulations of the tropospheric ozone cycle. In order to test the reliability of these models, they need to be validated by comparison with measurements, which can be obtained by different approaches (e.g., Stevenson et al. 2006).

It is beyond the scope of the paper to present any exhaustive discussion of tropospheric ozone trend analysis and its different causes, and we refer to Logan et al. (1999) and Oltmans et al. (1998, 2006) for a more complete discussion of the subject. Here we focus on the discussion of long-term global tropospheric ozone trends excluding local and regional scales. We will shortly describe the most important processes determining long-term tropospheric ozone trends (Sect. 2), followed by a short overview of available measurements (Sect. 3). Thereafter we will discuss ozone trends derived from measurements for different regions of the world (Sects. 4–6), and Sect. 7 contains conclusions.

2 Relevant Processes

Tropospheric ozone is produced in the troposphere by a complex series of photochemical reactions (Fig. 1). The ozone precursors include nitrogen oxides (NO_x: $NO + NO_2$), volatile organic compounds (VOC) and carbon monoxide (CO). At larger temporal and spatial scales, nitrogen oxides are the species that restrict ozone formation. The photochemical reactions shown in Fig. 1 lead to a gradual decrease in NO_x concentrations (by formation of HNO_3). If NO_x concentrations are very low, ozone is destroyed by photochemistry (not shown in Fig. 1). In addition to photochemical ozone production and destruction, the global tropospheric ozone cycle also includes the transport from the stratosphere and the destruction of ozone at the Earth's surface. In a recent study, Stevenson et al. (2006) used 26 numerical simulations (most of them set up as chemical transport models) to deduce the present day tropospheric ozone budget (the numbers of the results of the numerical simulations are given in teragrams ozone per year, those in brackets describe the standard deviation):

Global photochemical ozone production from anthropogenic and natural precursors: 5056 (±571)
Global photochemical ozone destruction: 4561 (±722)
Surface destruction: 1014 (±219)
Import from stratosphere: 519 (±195)

The results of this study illustrate that the uncertainties in the individual budget terms are still considerable, particularly with regard to ozone import from the stratosphere.

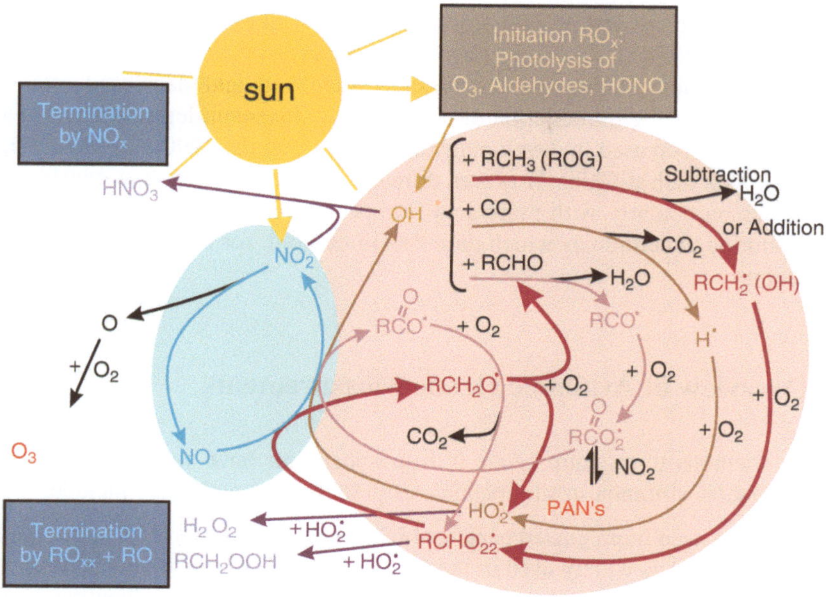

Fig. 1 Overview of photochemical tropospheric ozone production (from Staehelin et al. 2000). Copyright Springer 2000. With kind permission of Springer Science and Business Media.

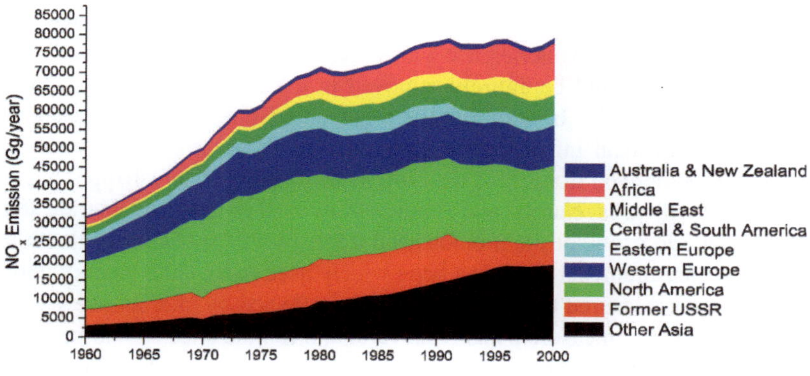

Fig. 2 Fossil-fuel related NO_x emissions from continents (TEAM (TNO emission assessment model), RETRO, Pulles et al. 2007). Note that emissions from ships and aircraft are not covered in this emissions inventory.

To understand tropospheric ozone trends, they need to be compared with changes in anthropogenic ozone precursor emissions. Emission changes are described by emission models. They rely on the description of anthropogenic activities leading to respective emissions and emission factors, that describe the type of technology (e.g., temporal changes in technology for road traffic vehicles by the introduction of catalytic converters in gasoline driven vehicles depending on national legislation). Emissions of society such as road traffic and industry also depend on economy, and all emission models show large increases in the period following World War II due to the large economic growth (cf. Fig. 2). In the industrialized countries anthropogenic emissions stabilized and decreased after the 1980s because of the introduction of air pollution abatement legislation while the emissions still increased in developing countries during the 1990s, for example, in Southeast Asia such as China. However, different emission models show considerable differences particularly with respect to European emission changes in the 1990s demonstrating uncertainties which data should be used, for example, in numerical simulations.

3 Overview of Available Ozone Measurements

Various methods to measure ozone in tropospheric air have been developed over time. The most common in situ methods used in the past and today include:

- Schönbein used impregnated papers which change color when exposed to ozone. This method was widely used in the 19th century. However, it now appears that this method is not reliable and should therefore not be used in a quantitative way (neither for trend analysis nor for comparison with numerical simulations) (cf. Staehelin et al. 1994).
- In the observatory of Montsouris close to Paris ozone was measured from 1876 to 1911 by a chemical method using arsenite (Volz and Kley 1988). However, this method is susceptible to interference with sulfur dioxide, and it is not clear whether this interference could be entirely removed in the measurements presented by Volz and Kley (1988). In addition, the representativeness of the measurements with respect to local influences is questionable. Therefore, the average ozone concentration derived from these measurements (approximately 10 ppb) has to be considered with great caution and should not be used in model validation studies of tropospheric ozone of the preindustrial troposphere (Staehelin et al. 1994).
- Chemical measurements (KI) have been widely used in the 20th century until the 1970s. Ozone measured by this method is also affected by interference with SO_2, and therefore only measurements from rural and remote sites should be used for trend analysis (Staehelin et al. 1994). The same method using electrochemical detection is also applied for ozone profile measurements from light balloons (ozone sondes). A network of ozone sonde stations has been operating since the late 1960s providing a unique data set to document long-term tropospheric ozone changes (e.g., Logan et al. 1999). However, note that different sensors have been developed over time (e.g., Brewer Mast and Electro Chemical sondes (ECC)) which show systematic differences. Furthermore, the evaluation of the data quality of old ozone sonde measurements (such as in the 1970s) is a difficult task.
- Reliable ozone measurements can be obtained by UV absorption using the Hg emission line at 253.7 nm, which is presently commonly applied in ground-based ozone monitoring (e.g., Klausen et al. 2003). Continuous ozone measurements

from air pollution monitoring are available from an increasing number of stations. However, reliable ozone measurements useful for long-term trend analysis from remote sites are still restricted. The method is also used to measure ozone from air planes, as well as from ships. Measurements from regular aircraft are particularly valuable for trend analysis if they provide extended data sets such as in the MOZAIC program (Measurement of Ozone and Water Vapour by Airbus In-Service Aircraft, since 1994, Marenco et al. 1998) and the earlier GASP program (Global Atmospheric Sampling Program, 1975–1979) (Nastrom 1979; Schnadt Poberaj et al. 2007).

It is important to note that surface measurements at high mountain sites as discussed in the following text, are mostly indicative of free tropospheric air masses (e.g., Zellweger et al. 2003). Hence, ozone trends derived from these stations are largely representative for changes in the free troposphere. Similarly, surface ozone measurements at remote sites such as, for example, available in the extratropical southern hemisphere (Sect. 6) also largely reflect free tropospheric conditions.

4 Ozone Trends in the Northern Extratropics

While surface ozone measurements of the 19th century, particularly those measured by Schönbein papers, have to be regarded with caution, the comparison of available ozone measurements performed before the end of the 1950s and those measured at the same sites or sites at similar altitudes show an increase by more than a factor two for August and September (Fig. 3) which is most probably attributable to the large increase in anthropogenic ozone precursors in this period (cf. Fig. 2). Note that a similar increase was documented from measurements of the High Alpine Observatory Jungfraujoch (3,560 m a.s.l.). These trends mainly reflect changes in the free troposphere, since the high alpine station samples free tropospheric air during more than half of the time (Zellweger et al. 2003). Most of the measurements shown in Fig. 3 were only performed during campaigns. However, the long series of surface ozone measurements of Arosa, a resort area in the Swiss Alps (Staehelin et al. 1994), confirms that the ozone increases reported in Fig. 3 took place during all seasons.

Measurements of the European ozone sonde stations show large increases in free tropospheric ozone from the early 1970s to the 1990s (5–25% per decade during 1970–1996; Logan et al. 1999). However, it is difficult to assess the data quality of the Brewer Mast sondes in the 1970s, and the temporal evolutions of ozone at 500 hPa are markedly different at the two nearby stations of Hohenpeissenberg, situated close to Munich in Northern Germany, and Payerne, located in the Swiss plateau (see, e.g., Logan et al. 1999), which seems be difficult to explain.

Since 1992 ozone, measured at high mountain sites, has increased further (Brönnimann et al. 2002; Ordóñez et al. 2007, see Fig. 4). These increases are more difficult to explain as anthropogenic ozone precursor emissions have decreased in Europe and North America since the early 1990s due to air pollution legislation

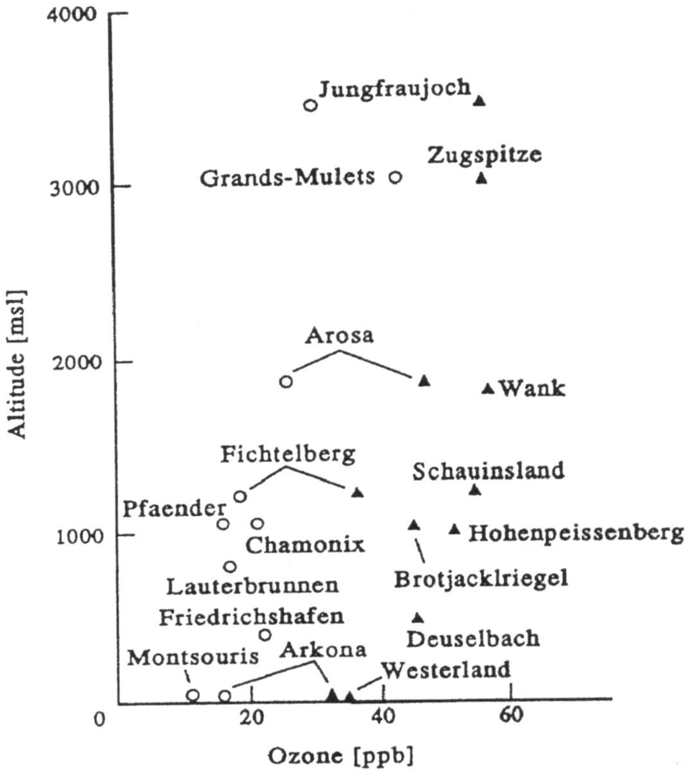

Fig. 3 Historical (circles) and recent (triangles) surface ozone concentrations of August/September from different locations in Europe as a function of altitude. The data of 1988–1991 are based on continuous monitoring measurements, whereas the historical measurements from the different sites mostly include measurements collected over short periods (from Staehelin et al. 1994). Copyright 1993 Pergamon Press Ltd.

(cf. Fig. 2). Since 1994 ozone has been continuously measured from regular aircraft within the MOZAIC project. The analysis of these measurements shows a considerable increase in upper tropospheric and lower stratospheric ozone (Thouret et al. 2006). Another study of the same data set shows that the tropospheric ozone column amount measured over airports has increased as well (Zbinden et al. 2006). Both investigations indicate that the increase extends over much of the northern extratropics. An ozone increase of similar magnitude has been found in the surface ozone measurements at Mace Head, a station at the west coast of Ireland (Simmonds et al. 2005). All measurements including background surface stations and measurements in the free troposphere show that ozone trends are most pronounced in winter. Note that European ozone sonde measurements do not support this increase, possibly because of lower sampling frequencies, lower precision of the sensors and possibly limitation in data quality (Ordóñez 2006). Different causes for this remarkable increase have been discussed in the literature: (i) effects of more frequent biomass burning,

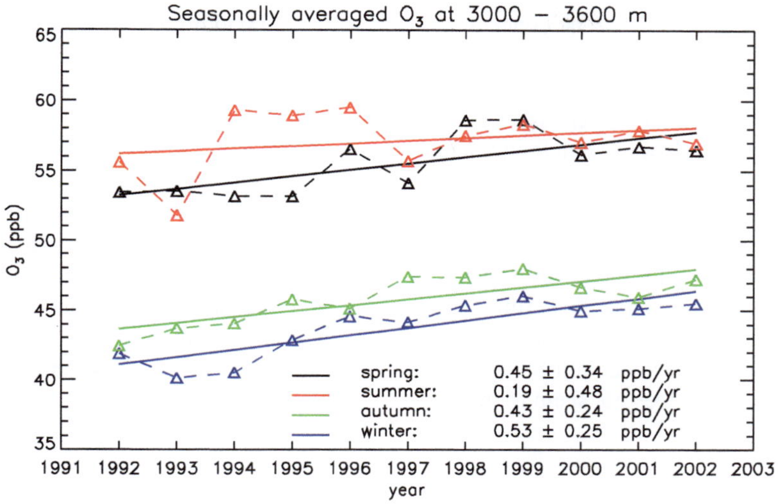

Fig. 4 Time series of seasonal mean ozone values of three European high mountains sites (Jungfraujoch, Zugspitze, and Sonnblick) and seasonal trends (from Ordóñez 2006).

(ii) changes in transport of air masses over the Atlantic in association with changes of the North Atlantic Oscillation (note that correlations between the North Atlantic Oscillation (NAO) and column ozone amount over Europe, as well as between the Arctic Oscillation (AO) and springtime ozone amount over the northeastern Atlantic for the period 1979–2000 were found using satellite data (Creilson et al. 2003, 2005 and references therein), (iii) influence of the strongly increasing emissions from South East Asia, and (iv) changes in the transport of ozone from the stratosphere, a parameter difficult to describe in numerical simulations of the global ozone cycle using state-of-the-art chemical transport models (cf. Sect. 2). Indeed, during the 1990s, ozone substantially increased in the lowermost stratosphere at northern midlatitudes. A recent study (Ordóñez et al. 2007) showed strong correlations between the ozone anomalies in the lowermost stratosphere and ozone concentrations measured at three European high mountain stations in the 1990s. This suggests a major contribution of an increased flux of ozone from the stratosphere to the troposphere to the well-documented ozone background increase in Europe which occurred during the 1990s.

Ozone balloon measurements from Japan, as well as surface and free tropospheric ozone measurements on Japanese islands in the Pacific show increasing values since the 1970s. Such increases have been attributed to the increasing ozone precursor emissions in continental Asia such as China (Naja and Akimoto 2004). However, it was noted that the ozone increases from the ozone balloon series have not longer been statistically significant since the early 1990 (Oltmans et al. 2006).

The longest ozone sonde record from USA was measured at Wallops Island (38°N): it shows a hardly significant increase in free tropospheric ozone for the period 1970–2003. The record from Boulder, Colorado (40°N) shows an insignificant decrease in free tropospheric ozone concentrations for 1985–2003 (Oltmans et al. 2004). The more northern Canadian stations Resolute (75°N) and Churchill (59°N) show significant decreases in tropospheric ozone in the period 1980–1990 extending from the planetary boundary layer up to the tropopause, while during 1991–2001 increasing ozone concentrations were found, which are significant at most tropospheric altitudes (Tarasick et al. 2005). This temporal evolution is difficult to explain by changes in anthropogenic ozone precursor emissions. However, the temporal change in upper tropospheric ozone was similar to the changes observed in the lowermost stratosphere suggesting that long-term tropospheric ozone trends are considerably influenced by changes in ozone in the lowermost stratosphere. At polar sites, ozone in the planetary boundary layer sometimes suddenly decreases in spring (Oltmans et al. 2006). This phenomenon has been attributed to biogenic emissions of halogen containing compounds, but it does not seem to influence long-term trends.

5 Ozone Trends in the Tropics

Available ozone measurements from the tropics suggest variable trends depending on regions. Upward trends were reported from long-term measurements from ship cruises over the Atlantic for 1978–2004 showing largest increases in the latitude bands of 20°–40°N (Fig. 5) (Lelieveld et al. 2004). Twice as large increases have been

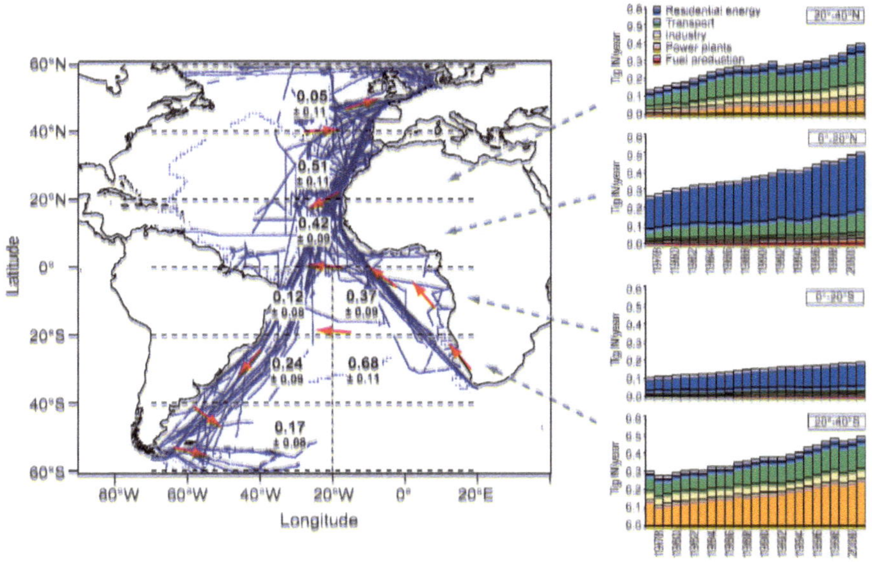

Fig. 5 Ozone trends derived from measurements from ships (Lelieveld et al. 2004). Ship tracks are shown in blue, derived trends for 20° latitude bands in ppb ozone per year. The increases in anthropogenic NO_x emissions in different regions are shown at the right side. Reprinted with permission from AAAS.

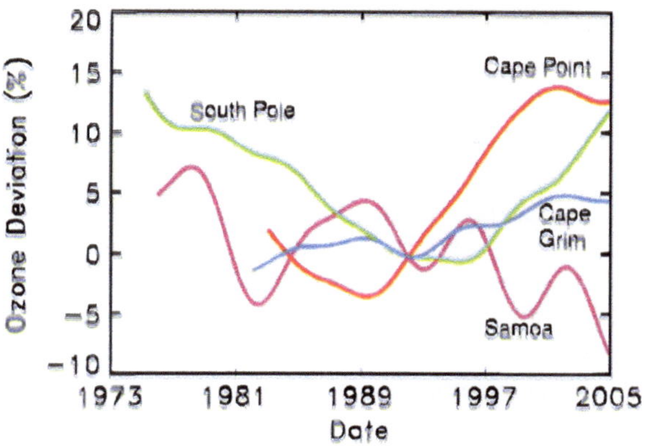

Fig. 6 Long-term ozone changes including southern extratropics measured at several surface sites (from Oltmans et al. 2006). The ozone deviations are derived from a statistical model described in Harris et al. (2001). Copyright 2006 Elsevier Ltd.

found for the free tropical troposphere over the Atlantic from the MOZAIC measurements starting in 1994 (Bortz et al. 2006). In both papers these large increases were attributed to increasing fossil-fuel emissions in the regions surrounding the Atlantic.

Surface-ozone measurements from the Mauna Loa Observatory at Hawaii available since 1974, and ozone sonde measurements from Hilo, available since 1982, show moderate increases. Long-term backward trajectory analysis have provided evidence that changes in the origin of air are the most likely primary factor for the observed increase (Oltmans et al. 2006).

6 Ozone Trends in the Southern Extratropics

The knowledge on long-term tropospheric ozone trends in the Southern extratropics is strongly limited because of the restriction of measurements. The ship cruise measurements of Fig. 5 also show upward near-surface trends in the Southern mid-latitude Atlantic region, although less pronounced than in the tropics. The only available sonde station operated at the remote site of Lauder in New Zealand shows a continued increase in the middle troposphere since 1986, which is qualitatively similar to the surface ozone increase reported from the long-term stations Cape Grim (Australia) and Cape Point (South Africa) (Fig. 6). Particularly interesting is also the temporal evolution of surface ozone at the South Pole showing a continuous decrease from the middle of the 1970s until the middle of the 1990s followed by an increase. However, recent studies have indicated that particular processes are important in Antarctica possibly influencing ozone trends (see Oltmans et al. 2006 including references).

7 Conclusions

Ozone in the troposphere is an important greenhouse gas, and numerical simulations show that its changes due to the emissions of anthropogenic precursors have significantly contributed to radiative forcing. While the concentrations of long-lived greenhouse gases such as carbon dioxide, methane, and others can be determined from ice cores, no reliable measurements of pre-industrial tropospheric ozone exist. According to our knowledge the oldest reliable measurements are available from European sites, which show a large increase on the order of a factor two from the late 1950s until the early 1990s, which can most probably be attributed to the large increase in anthropogenic emissions. Measurements from northern midlatitudes show a further increase in free tropospheric ozone since the early 1990s extending from North America to Europe (Zbinden et al. 2006) including North Canada, where increases can hardly be attributed to regional changes of anthropogenic ozone precursors. The scientific reasons of these upward trends are still discussed controversially. The simultaneous large increase in lowermost stratospheric ozone is consistent with the speculation that increased amounts of ozone imported from the stratosphere could be an important cause of the observed increase. This interpretation seems to be supported by recent data analyses (Tarasick et al. 2005; Ordóñez et al. 2007), but present chemical transport models have problems to adequately describe the influence of stratosphere–troposphere exchange. Over the tropical region of the Atlantic strong increases in ozone have been documented since the early 1980s, which have been attributed to the increase of anthropogenic fossil-fuel emissions. However, in many other tropical regions reliable long-term measurements are not available. Thus, it seems premature to draw overall conclusions for the tropical regions. The same is true for the southern midlatitudes, where also only very few long-term measurements are available. This shows the strong demand for reliable long-term observations, particularly for many remote sites. In addition, the very valuable monitoring from regular aircraft needs to be extended into other parts in the world.

References

Brönnimann, S., B. Buchmann, and H. Wanner, 2002: Trends in near surface ozone concentrations in Switzerland. *Atmos. Environ.*, **36**, 2841–2853.

Bortz, S. E., M. J. Prather, J.-P. Cammas, V. Thouret, and H. Smit, 2006: Ozone, water vapor, and temperature in the upper tropical troposphere: variations over a decade of MOZAIC measurements. *J. Geophys. Res.*, **111**, DO5305, doi:10.1029/2005JD006512.

Creilson, J. K., J. Fishman, and A. E. Wozniak, 2003: Intercontinental transport of tropospheric ozone: a study of its seasonal variability across the North Atlantic utilizing tropospheric residuals and its relationship to North Atlantic Oscillation. *Atmos. Chem. Phys.*, **3**, 2053–2066.

Creilson, J. K., J. Fishman, and A. E. Wozniak, 2005: Arctic Oscillation-induced variability in satellite-derived tropospheric ozone. *Geophys. Res. Lett.*, **32**, L14822, doi:10.1029/2005 GRL023016.

Haagen-Smith A. J., 1952: Chemistry and physiology of Los-Angeles Smog. *Ind. Eng. Chem.*, **44**, 1342–1346.

Harris, J. M., S. J. Oltmans, P. P. Tans, R. D. Evnas, D. L. Quincy, 2001: A new method for describing long-term changes in total ozone. *Geophys. Res. Lett.*, 28, 4535–4538.

IPCC, 2001: Climate Change: The Scientific Basis, Report from the working group I, Contribution of Working Group I to the Third Assessment Report of the Intergovernmental Panel on Climate Change. J. T. Houghton et al., eds., Cambridge University Press.

Klausen, J., C. Zellweger, B. Buchmann, and P. Hofer, 2003: Uncertainty and bias of surface ozone measurements at selected Global Atmosphere Watch sites. *J. Geophys. Res.*, **108**, doi:10.1029/2003JD003710.

Lelieveld, J., J. van Aardenne, H. Fischer, M. de Reus, J. Williams, and P. Winkler, 2004: Increasing ozone over the Atlantic Ocean. *Science*, **304**, 1483–1486.

Logan, J. et al., 1999: Trends in vertical distribution of ozone: a comparison of two analyses of ozone sonde data. *J. Geophys. Res.*, **104**, 26373–26399.

Marenco, A. et al., 1998: Measurement of ozone and water vapour by Airbus inservice aircraft: the MOZAIC program, an overview. *J. Geophys. Res.*, **103**, 25631–25642.

Naja, M. and H. Akimoto, 2004: Contribution of regional pollution and long-range transport to the Asian-Pacific region: analysis of log-term ozone sonde data over Japan. *J. Geophys. Res.*, **109**, doi:10.1029/2004JD004687.

Nastrom, G. D., 1979: Ozone in the upper troposphere from the GASP measurements. *J. Geophys. Res.*, **84**, 6383–6388.

Ordóñez, C., 2006: Trend analysis of ozone and evaluation of nitrogen dioxide satellite data in the troposphere over Europe, Doctoral thesis, 16544, ETHZ.

Ordóñez, C., D. Brunner, J. Staehelin, P. Hadjinicolaou, J. A. Pyle, M. Jonas, H. Wernli, and A. S. H. Prévôt, 2007: Strong influence of lowermost stratospheric ozone on lower free tropospheric ozone changes over Europe. *Geophys. Res. Lett.*, **34**, L07805, doi:10.1029/2006GL029113.

Oltmans, S. et al., 1998: Trends of ozone in the troposphere. *Geophys. Res. Lett.*, **25**, 139–142.

Oltmans, S. et al., 2004: Trends in tropospheric ozone. In: *Proc. Quadr. Ozone Symp., Kos, Greece, Atehns, Greece*, vol. 1, Zerefos CD, ed., 199–200.

Oltmans, S. et al., 2006: Long-term changes in tropospheric ozone. *Atmos. Environ.*, **40**, 3156–3173.

Pulles, T., M. van het Bolscher, R. Brand, and A. Visschedijk, 2007: Assessment of global emissions from fuel combustion in the final decades of the 20th century, Application of the Emission Inventory Model TEAM, TNO Built Environment and Geosciences, TNO report 2007-A-R0132/B.

Schönbein, C. F., 1844: Abhandlungen der Bayrischen Akademie der Wissenschaften Naturwissenschaftlich- mathematische Klasse, München **1844**, 257.

Schnadt Poberaj, C., J. Staehelin, D. Brunner, V. Thouret, and V. Mohnen, 2007:A UT/LS ozone climatology of the nineteen seventies deduced from the GASP aircraft measurement program. *Atmos. Chem. Phys. Discuss*, **7**, 3451–3517.

Simmonds, P. G., A. J. Manning, R. G. Derwent, P. Ciais, M. Ramonet, V. Kazan, and D. Ryall, 2005: A burning question. Can recent growth rate anomalies in the greenhouse gases be attributed to large-scale biomass burning events. *Atmos. Environ.*, **39**, 2513–2517.

Staehelin, J., 2002: Ozone measurements and trends (Troposphere). In: *Encyclopedia of Physical Sciences and Technology*, 3rd edn., vol. 11, Academic Press.

Staehelin, J., A. S. H. Prévôt, and J. Barnes, 2000: Photochemie der Troposphäre. In: *Handbuch der Umweltveränderungen und Ökotoxikologie, Band IA: Atmosphäre*, Guderian R, ed., Springer Verlag, 207–341.

Staehelin, J., J. Thudium, R. Bühler, A. Volz-Thomas, and W. Graber, 1994: Surface ozone trends at Arosa (Switzerland), *Atmos. Environ.*, **28**, 75–87.

Stevenson, D. S. et al., 2006: Multimodel ensemble simulations of present-day and near future tropophsheric ozone. *J. Geophys. Res.*, **111**, D08301, doi:10.1029/2005JD006338.

Tarasick, D. W., V. E. Fioletov, D. I. Wardle, J. B. Kerr, and J. Davies 2005: Changes in the vertical distribution of ozone over Canada from ozonesondes: 1980–2001. *J. Geophys. Res.*, **110**, doi:10.1029/2004JD004643.

Thouret, V., J.-P. Cammas, B. Sauvage, B. Athier, R. M. Zbinden, P. Nédélec, P. Simmon, and F. Karcher, 2006: Tropopause referenced ozone climatology and inter-annual variability (1994–2003) from MOZAIC programme. *Atmos. Chem. Phys.*, **6**, 1033–1051.

Volz, A. and D. Kley, 1988: Evaluation of the Montsouris series of ozone measurements made in the nineteenth century. *Nature*, **332**, 240–242.

Zbinden, R. M., J.-P. Cammas, V. Thouret, P. Nédélec, F. Karcher, and P. Simmon, 2006: Mid-latitude tropopheric ozone columns from the MOZAIC program: climatology and interannual variability. *Atmos. Chem. Phys.*, **6**, 1053–1073.

Zellweger, C., J. Forrer, P. Hoer, S. Nyeki, B. Schwarzenbach, E. Weingartner, A. Ammann, and U. Baltensperger, 2003: Partitioning of reactive nitrogen (NOy) and depenedence on meteorological conditions in the lower free troposphere. *Atmos. Chem. Phys.*, **3**, 779–796.

Simulation of Long-term Evolution of Stratospheric Dynamics and Chemistry: Role of Natural and Anthropogenic Forcings

M. Dameris and R. Deckert

Abstract In recent years numerical models describing dynamical, physical and chemical processes of the Earth's atmosphere have been significantly improved. Simulations with so-called Climate–Chemistry Models (CCMs) do not only allow investigations of single processes but also enable analyses of feedback mechanisms. Detailed inspections of model results with observations can help to identify gaps in our understanding and to improve our knowledge of atmospheric processes. This is the basis for model systems which are needed for estimates of the future evolution of atmospheric behaviour. In the following some examples are discussed which clearly indicate the need for such models. The role of natural as well as man-made forcings of the atmospheric system is examined. Moreover, the interaction of climate change effects and atmospheric chemistry, especially ozone, is demonstrated with results derived from multi-year CCM simulations.

1 Introduction

The importance of the coupling between dynamical, physical and chemical processes in the stratosphere has been recognised for many years. For example, investigations of the relations and feedbacks between ozone (O_3) depletion and climate-change processes that control them have demonstrated that it is not possible to achieve a complete understanding of recent ozone changes without the consideration of dynamical and physical processes and their interaction with chemical processes. An increase of well-mixed greenhouse gas concentrations (e.g., carbon dioxide: CO_2; methane: CH_4; nitrous oxide: N_2O) in the atmosphere leads to higher tropospheric temperatures (greenhouse effect) and lower stratospheric temperatures. The rates of many chemical reactions are temperature dependent, and these reaction rates affect the chemical composition of the atmosphere. Reduced stratospheric temperatures lead to a slowing of some gas-phase reactions that destroy ozone, but also lead to intensified depletion of ozone in the lower

Deutsches Zentrum für Luft- und Raumfahrt e.V., Institut für Physik der Atmosphäre, Oberpfaffenhofen, Germany
martin.dameris@dlr.de
rudolf.deckert@dlr.de

polar stratosphere due to increased activation of halogens (e.g., chlorine: Cl, bromine: Br) on polar stratospheric clouds (PSCs). Since climate change also influences the dynamics of the troposphere and the stratosphere, dynamically induced temperature changes could locally reinforce or oppose the temperature changes caused by radiative processes. The net effect of radiative, chemical and dynamical feedbacks is poorly understood and quantified at present because many of the interactions are non-linear. Assessments of future stratospheric composition and climate change are thus uncertain.

Numerical models are useful for investigations of chemical composition, and thermal and dynamical structure of Earth's atmosphere. They allow evaluations of various processes and mechanisms as well as feedbacks. Scientific progress can be achieved by understanding the discrepancies between observations and results derived from model simulations. Assessments of the future development of atmospheric dynamics and chemistry are typically based on scenario simulations and sensitivity studies. During recent years a number of Climate–Chemistry Models (CCMs) have been employed to address the questions how dynamical, physical and chemical processes interact, and how natural and anthropogenic forcings affect climate and atmospheric composition, especially the evolution of the ozone layer and ozone recovery (e.g., Eyring et al. 2006, 2007).

2 Climate–Chemistry Models (CCMs)

CCMs are adequate numerical tools to investigate the impact of climate change on atmospheric composition and vice versa, since these models do consider interactions of radiative, dynamical and chemical processes. They have been used for long-term simulations to study recent and future changes. In a CCM the simulated concentrations of radiatively active gases are used in the calculations of net heating rates. Changing abundances of these gases due to chemistry and advection, influence heating rates and, consequently, variables describing atmospheric dynamics such as temperature and wind. This gives rise to a dynamical-chemical coupling in which chemistry influences dynamics (via radiative heating) and vice versa (via temperature and advection). Not all CCMs consider the full coupling for all chemical constituents – some radiatively active gases are specified in either the climate or the chemistry modules. Ozone is always fully coupled as it represents the dominant radiative-chemical feedback in the stratosphere. In particular, transient CCM simulations consider observed or predicted gradual changes in concentrations of radiatively active gases and other boundary conditions (e.g., emissions). The temporal development of sea surface temperatures (SSTs) are prescribed for a specific episode (years to decades), since CCMs are not coupled to ocean models so far. For example, the CCM E39/C has been employed for an ensemble simulation covering the time period between 1960 and 2020. Three transient simulations have been performed for the years 1960–1999, four for 2000–2020 (Dameris et al. 2005, 2006). As boundary conditions serve observed and predicted changes with respect to SST and ice cover, concentrations of radiative active gases (e.g., CO_2, CH_4, N_2O) and chlorine compounds. Natural and anthropogenic nitrogen oxide (NO_X) emissions at

Earth's surface, and NO_x emission from lightning and air traffic are considered. Additionally, the effects of large volcanic eruptions (Agung in year 1963; El Chichón in year 1982; Pinatubo in year 1991) are considered in both, chemistry module and radiation scheme of E39/C. The solar cycle is taken into account and the quasi-biennial oscillation (QBO) is nudged (only between 10°S and 10°N) according to observations of the equatorial wind field.

3 Results

Figure 1 displays zonal mean temperature changes in the northern and southern polar stratosphere during the 60 model years. The lines represent means of the E39/C ensemble simulation separately for each decade during the period covered

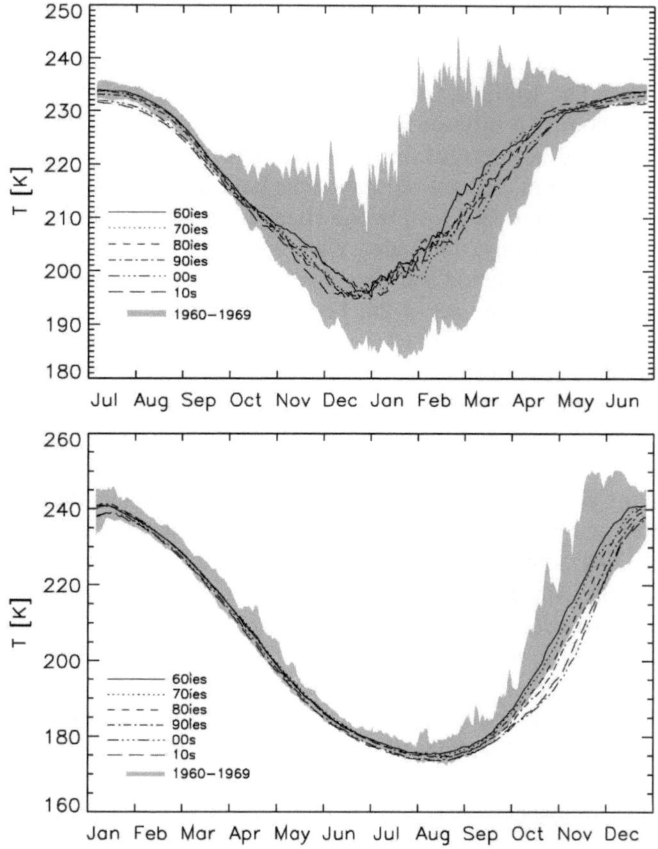

Fig. 1 Climatological mean curves of the annual cycle of zonal mean temperature (in K) as simulated with the CCM E39/C in 80°N (top) and 80°S (bottom) at 30 hPa. Different lines indicate decadal mean values for the six decades, i.e., the 1960s, 1970s, 1980s, 1990s, 2000s, and 2010s. The shaded areas indicate the range of minimum and maximum temperatures within the 1960s.

by transient simulations. The shaded area denotes the minimum and maximum values calculated for the 1960s. A comparison with respective long-term observations (e.g., analyses of the National Centre for Environmental Prediction: NCEP, see web page: http://code916.gsfc.nasa.gov/Data_services/met/nmc_stats.html) shows that the model is able to reproduce hemispheric differences in dynamical variability changes. While the annual cycle is well captured in both hemispheres, and the mean values in the northern hemisphere agree well with respective analyses derived from observations, the model shows deficiencies in the southern hemisphere (i.e., too cold lower stratosphere, understated inter-annual variability) which are due to a cold bias (Dameris et al. 2005). With regards to the high dynamical variability of the northern hemisphere during winter and spring, E39/C shows reduced inter-annual variability in early winter (December and January). Nevertheless, E39/C is able to reproduce stratospheric warmings in mid and late winter, but the number of major events is smaller than observed. In the northern hemisphere the climatological mean values for the six decades neither show systematic changes nor statistically significant differences, whereas in the southern hemisphere the lower stratosphere changes steadily towards colder conditions in late winter and early spring (i.e., more stable polar vortices). The lifetime (persistence) of the south polar vortex has been prolonged by about two to three weeks from the 1960s to the 1990s which is in agreement with analyses derived from observations (e.g., Zhou et al. 2000).

Short- and long-term changes of total ozone have been investigated by means of the ensemble simulation. Past total ozone changes are well simulated on both long (decadal) and short (monthly) timescales. There are many apparent features of episodic similarities between simulation and observation (for details see Dameris et al. 2005, 2006). For example, the impact of the 11-year solar cycle on stratospheric ozone has been investigated. Analyses of total ozone anomalies in the tropics (25°N–25°S), which are based on combined observations of the years from 1964 to 2005, suggest that decadal variations of tropical total ozone correspond with the solar activity. The maxima and minima are nearly in phase with the 11-year solar cycle (e.g., WMO 2003, 2007). Calculated ozone anomalies in the tropical belt derived from E39/C develop similarly to observations, with positive anomalies around solar maxima (1981 and 1991) and negative anomalies around solar minima (1986 and 1996). The E39/C simulations consider the effects of changing photolysis rates with solar activity. Since E39/C has an upper boundary which is centred at 10 hPa, solar induced chemical ozone changes are additionally forced by prescribing the time evolution of NO_Y at the model top. (Monthly mean concentrations are taken from results of the 2D model by Brühl and Crutzen (1993) which also considers solar cycle effects; see Dameris et al. 2005.) Interestingly, after the eruption of volcano Agung in 1963, the total amount of ozone was obviously reduced, particularly in the tropics. Although the chemical "forcing" is expected to be weak since the chlorine loading of the atmosphere was low at that time, the direct radiative effects of enhanced stratospheric aerosol abundance has an influence. Additionally, the solar cycle influences were of particular importance. E39/C reproduces the observed decrease of tropical ozone in the mid-1960s, when solar activity was minimal.

Global mean total ozone anomalies derived from observations and E39/C simulations for 1960–2020 are shown in Fig. 2. In the years between 1964 and 2003, the range of model results (red shading) is mostly congruent with observations. Not only simulated long-term changes match the observations, but also short-term variability is well reproduced. A notable example is the rapid change towards lower ozone values after the eruption of Pinatubo in June 1991 and the year-to-year fluctuations in the following years. It is obvious that volcanic eruptions strongly influence dynamics and chemistry, though only for a few years. Again, the impact of the 11-year solar cycle is clearly visible. Following a period of increasing total ozone between the mid-1990s and 2003, E39/C simulates an episode of about six more years with quite low global mean ozone. Minimum total ozone values are reached between 2004 and 2009, which is around the upcoming solar minimum. In the model the observed increase of ozone in the second half of the 1990s can be explained as a combined post-Pinatubo and solar cycle effect rather than the beginning of a sustainable recovery of the ozone layer caused by less stratospheric chlorine. The model results indicate that solar activity may delay the beginning of a sustained ozone recovery (Dameris et al. 2006).

It is not possible to achieve a complete understanding of stratospheric composition changes without the consideration of climate change and its feedback. The long-term evolution of stratospheric ozone concentrations depends not only on changes of many stratospheric constituents (including ozone depleting substances (ODSs), greenhouse gases, water vapour, and aerosols), but also on changes in the climate of the troposphere and stratosphere caused by natural variability and anthropogenic forcings. While it is expected that the reduction of ODSs in the next years to decades will lead to an increase in ozone, this increase could be affected by changes in temperature and in chemical composition and transport.

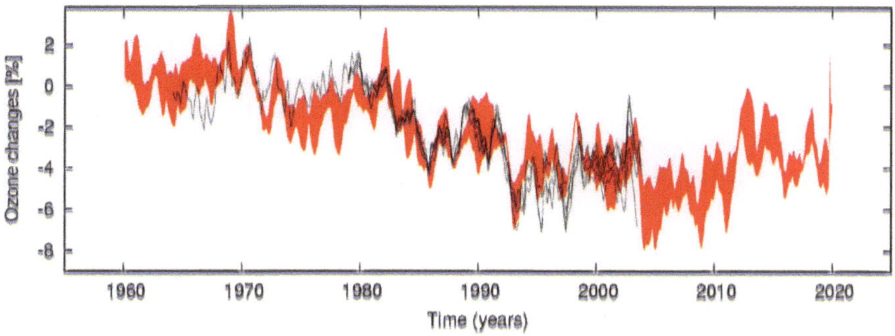

Fig. 2 The black curves show deseasonalised, area-weighted seasonal (3-month average) total ozone deviations, derived from different observations. The red shaded area indicates the minimum/maximum values derived from E39/C simulations. All data sets are deseasonalised with respect to the period 1979–1987, and deviations are expressed as percentages of the time average for 1964–1980. Results are shown for the entire globe (90°S–90°N) (from Dameris et al. 2006).

To identify and quantify the impact of climate change on the future evolution of the ozone layer the three CCMs E39/C, ULAQ (e.g., Pitari et al. 2002) and WACCM (e. g., Beres et al. 2005) have been used in a common way: In addition to so-called reference (REF) simulations which assume continuous changes in greenhouse gas concentrations (see Eyring et al. 2006), sensitivity simulations have been especially performed for WMO (2007) wherein concentrations of well-mixed greenhouse gases (CO_2, CH_4, and N_2O) are held constant after a given date (NCC). The evolution of halocarbon is based on the "Ab" scenario from WMO (2003) in both the reference and the sensitivity simulation. In the reference simulations the SSTs follow observations and predicted values from a coupled atmosphere ocean model, whereas in the sensitivity simulations with E39/C and ULAQ the SSTs are fixed according to observations of the years 1970–1979 and this SST data set is used again for every succeeding decade in the model simulation. A different approach was chosen in the WACCM sensitivity simulation, where SSTs from a coupled ocean-atmosphere model are used that was run in support of the IPCC assessment (2001). In all three CCMs ozone and water vapour are prognostic variables, that is, the radiative feedback of ozone and water vapour changes is considered. A direct comparison of results from the sensitivity simulations with results from reference simulations allows a qualitative estimation of the future impact of climate change on ozone (for details see Sect. 5.3.7 in WMO 2007).

The CCM results derived from the reference simulations all show an obvious cooling of the stratosphere because of increasing greenhouse gas concentrations whereas a nearly zero trend is calculated by all three CCMs when fixed greenhouse gas concentrations are assumed. The results with regard to long-term ozone changes demonstrate that the recovery of global mean column ozone is clearly accelerated in a changing climate (Fig. 3): Since the global stratosphere is under photochemical control (constrained by gas-phase reaction cycles), the ozone loss cycles in the stratosphere (primarily O_x) slow down with decreasing temperature. When climate change is considered, CCM reference simulations show a return to the 1980 global ozone amount several years earlier than without climate change (Chapter 5 in WMO 2007). The results derived from the CCM E39/C are not as clear in this sense, but here it must be considered that in contrast to ULAQ and WACCM, E39/C has a low uppermost layer which is centered at 10 hPa, and therefore neglects the impact of dynamical and photochemical effects on ozone in the upper part of the stratosphere. In any case, the E39/C simulations were not long enough for the difference between REF and NCC simulations to become apparent. CCM results show that around the year 2050 the middle to upper stratosphere is about 5–10 K colder in the reference runs than in corresponding sensitivity simulations (not shown). As a result of this cooling, middle and upper stratospheric ozone mixing ratios in the reference simulations exceeded those calculated in the sensitivity simulations (e.g., up to 15% below the stratopause). The most important impact is via the effect of temperature on the ozone loss rate. Below 10 hPa the model results indicate a slight delay of ozone recovery in the Antarctic lower stratosphere (not shown) due to intensified heterogeneous chemistry caused by reduced temperatures in the lower polar stratosphere. So, lower temperatures in the ozone loss regions of high latitudes have the opposite effect to lower temperatures in the ozone production region in the tropics.

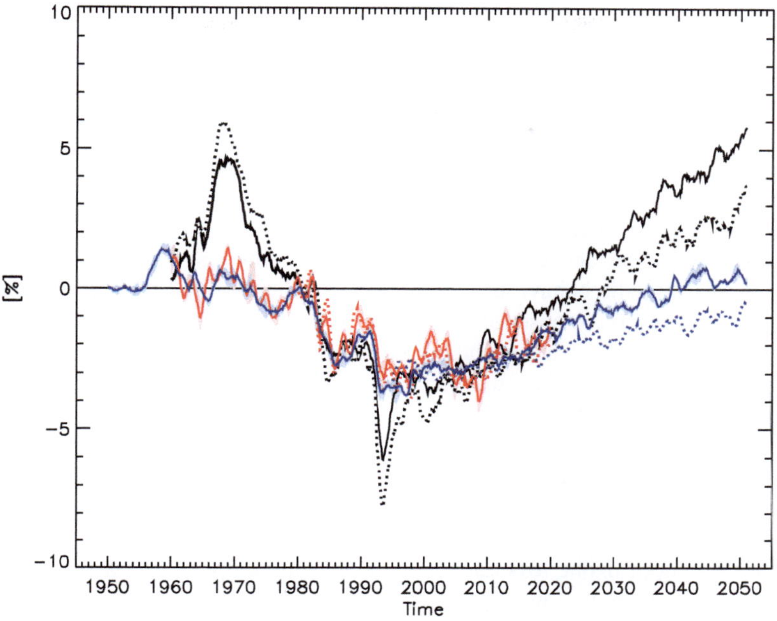

Fig. 3 Time series of zonally averaged near-global (60°N–60°S) total ozone deviations between 1960 and 2050 with regard to the year 1980 (in %). For each model, i.e., E39/C (red lines), ULAQ (black lines), and WACCM (blue lines), the solid curves show results derived from the reference simulations (REF). Since WACCM and E39/C have performed ensemble simulations, the envelope of results is shown for each model. The dotted curves indicate the results derived from the "no climate change runs" (NCC). All data are smoothed with a 13-month running mean; Fig. 5–25 in WMO (2007).

Interestingly, the analysis of all CCM simulations indicates that the Brewer-Dobson circulation will be intensified in a future climate with enhanced greenhouse gas concentrations. The results show an increased upwelling in the tropical upper troposphere and stratosphere which is in agreement with several other simulations of the climate response to increasing greenhouse gas concentrations (e.g., Butchart and Scaife 2001; Gillett et al. 2003; Langematz et al. 2003; Sigmond et al. 2004; Austin and Li 2006; Fomichev et al. 2007). Moreover, the increasing upwelling decreases lower tropical stratospheric ozone via the enhanced transport of ozone-poor air from the tropical tropopause region which is also apparent in the model results. In the winter extra-tropical stratosphere the poleward meridional circulation is increased as well as the downwelling at high latitudes. The transport of air with enhanced ozone mixing ratios to higher latitudes by an intensified Brewer–Dobson circulation seems to partly compensate the effect of enhanced Br/Cl catalysis of ozone in the polar region. A more detailed analysis including a quantification of the effects of single processes is needed before more reliable statements can be given.

4 Summary

An ensemble simulation with the CCM E39/C shows that the overall temporal and spatial development of observed atmospheric parameters can be reproduced. E39/C results confirm that it is necessary to consider major forcings (i.e., due to changes in greenhouse gas and halogen concentrations, SST and ice cover, the QBO, the solar cycle, major volcanic eruptions) to obtain a realistic short- and long-term variability of dynamical and chemical values and parameters. The model obviously shows a deterministic behaviour with regard to the prescribed forcing. Future estimates with regard to the onset of ozone recovery indicate that the solar cycle effect delays the beginning of a sustainable ozone increase for some years. Minimum ozone values are expected around the upcoming minimum of solar activity which is expected for 2007/08.

Sensitivity studies with different models including CCMs have shown that future increases of greenhouse gas concentrations will enhance the average cooling in the stratosphere. This involves a slow down of the ozone destruction rates via the major chemical loss cycles (HO_x, NO_x, ClO_x). CCM simulations consistently indicate that the slowing of ozone loss rates with decreasing temperatures will be offset by ozone losses in the polar lower stratosphere due to elevated concentrations of active halogen species. There are clear indications of an intensified Brewer–Dobson circulation in an atmosphere with enhanced greenhouse gas concentrations which lead to an increased transport of ozone rich air into the polar winter lower stratosphere. Altogether, the global amount of total ozone will recover more quickly by several years if stratospheric temperatures further decrease.

Acknowledgements The authors wish to acknowledge the European Commission for funding the SCOUT-O3 project. Special thanks go to Rolando Garcia and Doug Kinnison (NCAR), and Giovanni Pitari and Eva Mancini (University of L'Aquila) for the provision of WACCM and ULAQ data for the comparison of model results and for common interpretation of the climate change effects on the evolution of the ozone layer. Thanks also to the ECHAM team (in particular Veronika Eyring, Volker Grewe, Sigrun Matthes, Michael Ponater, Andrea Stenke) for supporting the E39/C analyses and for helpful discussions.

References

Austin, J. and F. Li, 2006: On the relationship between the strength of the Brewer-Dobson circulation and the age of stratospheric air. *Geophys. Res. Lett.*, **33**, L17807, doi:10.1029/2006GL026867.

Beres, J. H., R. R. Garcia, B. A. Boville, and F. Sassi, 2005: Implementation of a gravity wave source spectrum parameterization dependent on the properties of convection in the Whole Atmosphere Community Climate Model (WACCM). *J. Geophys. Res.*, **110**, D10108, doi:10.1029/2004JD005504.

Brühl, C. and P. J. Crutzen, 1993: MPIC two-dimensional model. *NASA Ref. Publ.*, **1292**, 103–104.

Butchart, N. and A. A. Scaife, 2001: Removal of chlorofluorocarbons by increased mass exchange between the stratosphere and troposphere in a changing climate. *Nature*, **410**, 799–802.

Dameris, M., V. Grewe, M. Ponater, R. Deckert, V. Eyring, F. Mager, S. Matthes, C. Schnadt, A. Stenke, B. Steil, C. Brühl, and M. A. Giorgetta, 2005: Long-term changes and variability in a transient simulation with a chemistry-climate model employing realistic forcing. *Atmos. Chem. Phys.*, **5**, 2121–2145.

Dameris, M., S. Matthes, R. Deckert, V. Grewe, and M. Ponater, 2006: Solar cycle effects delays onset of ozone recovery. *Geophys. Res. Lett.*, **33**, L03806, doi:10.1029/2005GL024741.

Eyring, V., N. Butchart, and D. W. Waugh, et al., 2006: Assessment of temperature, trace species and ozone in chemistry-climate model simulations of the recent past. *J. Geophys. Res.*, **111**, 22308, doi:10.1029/2006JD007327.

Eyring, V., D. W. Waugh, and E. Bodeker, et al., 2007: Multimodel projections of stratospheric ozone in the 21st century. *J. Geophys. Res.*, **112,** D16303, doi:10.1029/2006 JD008332, 2007.

Fomichev, V. I., A. I. Jonsson, J. de Grandpré, S. R. Beagley, C. McLandress, K. Semeniuk, and T. G. Shepherd, 2007: Response of the middle atmosphere to CO_2 doubling: results from the Canadian Middle Atmosphere Model. *J. Climate*, **20,** 1121–1141.

Gillett, N. P., M. R. Allen, and K. D. Williams, 2003: Modelling the atmospheric response to doubled CO_2 and depleted stratospheric ozone using a stratosphere-resolving coupled GCM. *Q. J. Roy. Meteor Soc.*, **129**, 947–966.

IPCC, 2001: Climate Change 2001: *The Scientific Basis*: Contribution of Working Group I to the Third Assessment Report of the Intergovernmental Panel on Climate Change, J. T. Houghton et al., ed., Cambridge University Press, Cambridge, 881 pp.

Langematz, U., M. Kunze, K. Krüger, K. Labitzke, and G. L. Roff, 2003: Thermal and dynamical changes of the stratosphere since 1979 and their link to ozone and CO_2 changes. *J. Geophys. Res.*, **108**, 4027, doi:10.1029/2002JD002069.

Pitari, G., E. Mancini, V. Rizi, and D. T. Shindell, 2002: Impact of future climate and emission changes on stratospheric aerosols and ozone. *J. Atmos. Sci.*, **59**, 414–440.

Sigmond, M., P. C. Siegmund, E. Manzini, and H. Kelder, 2004: A simulation of the separate climate effects of middle-atmospheric and tropospheric CO_2 doubling. *J. Climate*, **17**, 2352–2367.

WMO (World Meteorological Organization), 2003: Scientific Assessment of Ozone Depletion: 2002. Global Ozone Research and Monitoring Project-Report, **47**, Geneva, Switzerland.

WMO (World Meteorological Organization), 2007: Scientific Assessment of Ozone Depletion: 2006. Global Ozone Research and Monitoring Project-Report, **50**, Geneva, Switzerland.

Zhou, S., M. E. Gelman, A. J. Miller, and J. P. McCormack, 2000: An inter-hemispheric comparison of the persistent stratospheric polar vortex. *Geophys. Res. Lett.*, **27**, 1123–1126.

Dynamical Changes in the Arctic and Antarctic Stratosphere During Spring

U. Langematz and M. Kunze

Abstract Short- and long-term changes in the intensity and persistence of the Arctic and Antarctic stratospheric polar vortices during spring have been analyzed, using NCEP/NCAR (National Centers for Environmental Prediction/National Center for Atmospheric Research) reanalyses. For the Arctic the results confirm the existence of low frequency variability in the winter stratosphere. During the 1980s and early to mid-1990s the northern hemisphere (NH) polar vortex was intensified in spring and broke up late. Since the late 1990s however, major stratospheric warmings occurred more frequently, so that the polar vortex in spring still intensified in March but with a smaller magnitude. As some of the major warmings occurred early in winter, the polar vortex was able to recover leading to late breakup dates in spite of the dynamical disturbances. In the long-term, there is no statistically significant change in Arctic vortex intensity or lifetime. In the Antarctic, the significant intensification of the polar vortex found in the 1980s and 1990s has been considerably reduced due to an unexpected enhancement of dynamical activity in southern hemisphere (SH) winter since 2000, masking the significant increase in polar vortex persistence found for the period 1979–1999. Still on the long-term, the Antarctic vortex shows a significant deepening and shift towards later spring transitions.

1 Introduction

Since the 1980s temperature measurements revealed a tendency of the Arctic and Antarctic stratosphere to cool during spring (e.g., Pawson and Naujokat 1997; Randel and Wu 1999; Ramaswamy et al. 2001). Rex et al. (2004) report that the minimum temperatures in cold, dynamically undisturbed Arctic winters have become lower over the past decades. These thermal changes have been accompanied by changes in stratospheric dynamics, like an enhancement of the Arctic polar night jet in the middle stratosphere during late winter and spring associated with an

Institut für Meteorologie, Freie Universität Berlin, Germany
Ulrike. Langematz@met.fu-berlin.de
Markus.Kunze@met.fu-berlin.de

S. Brönnimann et al. (eds.), *Climate Variability and Extremes during the Past 100 Years.*

intensification of the Arctic polar vortex (Kodera and Koide 1997; Langematz et al. 2003). In spring, both the Antarctic and the Arctic polar vortex broke down later than in earlier decades which was explained by a stabilizing effect of enhanced cooling associated with chemical ozone depletion in the Antarctic (Zhou et al. 2000; Shine 1986). In the Arctic, an additional decrease of tropospheric wave forcing was found (Coy et al. 1997; Waugh et al. 1999) leading to an unusual clustering of nine consecutive winters without major stratospheric warmings (Labitzke et al. 2002; Manney et al. 2005).

In recent years however, the winter stratosphere was dynamically more disturbed than during the 1980s and 1990s. Since 1998/99, major stratospheric sudden warmings took place in most Arctic winters, with a very strong one occurring in January 2006. Moreover, the Antarctic polar vortex featured unusual dynamic activity culminating in 2002 with the first Antarctic major warming ever observed (Naujokat and Roscoe 2005). Langematz and Kunze (2006) showed that the trends in the polar vortices and circulation, calculated for the 1980s to mid-1990s, were considerably reduced when adding the most recent years to the time series. The deepening of the Arctic polar vortex since 1979 still exists however with smaller magnitude. Surprisingly, the trend towards a later breakdown of the Arctic vortex was enhanced in spite of the increased frequency of major stratospheric warmings. This could be explained by the early timing of the major warmings in December and January, thus allowing the vortex to build up again in late winter and persist into spring. In the Antarctic, the trend towards an enhanced polar vortex in spring and later breakdown dates was severely reduced due to the accumulation of three dynamically active winters in the Antarctic since 2000.

The above developments are of scientific interest for two reasons:

- They occur at a time when measurements indicate for the first time a stagnation of stratospheric halogen concentrations thus suggesting a possible reduction of chemical ozone depletion in the near future. This leads directly to the question if the observed weak "ozone holes" in the years 2000, 2002 and 2004 are a first sign of polar ozone recovery or rather due to the enhanced dynamical activity leading to higher stratospheric temperatures in the Antarctic polar vortex and mixing of vortex with midlatitude air thus causing weaker ozone depletion.
- What is the reason for the unexpectedly high dynamical variability in the southern winter? The intensity of the polar vortex is determined by different factors, with planetary wave forcing from the troposphere being the major agent. Labitzke et al. (2006) showed that the state of the polar vortex also depends on the phase of the Quasi-Biennial Oscillation of the zonal wind in the tropical stratosphere and the level of solar activity. While for the northern hemisphere long-term oscillations of the occurrence frequency for stratospheric warmings are well known (Labitzke et al. 2002), a dynamically active period like the current one seems to be unusual for the Antarctic.

To address the above issues it is of particular importance to continuously monitor the dynamical development during stratospheric winter and spring. We focus here on changes of the polar vortex characteristics during spring, including the Arctic and Antarctic spring seasons 2006. In Sect. 2, we describe the data used for the analysis; results are presented in Sect. 3 for the Arctic spring and in Sect. 4 for the Antarctic spring. A summary follows in Sect. 5.

2 Data

In this study we use NCEP/NCAR reanalyses for the lower and middle stratosphere between 100 and 10 hPa (~16 and 30 km) from 1965 to 2006 (Kalnay et al. 1996). Trends are calculated by linear regression for three periods of interest: 1979–1999 (21 years, a frequently studied period due to the availability of satellite measurements and the increasing stratospheric ozone depletion), 1979–2006 (28 years, like the previous period but updated to 2006), and 1965–2006 (42 years, the longest period available with useful global stratospheric analyses). To exclude an impact on the derived trends of the three major volcanic eruptions of Mount Agung in 1963, El Chichòn in 1982 and Mount Pinatubo in 1991 we removed data for 2 years following the eruptions from the time series. The statistical significance of the linear trends was tested using a *Student's t*-test (Brownlee 1965).

Langematz and Kunze (2006) discussed in detail the reliability of reanalyses for the period before the assimilation of satellite measurements in 1979, in particular for long-term trend calculations. They found by comparison with the radiosonde-based FUB analyses (Labitzke et al. 2002) that in spring trends at middle and high latitudes are in reasonable agreement also for the extended period. Note however that measurements in the Antarctic during the early years have been sparse thus limiting the credibility of the derived trends for the Antarctic in the extended period.

3 Arctic Spring

Figure 1 shows the decadal change of the polar vortex intensity in terms of geopotential height at 30 hPa (~25 km) in March. Three periods are compared: 1979–1999, 1979–2006 and 1965–2006. Since 1979 the Arctic polar vortex has intensified. The magnitude of the intensification strongly depends on the considered period. During the 1980s and 1990s, when only few major stratospheric warmings took place, the polar vortex was up to 28 dam/decade stronger in the middle atmosphere in March. This polar vortex enhancement was accompanied by a cooling of the polar cap in the lower stratosphere due to a decrease in total eddy heat flux entering from the troposphere in January and February, as shown in Langematz and Kunze (2006, their Fig. 7). The connection between tropospheric forcing in midwinter and stratospheric temperature and polar vortex evolution in spring was discussed in detail by Newman et al. (2001).

The intensification of the polar vortex is still evident but weaker for the extended period until 2006 due to the enhanced dynamical variability in the recent years.

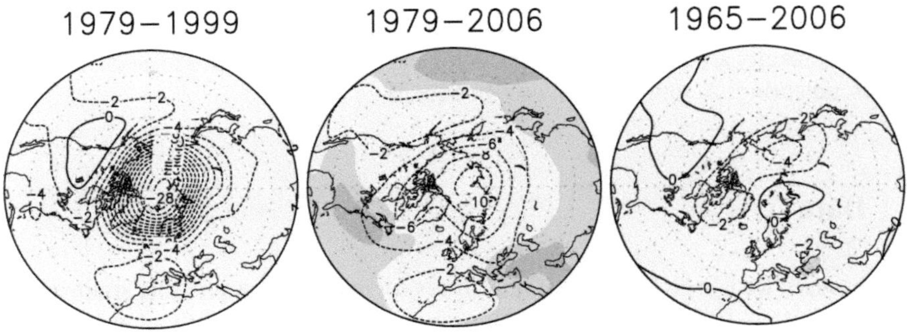

Fig. 1 Decadal trends in geopotential height (dam/decade) at 30 hPa in Arctic spring (March) between 20°N and 90°N calculated from NCEP/NCAR reanalyses for the periods 1979–1999, 1979–2006, and 1965–2006. Contour interval: 2 dam/decade. Shadings denote the 95% and 99% significance levels (*Student's t*-test).

However, as some of the recent major warmings occurred rather early in winter, as for example, in 2006, the polar vortex was able to recover again leading to well developed polar vortices in March. Thus the extended period 1979–2006 still displays an intensification of the polar vortex in spring. In the full time period (1965–2006) the magnitude of these changes is further reduced, and the changes are statistically not significant as long-term decadal oscillations exist in the stratosphere which lead to periods with opposite trends and thus partly offset the strongly negative trends of the 1980s–1990s.

Strong interannual variations combined with long-term oscillations in Arctic spring are also evident in Fig. 2 that shows the time series of the geopotential height in March from 1965 to 2006 averaged over the polar cap (60°–90°N) in the middle stratosphere (30 hPa, ~25 km). Superimposed are linear trends for the periods 1979–2006 and 1965–2006. Neither the decrease in geopotential height since 1979 nor the zero change since 1965 is statistically significant due to the large variations in single years.

The date of the Arctic polar vortex spring breakup is displayed in Fig. 3. The breakup date is defined as the day of the year when the zonal mean westerlies at 65°N or S at 50 hPa decrease below a threshold value of 10 m/s. This definition is based on the observed change of zonal winds from westerlies to easterlies when the polar vortex breaks up. The specific threshold value, pressure level and latitude were chosen to approximate best the more sophisticated methods applied in Waugh et al. (1999). More details can be found in Langematz and Kunze (2006).

The interannual variability of the breakup date is high in the Arctic. The transition date varies over more than 2 months between day 56 (February 25) in 1999 and 121 (May 1) in 1968. The early years until 1970 showed more late transitions, while the 1970s to mid-1980s showed more early transitions. Since the mid-1990s the interannual variability in the Arctic is enhanced including the second latest transition in 1997 and the record early transition in 1999. In spring 2006, the Arctic vortex broke up late, on day 111 (April 21).

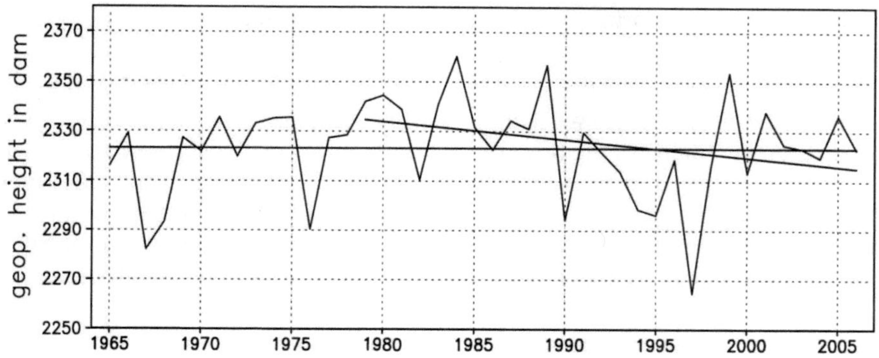

Fig. 2 Geopotential height (dam) in the Arctic (60°–90°N) at 30 hPa in March, calculated from NCEP/NCAR reanalyses from 1965 to 2006. Superimposed are linear trends for the periods 1979–2006 and 1965–2006.

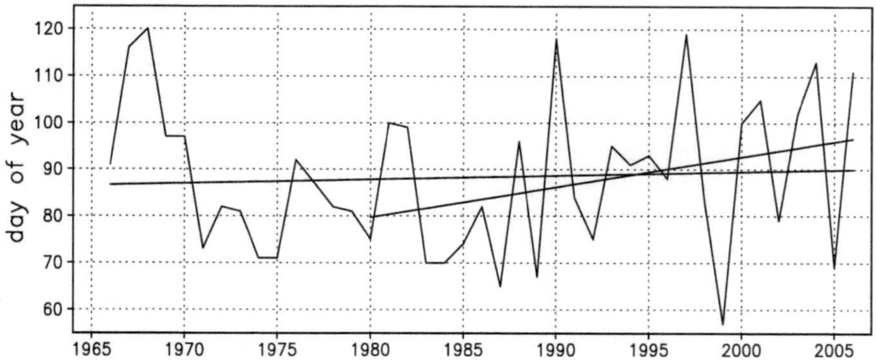

Fig. 3 Date of spring breakup (day of year) of the Arctic polar vortex at 50 hPa calculated from NCEP/NCAR reanalyses for the period 1965–2006 according to the criterion explained in the text. The straight lines indicate the trends in the breakup dates for the springs 1980–2006 and 1966–2006.

A comparison of the calculated breakup dates with updated values from Waugh et al. (provided by Paul Newman) shows a very good agreement for Antarctic spring while for the Arctic, larger discrepancies occur in single years. This is an indication for the sensitivity of both methods to the dynamical variability of the polar vortex which is larger in the Arctic than in the Antarctic. For example, according to the Waugh et al. method, the polar vortex broke up very early in February 2006, when a strong major warming occurred in the stratosphere, while in this study the vortex breakup was calculated for mid-April 2006, due to the temporal recovery of the vortex after the warming.

The trend in polar vortex breakup date calculated for the period 1979–2006 reveals an increase of 6.5 ± 1.9 days/decade, which is however only significant at the 87% level due to the interannual variations (Fig. 3). There seems to be a contradiction between the enhanced dynamical variability since 1999 resulting in more major stratospheric warmings on the one side and a more persistent Arctic polar vortex in spring on the other side. However, the intensity of the polar vortex in spring strongly depends on the timing of major warmings: early major midwinter warmings, such as in 2006, may lead to an even stronger polar vortex in spring, while late major warmings may weaken the polar vortex in a way that in conjunction with the increasing radiative heating in spring, the polar vortex is no more able to recover. Such late major warmings are usually connected with early vortex breakup dates.

4 Antarctic Spring

In the Antarctic the polar vortex significantly intensified during the period 1979–1999. Figure 4 shows the change of the geopotential height in December at the 50 hPa pressure level, where the deepening of the Antarctic polar vortex is most pronounced. In contrast to the deepening of the polar vortex in Arctic spring (Fig. 1), the Antarctic changes of the polar vortex are statistically highly significant. Note that the polar vortex is stronger during winter in the Antarctic than in the Arctic leading to a longer lifetime of the Antarctic vortex and later transitions from spring to summer circulation. Thus the month of December is still representative for spring conditions. In the updated period 1979–2006 the Antarctic polar vortex still shows an intensification however with reduced magnitude. In contrast to the northern hemisphere, the negative trend remains negative in the long period 1965–2006. The sign of the trend is thus a robust feature and not an artefact of a limited time series, while its magnitude varies depending on the length of the considered period. This decrease of geopotential height is directly related to the cooling of the lower stratosphere (e.g., Randel and Wu 1999) that is driven by the Antarctic ozone losses during the August–October period.

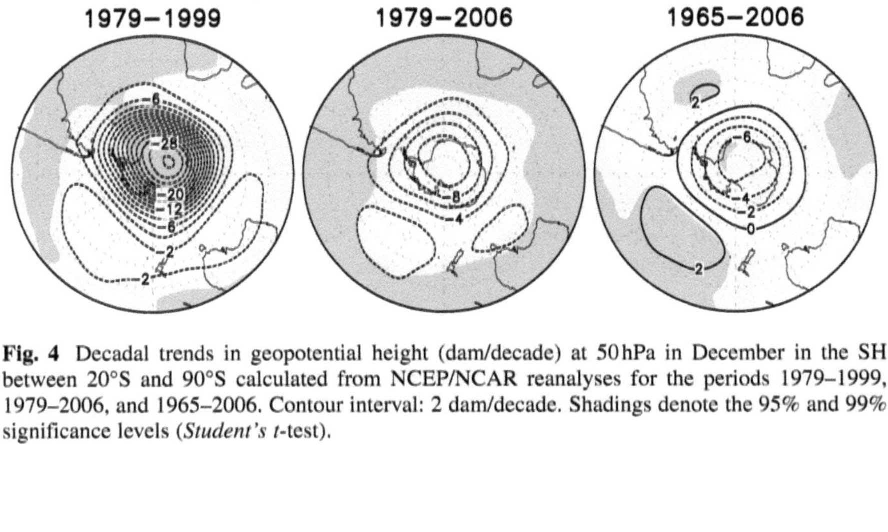

Fig. 4 Decadal trends in geopotential height (dam/decade) at 50 hPa in December in the SH between 20°S and 90°S calculated from NCEP/NCAR reanalyses for the periods 1979–1999, 1979–2006, and 1965–2006. Contour interval: 2 dam/decade. Shadings denote the 95% and 99% significance levels (*Student's t*-test).

The breakup date for the Antarctic polar vortex reveals considerable interannual variability since the mid-1980s with alternating early and late transition dates (Fig. 5). However both, early and late transition dates showed a tendency towards later days of the year, so that for the period 1979–1999 a highly significant increase of the polar vortex lifetime of 9.3 ± 1.8 days/decade could be found. Since the turn of the century, however, three winters with early vortex breakups occurred, including the unexpected major stratospheric warming in 2002 reducing the increase in polar vortex lifetime for the period 1979–2006. In 2006, the Antarctic polar vortex was again very well developed until beginning of December when it started to be eroded by increasing planetary wave activity. The vortex breakup occurred on December 6, and was late compared to the long-term mean of day 329 (November 25). Despite of the two recent dynamically active winters, the long-term trend in breakup date shows an increase of 3.0 ± 0.7 days/decade with a probability of 94%.

In the period 1979–2006 the southern hemisphere total meridional heat flux at 100 hPa, averaged for September and October and between 40°S and 80°S, was well correlated with the polar vortex breakup date (correlation coefficient r = 0.44, not shown). This allows estimating the impact of tropospheric dynamical variability on the breakup date of the Antarctic polar vortex. We determined the delay in vortex breakup date per heat flux unit and "corrected" the breakup dates as a function of the heat flux for each individual winter. The result is a delay of the breakup date, with the new date indicating the theoretical vortex breakup occurring without any tropospheric forcing. In the year 2000, for example, when the heat flux was strong in September and October, the Antarctic vortex broke up on day 322. Without heat flux correction the breakup would have occurred on day 360. Thus tropospheric variability caused the vortex to break up about one month earlier. In contrast, the impact of tropospheric variability on the breakup date was low in 1987: the vortex

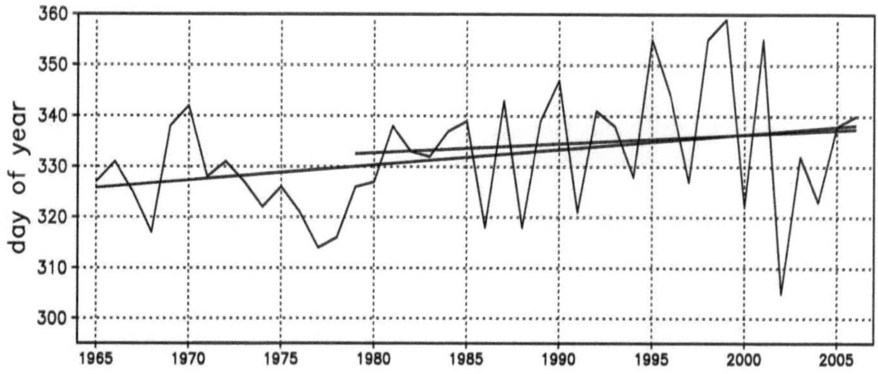

Fig. 5 Date of spring breakup (day of year, doy) of the Antarctic polar vortex at 50 hPa calculated from NCEP/NCAR reanalyses for the period 1965–2006 according to the criterion explained in the text. The straight lines indicate the trends in the breakup dates for the springs 1979–2006 and 1965–2006.

broke down only 14 days earlier due to tropospheric forcing. The corrected time series of breakup dates shows a slightly increased tendency towards later breakup dates indicating the influence of stratospheric ozone depletion on vortex intensity. The corrected time series displays however still considerable interannual variability possibly influenced by the tropical QBO or decadal variability. A quantitative assessment requires a more detailed analysis which is beyond the scope of this paper.

5 Summary

We analyzed changes in the intensity and persistence of the Arctic and Antarctic stratospheric polar vortices in spring, based on NCEP/NCAR reanalyses for the period 1965–2006. Our results for three different periods (1979–1999, 1979–2006, and 1965–2006) emphasize the uncertainties associated with deriving trends from short data records. Different trends for the different periods arise due to the existence of low-frequency dynamical variability in the winter stratosphere. In the Arctic, stratospheric winters were dynamically rather undisturbed during the 1980s and early to mid-1990s; the polar vortex intensified in spring during that period with a delay in the breakup date. This was related to a concurrent decrease in dynamical forcing from the troposphere in midwinter (Langematz and Kunze 2006). Since the late 1990s however, major stratospheric warmings occurred in almost every winter, so that the polar vortex in spring still intensified in March but with a smaller magnitude. Nevertheless, as some of the major warmings took place rather early in winter (e.g., in 2006), the polar vortex was able to recover leading to a comparatively late breakup in spite of the dynamical disturbance. In the long-term, there is no statistically significant change in Arctic vortex intensity or lifetime.

The Antarctic had been characterized by a significant deepening of the polar vortex during the 1980s and 1990s. This had been explained as a response to the radiative cooling caused by stratospheric ozone depletion since the 1980s (Shine 1986; Jones and Shanklin 1995; Langematz et al. 2003). Zhou et al. (2000) argued that the delay in Antarctic polar vortex breakup during that time was mainly due to ozone depletion as it was strongest in the lower stratosphere where ozone is chemically destroyed by active chlorine. However in the years since 2000, the deepening of the polar vortex has been considerably reduced due to an unexpected enhancement of dynamical activity in southern hemisphere (SH) winter. The first major stratospheric warming ever observed in the SH (Naujokat and Roscoe 2005) in 2002 and two early vortex breakups in 2000 and 2004 weakened the robust trend towards a later vortex breakup date of the period 1979–1999. They thus obscure a direct detection of an effect on polar ozone depletion of a decrease in ozone depleting substances. Whether the most recent dynamical winters in the Antarctic are an indication of a long-term climate change can not yet be determined. During the winter 2006, the Antarctic polar vortex was very strong and persisted into December, confirming that longer time series of observations are required for deriving significant trend estimates in the stratosphere in winter and spring.

Acknowledgments The authors wish to thank Dr. Paul Newman for providing data and instructive comments. MK was supported by the European Community in the SCOUT-O3 project (505390-GOCE-CT-2004). The NCEP/NCAR data were downloaded from http://wesley.wwb.noaa.gov/reanalysis.html.

References

Brownlee, K. A., 1965: *Statistical Theory and Methodology in Science and Engineering*. Wiley, New York, 334–346.

Coy, L., E. R. Nash, and P. A. Newman, 1997: Meteorology of the polar vortex: spring 1997. *Geophys. Res. Lett.*, **24**, 2693–2696.

Jones, A. E. and J. D. Shanklin, 1995: Continued decline of total ozone over Halley, Antarctica since 1985. *Nature*, **376**, 409–411.

Kalnay, E. et al., 1996: The NCEP/NCAR 40-year reanalysis project. *Bull. Am. Meteorol. Soc.*, **77**, 437–471.

Kodera, K. and H. Koide, 1997: Spatial and seasonal characteristics of recent decadal trends in the northern hemispheric troposphere and stratosphere. *J. Geophys. Res.*, **102**, 19433–19447.

Labitzke, K. et al., 2002: *The Berlin Stratospheric Data Series*. CD from Institut für Meteorologie, Freie Universität Berlin.

Labitzke, K., M. Kunze, and S. Brönnimann, 2006: Sunspots, the QBO, and the stratosphere in the north polar region – 20 years later. *Meteorol. Z.*, **15**, 355–363.

Langematz, U. and M. Kunze, 2006: An update on dynamical changes in the Arctic and Antarctic stratospheric polar vortices. *Clim. Dynam.*, **27**, 647–660, DOI 10.1007/s00382-006-0156-2.

Langematz, U., M. Kunze, K. Krüger, K. Labitzke, and G. L. Roff, 2003: Thermal and dynamical changes of the stratosphere since 1979 and their link to ozone and CO2 changes. *J. Geophys. Res.*, **108**, 4027, doi:10.1029/2002JD002069.

Manney, G. L., K. Krüger, J. L. Sabutis, S. A. Sena, and S. Pawson, 2005: The remarkable 2003–2004 winter and other recent warm winters in the Arctic stratosphere since the late 1990s. *J. Geophys. Res.*, **110**, D04107, doi:10.1029/2004JD005367.

Naujokat, B. and H. K. Roscoe, 2005: Evidence against an Antarctic stratospheric vortex split during the periods of pre-IGY temperature measurements. *J. Atmos. Sci.*, **62**, 885–889.

Newman, P. A., E. R. Nash, and J. E. Rosenfield, 2001: What controls the temperature of the Arctic stratosphere during the spring? *J. Geophys. Res.*, **106**, 19999–2010.

Pawson, S. and B. Naujokat, 1997: Trends in daily wintertime temperatures in the northern stratosphere. *Geophys. Res. Lett.*, **24**, 575–578.

Ramaswamy, V., M. L. Chanin, J. Angell, J. Barnett, D. Gaffen, M. Gelman, P. Keckhut, Y. Koshelkov, K. Labitzke, J. J. R. Lin, A. O'Neill, J. Nash, W. Randel, R. Rood, K. Shine, M. Shiotani, and R. Swinbank, 2001: Stratospheric temperature trends: observations and model simulations. *Rev. Geophys.*, **39**, 71–122.

Randel, W. J. and F. Wu, 1999: Cooling of the Arctic and Antarctic polar stratospheres due to ozone depletion. *J. Climate*, **12**, 1467–1479.

Rex, M., R. J. Salawitch, P. von der Gathen, N. R. P. Harris, M. P. Chipperfield, and B. Naujokat, 2004: Arctic ozone loss and climate change. *Geophys. Res. Lett.*, **31**, L04116, doi:10.1029/2003GL018844.

Shine, K. P., 1986: On the modeled thermal response of the Antarctic stratosphere to a depletion of ozone. *Geophys. Res. Lett.*, **13**, 1331–1334.

Waugh, D. W., W. J. Randel, S. Pawson, P. A. Newman, and E. R. Nash, 1999: Persistence of lower stratospheric polar vortices. *J. Geophys. Res.*, **104**, 27191–27201.

Zhou, S., M. E. Gelman, A. J. Miller, and J. P. Mc Cormack, 2000: An inter-hemisphere comparison of the persistent stratospheric polar vortex. *Geophys. Res. Lett.*, **27**, 1123–1126.

Creating Knowledge from the Confrontation of Observations and Models: The Case of Stratospheric Ozone

G. P. Brasseur

Among the many interesting bodies which the researches of modern chemistry have brought to light, few are more remarkable than the substance to which the name of ozone has been given.

T. Andrews, Professor of Chemistry,
Queens College, Belfast, UK, 1855

Abstract Through three examples taken from the history of ozone research, this paper illustrates that knowledge has been created by using simultaneously different approaches and methodologies, and by confronting the information resulting from laboratory studies, observational programs and modeling activities. New knowledge on chemical and dynamical processes in the atmosphere has been produced from detailed studies of the vertical and meridional ozone distributions in the stratosphere, and from investigations on the cause of the formation of the Antarctic ozone hole.

1 Introduction

In April 1840, Christian Fredrich Schönbein, a German chemist working at the University of Basle in Switzerland, published a paper (Schönbein 1840) in which he reported the presence of a peculiar odor produced during the electrolysis of acidulated water. He did not identify the cause of this odor, but suggested to name this "smelling principle" (riechende Princip) *ozone*, after the Greek word ὄζειν (ozein, to smell). This discovery led to much speculation about the nature of this odor (de la Rive 1845; Marignac 1845; Andrews 1856), and it was only in 1868 that Jean-Louis Soret, also in Basle, provided an experimental proof that ozone is an allotropic form of oxygen [OOO or O_3] (Soret 1868), and so confirmed the suggestion made by W. Olding in England (Olding 1861) a few years earlier.

National Center for Atmospheric Research, Boulder, CO, USA
brasseur@ucar.edu

The presence of ozone in the air was detected in 1858 by André Houzeau in Rouen, France (Houzeau 1858). The French agronomist and chemist developed a quantitative method involving a mixture of iodine and arsenic to measure the abundance of ozone in air. During the last decades of the 19th century, the atmospheric concentration of this molecule became measured routinely at the surface in Europe and elsewhere, using in most cases a measurement technique developed by Schönbein himself. This method was based on a paper soaked in potassium iodide and a starch solution whose color changes with the level of the ozone concentration. A French chemist working at the Municipal Observatory of Parc Montsouris in Paris, Albert Levy, performed systematic observations of surface ozone between 1877 and 1907, and therefore established the first long-term ozone record. His measurements, which have been recently reanalyzed by Volz and Kley (1988), show that the surface ozone mixing ratio around 1900 was of the order of 5–15 ppbv, that is, a factor two or three lower than what is currently observed in the urban and industrial regions of the northern hemisphere (Fig. 1).

The history of ozone research is interesting because it illustrates how scientific knowledge can be acquired from the systematic confrontation of experimental evidence with theoretical approaches. Critical review of observations and their analyses with the aid of models are a powerful way of generating new information, and of improving our understanding of complex processes. Laboratory measurements of chemical kinetic parameters and molecular absorption cross sections have been key for calculating reaction and photolysis rates, and hence for estimating the atmospheric concentration of ozone and of related chemical species. In this paper, we provide three examples that illustrate how the simultaneous use of theoretical and experimental approaches, and specifically the use of observational and modeling techniques has led to breakthroughs on questions related to stratospheric chemistry, stratospheric dynamics, and the impact of human activities on the chemical composition of the stratosphere.

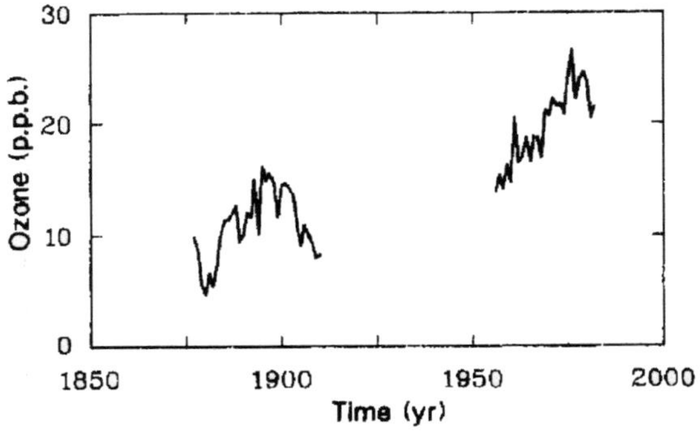

Fig. 1 Annual mean ozone concentration mixing ratio (ppbv) observed at Parc Montsouris (Paris, France) between 1876 and 1910, and in Arkona (Island of Rügen, Baltic Sea, northern Germany) between 1956 and 1983 (from Volz and Kley 1988, reprinted by permission from Macmillan Publishers Ltd: Nature, copyright 1988).

2 Understanding Stratospheric Chemistry: The Vertical Ozone Profile

Establishing the vertical distribution of the ozone concentration in the atmosphere at a time when experimental techniques were in their infancy was a real challenge for the scientists at the end of the 19th and the beginning of the 20th centuries. As early as 1879, Marie Alfred Cornu, Professor at Ecole Polytechnique in Paris (Cornu 1879), had observed a sharp cutoff (300 nm) in the solar ultraviolet radiation reaching the Earth's surface, a phenomena, which two years later was attributed by Sir Walter Noel Hartley to the absorption by atmospheric ozone (Hartley 1881a, b). The knowledge of the ozone absorption properties led Charles Fabry and Henri Buisson who had measured the optical properties of this gas in their laboratory in Marseilles, France, to derive in 1912, a thickness of about 5 mm STP for the total atmospheric ozone column (Fabry and Buisson 1913). They suggested that ozone is formed from oxygen in the upper atmosphere. Using a special spectrograph to observe the intensity of solar ultraviolet, they revised their number in 1920, and derived an ozone column abundance of 3 mm STP (Fabry and Buisson 1921). In 1928, Jean Cabannes and Jean Dufay, who were working also in Marseilles, highlighted that the measurements of ozone at the surface and the observations of the vertically integrated ozone concentration (ozone column abundance) made by Fabry and Buisson, as well as by Gordon Dobson at the University of Oxford (UK), could only be explained if the ozone density at high altitudes was considerably higher than near the surface (Cabannes and Dufay 1925, 1927). Cabannes and Dufay located an ozone layer near 50 km altitude. With a spectrophotometer built following the approach adopted by Fabry and Buisson, Dobson (1929) made systematic observations of the ozone column in different parts of the world. His spectrophotometer remained for many years the only instrument able to provide very accurate measurements of the ozone column abundance in the atmosphere. Using this instrument in Arosa, Switzerland, and measuring the incoming solar radiation at different wavelengths and for different zenith angles of the Sun, Paul Götz, a Swiss scientist working at the Lichtklimatisches Observatorium in Arosa together with Dobson concluded that the ozone layer should

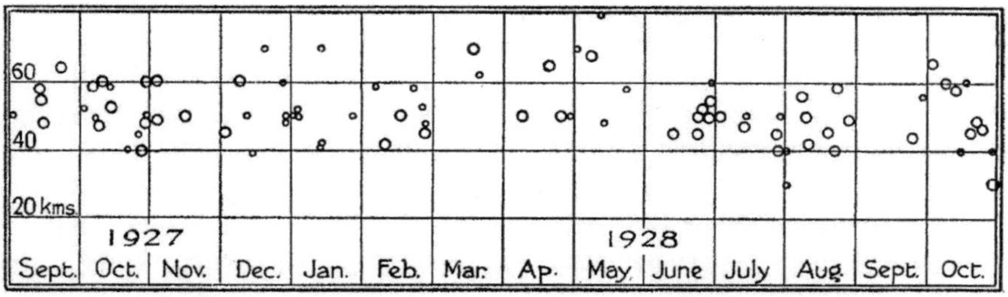

Fig. 2 Height of the ozone layer (km) over Arosa at different times of the year determined by Götz and Dobson (1929) between September 1927 to October 1928. Subsequent papers (e.g., Götz, 1931) suggest that the layer be located at much lower altitude (i.e., 20 km in Arosa).

be located between 40 and 50 km altitude (Fig. 2), in reasonable agreement with the earlier suggestion of Cabannes and Dufay. In their paper, Götz and Dobson (1928) noted, however, "that it is remarkable that the ozone situated at so great a height should be so closely connected with variations in pressure much lower down."

The first international ozone conference organized by Fabry in Paris in 1929 (at the suggestion of Dobson) gave geophysicist Sydney Chapman the opportunity to propose a photochemical theory (Chapman 1930) to explain the presence of an ozone layer at high altitude in the stratosphere. In his paper, Chapman reviewed different possible ozone production mechanisms. One of them was the dissociation of molecular oxygen by corpuscular radiation, a process that could be particularly intense in the polar regions. This idea had been proposed earlier to explain the greater ozone content observed by Dobson at high latitudes (see Sect. 3). Chapman rejected, however, this idea corpuscular radiation does not penetrate in the atmosphere below 90 km altitude (the heights where aurora are observed), and it would take a considerable amount of time for the ozone to descend to 45 km altitude, the height at which, according to Chapman, the ozone layer was located. Rather, he suggested that the formation mechanism is due to the photolysis of molecular oxygen by solar ultraviolet radiation (in the spectral region of 130–185 nm). According to Fabry, this part of the solar ultraviolet spectrum was penetrating precisely to about 45 km altitude. And Chapman stated in his paper that "the height of ozone is slightly favorable to the ultraviolet theory." He highlighted the role of atomic oxygen, produced by the O_2 photodissociation, which recombines with an oxygen molecule to produce the ozone molecule. The photolysis of ozone also had to be considered, but the net loss mechanism was believed to be due to the reaction between O and O_3. Using a simple photochemical model with reaction coefficients and absorption cross sections measured in the laboratory, Chapman could explain the most recent observations. He calculated the equilibrium ozone concentrations and determined that the maximum ozone concentration should indeed be located near 50 km altitude.

By the end of the 1920s, the problem of the vertical ozone profile seemed thus to be solved. However, a surprise came from Paul Götz in the early 1930s. During an expedition in Spitzbergen in 1929, Götz (1931) observed spectra of the zenith skylight, and noted an anomaly in the values of the relative radiances of scattered sunlight at certain wavelengths in the ultraviolet as the Sun approaches the horizon. He established that this anomaly was due to the presence of the ozone layer, and used this effect, that he called the *Umkehr Effect*, to estimate the vertical distribution of the ozone concentration. He deduced from his measurements that the ozone maximum should be located near 20–25 km altitude (Götz et al. 1933, 1934a, b) rather than 50 km (Fig. 3). These new results were confirmed in Germany by Erich Regener and his son Victor, who measured, in 1934, the solar ultraviolet absorption at different altitudes from an unmanned stratospheric balloon (Regener and Regener 1934). Additional evidence was provided a year later (1935) during the Explorer II balloon flight, which took place from Rapid City, South Dakota, and reached the record height of 72,000 ft with Captain Albert W. Stevens and Captain Orvil A. Anderson on board of an instrumented gondola.

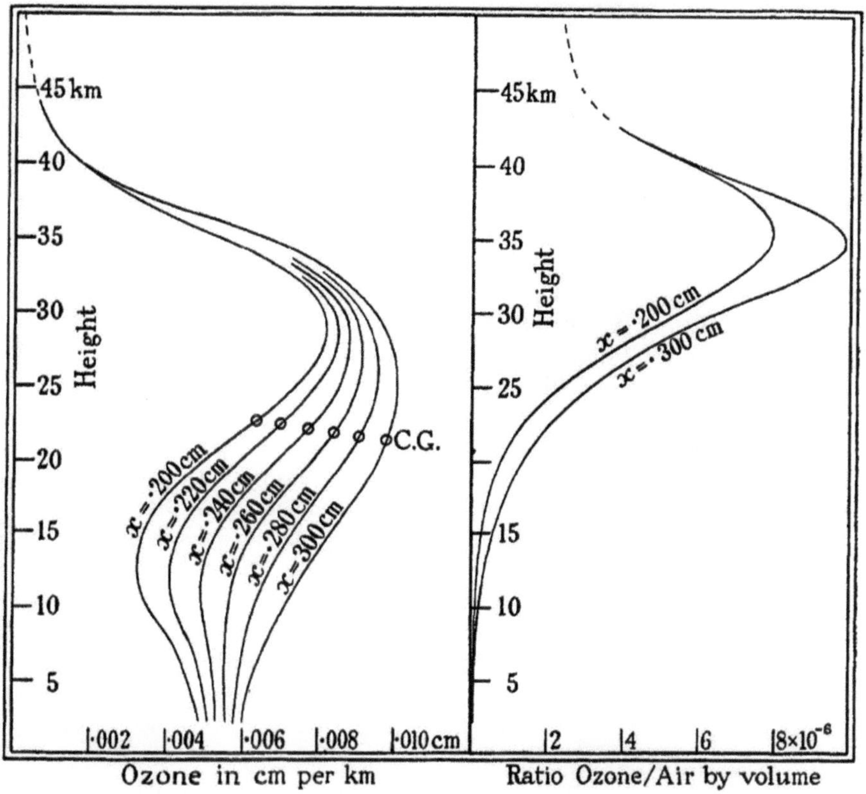

Fig. 3 Determination by Götz et al. (1934a) of the vertical distribution of atmospheric ozone in Arosa by the Umkehr method. Let panel: ozone column per km altitude; right panel: ozone volume mixing ratio.

The theoretical estimate by Chapman of an ozone concentration peak near 45 or 50 km altitude was corrected after 1953 when P. Brix and G. Herzberg at the National Research Council in Canada derived in the laboratory the absorption properties of molecular oxygen at wavelengths larger than 200 nm (Brix and Herzberg 1953). They established that the photolysis of molecular oxygen and therefore the photochemical production of ozone are occurring as a result of solar absorption between 200 and 242 nm, that is, at wavelengths that penetrate down to the lower stratosphere. This mechanism explains therefore the existence of an ozone maximum near 25 km altitude. This finding highlights the importance of laboratory measurements of fundamental properties of molecules. Without the accurate estimate of such parameters, both models of atmospheric ozone and inversion algorithms cannot provide quantitative information on the chemical composition of the atmosphere.

New surprises came with new observations. In 1960, S.V. Venkateswaran and his colleagues at the University of California in Los Angeles measured the solar light reflected by the Iota-1 communication satellite near sunset and sunrise, and were able to deduce the vertical profile of ozone (Venkateswaran et al. 1961). This first measurement of ozone using a spacecraft showed that, in addition to the well-documented ozone maximum at 20–25 km, another maximum was probably located at 55 km. This result reopened a debate in the scientific community since these last findings supported the early suggestions of a high-altitude ozone maximum. However, measurements made two years later by R. D. Rawcliffe and his coworkers at the Aerospace Corporation using a radiometer installed on a US Air Force satellite, showed unambiguously that no secondary ozone maximum was present near the stratopause (Rawcliffe et al. 1963). In fact, the presence of a secondary maximum was identified in 1973 by Paul Hays at the University of Michigan and by Ray Roble at the National Center for Atmospheric Research (NCAR) in Boulder, Colorado, but it is located at higher altitudes near the mesopause (Hays and Roble 1973).

Until the middle of the 20th century, the only theoretical estimate of the concentration of ozone in the stratosphere and mesosphere was provided by the Chapman model. In 1950, Sir David Bates, an applied mathematician at the University of Belfast, and Baron Marcel Nicolet, an atmospheric scientist from Belgium, who were working together at the California Institute for Technology in Pasadena, California and were studying the mechanisms responsible for the radiative emissions in the Meinel bands in the upper atmosphere, noted that reactions involving hydrogen atoms (H), hydroxyl (OH) and hydroperoxy radicals (HO_2) provide important catalytic destruction mechanisms for ozone in the mesosphere (Bates and Nicolet 1950). As a result, the ozone concentration predicted by the Chapman theory had to be reduced considerably above 50 km altitude. In spite of this new finding, it appeared during the 1960s that a discrepancy of nearly a factor of two existed between the concentrations derived from the photochemical theory accepted at that time and the ozone observations. John Hampson, a scientist working at the Canadian Armaments Research and Development Establishment in Quebec, was the first to show how HOx could catalytically destroy ozone in the stratosphere, and so linked the destruction of stratospheric ozone to the presence of water vapor in this layer of the atmosphere (Hampson 1964, 1966). These effects needed to be added in numerical models. But, it was the Dutch scientist Paul Crutzen working at that time at Oxford University, who identified in 1970 that a catalytic cycle involving nitric oxide (NO) is the most efficient ozone destruction mechanism in the stratosphere (Crutzen 1970). Harold Johnston from the University of Berkeley indicated that the release of considerable amounts of nitrogen oxides by the projected fleet of high-altitude commercial aircraft would destroy considerable amounts of ozone in the stratosphere (Johnston 1971). The Climatic Impact Assessment Program (CIAP) initiated by the US Department of Transportation to address this question and similar programs in the UK and in France became a catalyst for stratospheric research. It led to considerable progress in our understanding of dynamical and chemical processes based on advances in theory, observations and laboratory measurements. A few years later, Richard Stolarski and Ralph Cicerone, then at the University of Michigan, invoked a similar ozone destruction cycle, but with the chlorine atom replacing nitric oxide (Stolarski and Cicerone 1974). It was shown in 1974 by Mario Molina and Sherry Rowland from the University of California at Irvine that the major source of chlorine atoms in the stratosphere was the photolysis of industrially manufactured chlorofluorocarbons (Molina and Rowland 1974). Steven Wofsy from Harvard University suggested that bromine compounds could contribute to additional catalytic ozone destruction (Wofsy et al. 1975). Halons and a fraction of the metyl bromide present in the atmosphere are of human origin. These findings highlighted that human activity could become a serious threat to the ozone layer, and that ozone depletion could become a global environmental problem.

In spite of these major steps toward our understanding of atmospheric chemistry, the stratospheric models of the 1980s and 1990s that accounted for the key ozone-destroying catalytic cycles were systematically underestimating by 20–50% the ozone concentration in the upper stratosphere between 35 and 50 km altitude. At these heights, ozone is controlled almost exclusively by chemical reactions, and the effect of transport is therefore of secondary importance. In the 1980s, very few observations were available in this atmospheric layer until space observations of chemical compounds became available. Observations by the Halogen Occultation Experiment (HALOE) and the Microwave Limb Sounder (MLS) on board the Upper Atmosphere Research Satellite (UARS) launched by NASA in 1991, showed that by constraining the concentration of chlorine compounds to their observed values, no significant ozone deficit remained in the upper stratosphere (Crutzen et al. 1995; Khosravi et al. 1998). This led to new laboratory investigations that revealed that the rate of conversion of ozone-destroying chlorine monoxide (ClO) by reaction with the OH radical to form the less-reactive hydrogen chloride (HCl) was overestimated, since in 6% of the cases, this reaction was producing a chlorine atom (Cl), which contributes to the destruction of ozone. Today some uncertainties remain about the role of hydrogenated radicals and their effects on ozone in the mesosphere.

The progress towards a full understanding of the mechanisms that govern the vertical distribution of stratospheric ozone has been key for elucidating the role of different chemical processes. It has required a constant synthesis between different research methodologies including in situ observations, remote sensing, laboratory investigations, data analysis and modeling. The development of new observing technologies, the access to increasingly powerful computers and the continuous efforts of the international scientific community have been important conditions for making decisive progress.

3 Understanding Stratospheric Dynamics: The Meridional Ozone Distribution

The photochemical theory presented by Chapman in 1929, which recognized that the ozone production is driven by the intensity of solar ultraviolet radiation led to the conclusion that the ozone concentration should be larger in the tropics than at midlatitudes and high latitudes. Furthermore, in the extratropical regions, the ozone abundance should be largest during summertime, and very little ozone should be present in the atmosphere during winter. The observations conducted by Dobson suggested, however, the opposite.

Dobson had installed spectrophotometers in many regions of the world (Fig. 4), so that he could derive the latitudinal distribution and seasonal evolution of the ozone column (Dobson et al. 1930). He reported that the ozone column abundance is smallest in the tropics at all seasons and largest at high latitudes at the end of the winter and in early spring. He also noted an asymmetry between northern hemisphere and southern hemisphere ozone: The spring maximum was located very near the pole in the northern hemisphere (March–April), while it was observed near 60° latitude in the southern hemisphere (September–October) with a slight local minimum

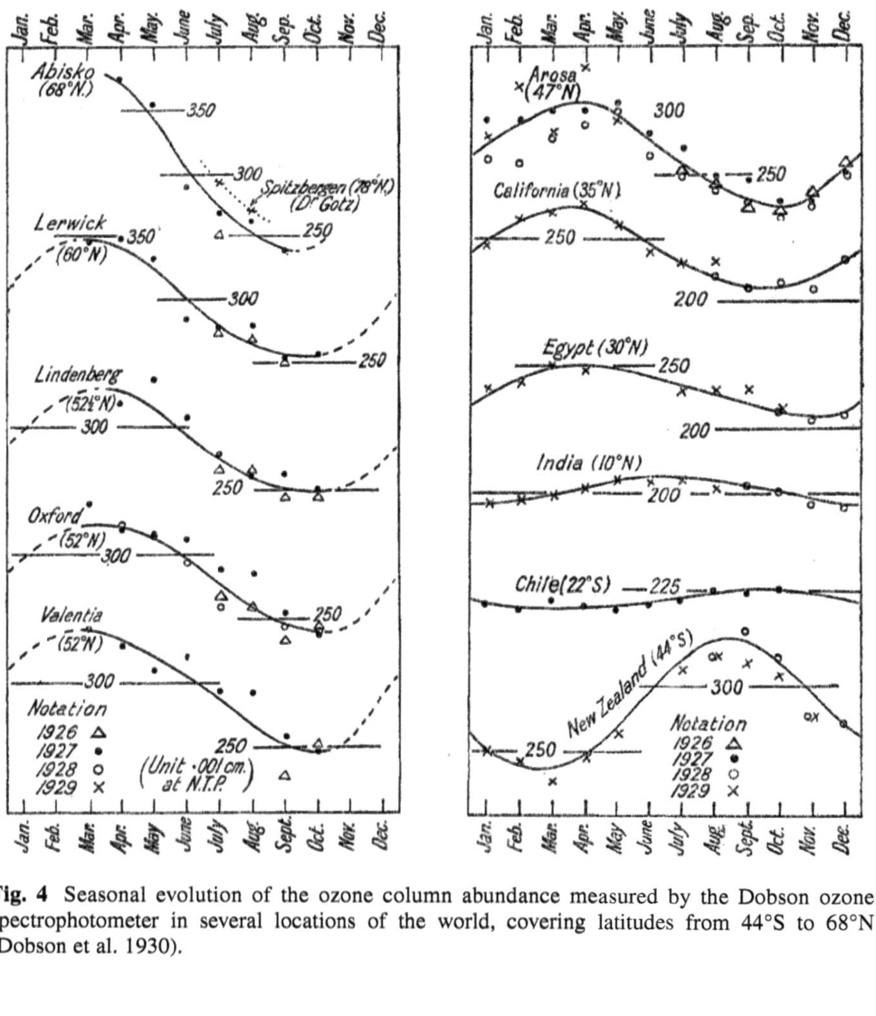

Fig. 4 Seasonal evolution of the ozone column abundance measured by the Dobson ozone spectrophotometer in several locations of the world, covering latitudes from 44°S to 68°N (Dobson et al. 1930).

at the South Pole. Dobson also noticed a strong day-to-day variability in the observed ozone column, which is related to the meteorological situation.

The discrepancy between the photochemical theory and the observational evidence needed to be resolved. Although incomplete, the Chapman theory showed that the photochemical lifetime of ozone is increasing with decreasing altitude, and is becoming long enough in the lower stratosphere for transport to govern the meridional distribution of this gas. Both the ozone observations by Dobson and the measurements of water vapor by Alan Brewer could only be explained by a circulation directed from the equator to the poles with upward motions in the tropics and slow descent at high latitudes (Brewer 1949; Dobson 1956). This deduction was, however, inconsistent with the *Eulerian zonal mean* circulation derived from meteorological observations and general circulation models. Measurements of winds suggested indeed the existence of upward motions in the tropics, but showed also the existence of downward motions at midlatitudes rather than near the poles. The solution to this apparent discrepancy was found later. It required a better understanding of the zonal mean averaging procedures that were used. The existence of planetary waves in the stratosphere (which introduce longitudinal departures from the zonal mean circulation), and their upward propagation during wintertime, noted by Charney and Drazin (1961) and Matsuno (1970), suggests that the mean meridional transport of heat and mass (e.g., ozone), should be expressed as the sum of two contributions: one from the Eulerian mean air flow and one from an eddy component produced by these waves. A large degree of cancellation exists between these two components, and it is the net meridional transport (also called the residual transport) that is derived from the observation of tracers like ozone and water vapor. The theoretical concept of the *transformed Eulerian mean* circulation, a mathematical formulation of the residual circulation that is more appropriate to represent the *net* mean meridional transport of atmospheric compounds, was shown to be consistent with the mean meridional circulation derived from the observations of Brewer and Dobson. The transformed Eulerian mean circulation is driven primarily by the dissipation of planetary waves in the stratosphere and of gravity waves in the mesosphere.

Observations of ozone and other chemical compounds from spaceborne instrumentation, and particularly from the LIMS instrument (Nimbus 7 satellite in the late 1970s), provided a clear picture of the role of planetary waves for the transport of ozone and other long-lived chemical compounds. The occurrence of wave-breaking events in the middle atmosphere, and the cascade of energy towards smaller spatial scales associated with irreversible mixing processes in a region named the "*surf zone*" by Cambridge dynamicist Michael McIntyre was highlighted by these systematic space observations of atmospheric tracers at the global scale (McIntyre 1993).

Satellite observations conducted in the following decades led to important progress in our understanding mechanisms that govern the dynamics of the middle atmosphere. For example, they provided information on the occasional occurrence of sudden warming in the polar middle atmosphere during winter, on the effects of tropical and gravity waves, and on the role of the quasi-biennial oscillation. They highlighted the importance of dynamical barriers such as the tropopause, the polar vortex, and the boundaries between tropical and extratropical air. In these areas, key contributions were made by James Holton, Michael McIntyre, and Alan Plumb, and many others.

Here again, progress has resulted from the continuous interactions between groups developing space experiments, those analyzing the observational data, and the atmospheric modeling community.

4 Understanding the Impact of Human Activities: The Ozone Hole

The report in 1985 by Joseph Farman, Brian Gardiner and Jonathan Shanklin from the British Antarctic Survey that showed (Fig. 5) a substantial decrease since the late 1970s in the ozone-column abundance at the Antarctic Station of Halley Bay, came as a surprise to the scientific community (Farman et al. 1985). This observed trend seemed to confirm the ozone decrease detected at the Japanese station of Syowa by Chubachi, and reported a few months earlier at the Quadriennal Ozone Symposium held during the summer of 1984 in Greece (Chubachi 1985). Modelers, who had been calculating the expected evolution of ozone in response to future emissions of chlorofluorocarbons, had not predicted this large reduction of polar ozone. No existing theory could explain the observations of the British and Japanese experimenters.

Different hypotheses were presented to explain this surprising finding. K. K Tung from the University of Washington (Tung et al. 1986), suggested that the atmospheric circulation could have changed in the previous decade, producing uplifting of ozone-poor air masses from the troposphere to the lower stratosphere at high latitudes (change in the Transformed Eulerian Mean circulation). Linwood Callis and M. Natarajan from NASA/Langley invoked a possible enhancement in the production of nitrogen oxides in the upper atmosphere in relation with high solar activity, and rapid descent of these compounds to the lower stratosphere (Callis and Natarajan 1986). Susan Solomon from the NOAA Aeronomy Laboratory in Boulder Colorado, together with several colleagues, proposed a mechanism by which slowly reactive inorganic chlorine reservoirs such as $ClONO_2$ and HCl are activated on the surface of tiny particles present in the coldest layers of the lower stratosphere, and produce ClO radicals that can rapidly destroy ozone (Solomon et al. 1986). The presence of polar stratospheric clouds, which could provide sites for these heterogeneous reactions to occur, had been observed earlier by Patrick McCormick at NASA/Langley using the SAGE satellite (McCormick et al. 1982).

In situ chemical measurements made from the NASA high-latitude ER-2 aircraft and space observations revealed that, as suggested by Solomon et al. (1986), chlorine activation on polar stratospheric clouds was taking place in the coldest regions of the Antarctic stratosphere, and that ozone was considerably depleted in precisely the same regions. Thus, it became clear that the ozone hole was produced as a result of the emissions of industrially manufactured chlorofluorocarbons. New investigations were urgently needed to identify more specifically the physical and chemical processes that were missing even in the most advanced models. Luisa Molina and

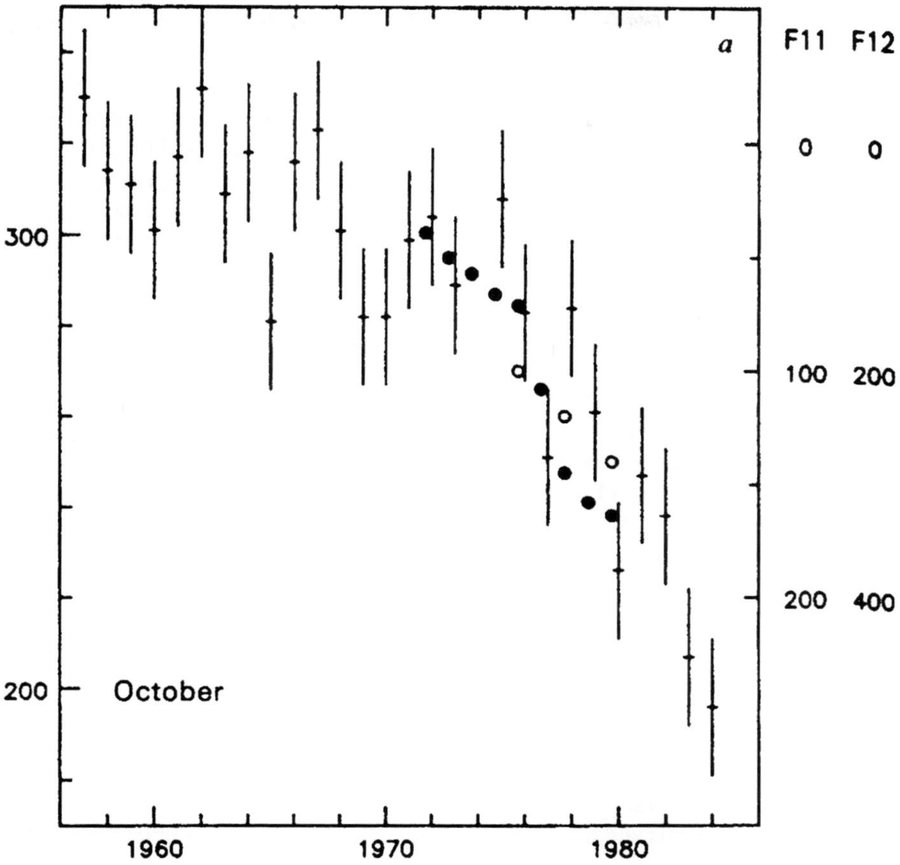

Fig. 5 October mean Ozone column abundance measured at the British Halley Base in Antarctica between 1958 and 1984. A large decrease is noticed, starting in the 1970s. The corresponding atmospheric mixing ratios of chlorofluorocarbons CFC-11 and –12 (pptv) are indicated by the full and open dots, respectively. The amounts of CFC-11 and –12 increase down the figure (from Farman et al. 1985, reprinted by permission from Macmillan Publishers Ltd: Nature, copyright 1985).

Mario Molina at MIT derived from laboratory studies the most efficient catalytic cycle that destroys ozone in the lower stratosphere in the presence of elevated concentrations of ClO, and highlighted the role of the Cl_2O_2 dimer (Molina and Molina 1987). It also soon became apparent that reactions involving bromine contribute to the formation of the Antarctic ozone hole. A large number of laboratory studies and in situ observations were also performed to characterize the nature and properties of atmospheric liquid and solid particles, which are responsible for chlorine activation. These studies showed that heterogeneous reactions could also occur on the surface of sulfate aerosol particles outside the Polar Regions, so that significant amounts of ozone should be destroyed by anthropogenic chlorine following large volcanic eruptions. Significant ozone depletion was indeed observed after the eruption of Mt Pinatubo in 1991 (see, e.g., Hofmann et al. 1994), which injected about 20 Tg of sulfur in the lower stratosphere.

All these experimental studies provided the information needed to improve chemical transport models of the middle atmosphere. When the newly discovered processes were included in the model formulation, the formation and the fate of the Antarctic ozone hole could be simulated numerically in good agreement with observations. These provided the necessary support to help in the interpretation of the observations conducted during extensive field campaigns in the Antarctic and in the Arctic.

The story of the discovery of the ozone hole provides another example that illustrates the need for using in a coherent manner, different approaches to solve a complex environmental problem. Here again, knowledge was incomplete when the fast reduction in Antarctic ozone was noticed, but it increased rapidly and dramatically as a result of the integration of experimental, observational and modeling results. The large perturbation produced as a consequence of human activities has triggered active research that has led to a spectacular improvement in our understanding of fundamental chemical, microphysical and dynamical processes that characterize the polar stratosphere.

5 Conclusions

The three examples presented here illustrate major milestones in the history of ozone research. They clearly show that substantial progress in our understanding of atmospheric processes has resulted from the interplay between observations, laboratory experiments and modeling. Ozone studies have been key in learning more about the chemistry, physics and dynamics of the atmosphere. It is therefore important that research programs include these different components and emphasize the need for critically confronting experimental results with theoretical work and the results of model simulations.

References

Andrews, T., 1856: On the constitution and properties of ozone. *Proc. Roy. Soc.*, **7**, 498.

Bates D. R. and M. Nicolet, 1950: The photochemistry of atmospheric water vapor. *J. Geophys. Res.*, **55**, 301.

Brewer, A. W., 1949: Evidence of a world circulation provided by measurements of helium and water vapor distribution in the stratosphere. *Quart. J. Meteorol. Soc.*, **75**, 351.

Brix, P. A. and G. Hezrberg, 1953: The dissociation energy of oxygen. *J. Chem. Phys.*, **21**, 2240.

Cabannes, J. and J. Dufay, 1925: Mesure de l'altitude de la couche d'ozone dans l'atmosphère. *Comptes Rendus, Acad. Sci., Paris*, **181**, 302.

Cabannes, J. and J. Dufay, 1927: Mesure de l'altitude et de l'épaisseur de la couche d'ozone dans l'atmosphère. *J. Phys. Rad.*, **8**, 125.

Callis, L. B. and M. Natarajan, 1986: The Antarctic ozone minimum:relationship to odd nitrogen, odd chlorine, the final warming and the 11-yeqar solar cycle. *J. Geophys. Res.*, **91**, 10771.

Chapman, S., 1930: A theory of upper-atmosphere ozone. *Memoirs of the R. Meteorol. Soc*, vol III(26), 103.

Charney, J. G. and P. G. Drazin, 1961: Propagation of planetary-scale disturbances from the lower into the upper atmosphere. *J. Geophys. Res.*, **66**, 83.

Chubachi, S., 1985: A special ozone observation at Syowa station, Antarctica, from February 1982 to January 1983. In: *Atmospheric Ozone. Proc. Quadrennial Ozone Symposium*, C. S. Zerfos and A. Ghazi, eds., 285–289, D. Reidel Publishing Co., Dordrecht.

Cornu, A., 1879: Sur la limite ultra-violette du spectre solaire a diverses altitudes. *Comptes Rendus, Acad. Sci., Paris*, **88**, 1101, 1285.

Crutzen, P. J., 1970: The influence of nitrogen oxides on the atmospheric ozone content. *Q. J. R. Meteorol. Soc.*, **96**, 320.

Crutzen, P. J., J. U. Gross, C. Bruehl, R. Mueller, and J. M. Russell, 1995: Reevaluation of the ozone budget with HALOE URAS data – No evidence for the ozone deficit. *Science*, **268**, 705.

de la Rive, M., 1845: Sur l'ozone. *Comptes Rendus Acad. Sci. Paris*, **20**, 1291.

Dobson, G. M. B., 1929: Atmospheric ozone. *Gerlands Beitr. Geophys.*, **24**, 8.

Dobson, G. M. B., 1956: Origin and distribution of the polyatomic molecules in the atmosphere. *Proc. Roy. Soc. London, A*, **236**, 187.

Dobson, G. M. B., H. H. Kimball, and E. Kidson, 1930: Measurements of the amount of ozone in the earth's atmosphere and its relation to other geophysical conditions, Part IV. *Proc. R. Soc., London, Ser. A*, **129**, 411.

Fabry, C. and H. Buisson, 1913: L'apsorption de l'ultraviolet par l'ozone et la limite du spectre solaire. *J. Phys.*, **3**, 196.

Fabry, C. and H. Buisson, 1921: Etude de l'extrémité ultraviolette du spectre solaire. *J. Phys. Radium*, **6**(2), 197.

Farman, J. C., B. G. Gardiner, and J. D. Shanklin, 1985: Large losses of total ozone in Antarctica reveal seasonal ClOx/NOx interaction. *Nature*, **315**, 207.

Götz, F. W. P., 1931: Zum Strahlungsklima des Spitzbergensommers. Strahlungs- und Ozonmessungen in der Konigsbucht 1929. *Gerlands Beitr. Geophys.*, **31**, 119.

Götz, F. W. P., 1934b: Über die Deutung der Umkehreffektes bei Messungen des atmospharischen Ozons. *Zeit. Astrophys.*, **8**, 267.

Götz F. W. P. and G. M. B. Dobson, 1928: Observations of the height of the ozone in the upper atmosphere. *Proc. Roy. Soc. London, Series A*, **120**(785), 251–259.

Götz F. W. P. and G. M. B. Dobson, 1929: Observations of the height of the ozone in the upper atmosphere. Part II. *Proc. Roy. Soc. London, Series A*, **125**(797), 292–294.

Götz, F. W. P., G. M. B. Dobson, and A. R. Meetham, 1933: Vertical distribution of ozone in the atmosphere. *Nature*, **132**, 281.

Götz, F. W. P., A. R. Meetham, and G. M. B. Dobson, 1934a: The vertical distribution of ozone in the atmosphere. *Proc. Roy. Soc. London, A*, **145**, 416.

Hampson, J., 1964: Photochemical behavior of the ozone layer, Tech. Note 1627/64, CARDE, Valcartier, Quebec.

Hampson, J., 1966: Chemiluminescent emissions observed in the stratosphere and mesosphere. In: *Les Problemes Météorologiques de la Stratosphere et de la Mesosphere*, M. Nicolet, ed., Presses Universitaires de France, Paris, 393 pp.

Hartley, W. N., 1881a: On the absorption spectrum of ozone. *J. Chem. Soc.*, **39**, 57.

Hartley, W. N., 1881b: On the absorption of solar rays by atmospheric ozone. *J. Chem. Soc.*, **39**, 111.

Hays, P. B. and R. G. Roble, 1973: Observation of mesospheric ozone at low latitudes. *Planet. Space Sci.*, **21**, 273.

Hofmann, D. J., S. J. Oltmans, W. D. Komhyr, J. M. Harris, J. A. Lathrop, A. O. Langford, T. Deshler, B. J. Johnson, A. Torres, and W. A. Matthews, 1994: Ozne loss in the lower stratosphere over the United States in 1992–1993: evidence for heterogeneous chemistry on the Pinatubo aerosol. *Geophys. Res. Lett.*, **21**, 65.

Houzeau, A., 1858: Preuve de la présence dans l'atmosphère d'un nouveau principe gazeux, l'oxygène naissant. *Comptes Rendus Acad. Sci. Paris*, **46**, 89.

Johnston, H. S., 1971: Reduction of stratospheric ozone by nitrogen oxide catalysts from supersonic transport exhaust. *Science*, **173**, 517.

Khosravi, R., G. Brasseur, A. Smith, D. Rusch, J. Water, and J. Russell III, 1998: Significant reduction in the ozone deficit: a 3-D model study using UARS data. *J. Geophys. Res.*, **103**(16), 203.

Marignac, C. 1845: Sur la nature et la production de l'ozone. *Arch. Electricité*, Geneva, **5**, 5.

Matsuno, T., 1970: Vertical propagation of stationary planetary waves in the northern hemisphere. *J. Atmos. Sci.*, **27**, 871..

McCormick, M. P., H. M. Steele, P. Hamill, W. P. Chu, and T. J. Swissler, 1982: Polar stratospheric cloud sightings by SAM II. *J. Atmos. Sci.*, **39**, 1387.

McIntyre, M. E., 1993: On the role of wave propagation and wave breaking in atmosphere-ocean dynamics. In: *Theoretical and Applied Mechanics 1992*, S. R. Bodner et al., eds., Elsevier, 281–304.

Molina, L. T. and M. J. Molina, 1987: Production of Cl_2O_2 from the self reaction of ClO radical. *J. Phys. Chem.*, **91**, 433.

Molina, M. J. and F. S. Rowland, 1974: Stratospheric sink for chlorofluoromethanes: chlorine atom-catalyzed destruction of ozone. *Nature*, **249**, 810.

Olding, W., 1861: Manual of Chemistry, descriptive and theoretical, Part 1, Longman Green, London.

Rawcliffe, R. D., G. E. Meloy, R. M. Friedman, and E. H. Rogers, 1963: Measurement of vertical distribution of ozone from polar orbiting satellite. *J. Geophys. Res.*, **68**, 6425.

Regener, E. and V. H. Regener, 1934: Aufnahme des ultravioletten Sonnenspektrums in der Stratosphäre und die vertikale Ozonverteilung. *Phys. Zeit.*, **35**, 788.

Schöenbein, 1840: Recherches sur la nature de l'odeur qui se dégage de certaines réactions chimiques. *Comptes Rendus, Acad. Sci., Paris*, **10**, 706.

Solomon, S., R. R. Garcia, F. S. Rowland, and D. J. Wuebbles, 1986: On the depletion of Antarctic ozone. *Nature*, **321**, 755.

Soret, J. L., 1868: Recherches sur la densité de l'ozone. *Ann. Chim. Phys., Paris*, **13**, 257.

Stolarski, R. S. and R. J. Cicerone, 1974: Stratospheric chlorine: a possible sink for ozone. *Can. J. Chem.*, **52**, 1610.

Tung, K. K., M. K. W. Ko, J. M. Rodriguez, and N. D. Sze, 1986: Are Antarctic ozone variations a manifestation of dynamics or chemistry? *Nature*, **333**, 811.

Venkateswaran, S. V., Krueger, A. J., and Moore, J. G., 1961: Determination of the vertical distribution of ozone by satellite photometry. *J. Geophys. Res.*, **66**, 1751.

Volz, A. and D. Kley, 1988: Evaluation of the Montsouris series of ozone measurements made in the nineteenth century. *Nature*, **332**, 240.

Wofsy, S. C., M. B. McElroy, and Y. L. Yung, 1975: The chemistry of stratospheric bromine. *Geophys. Res. Lett.*, **2**, 215.

Response of the Earth's Atmosphere to the Solar Irradiance Variability

E. Rozanov[1,2], T. Egorova[1], and W. Schmutz[1]

Abstract Recent satellite observations show that the solar ultraviolet irradiance is much more variable than the total solar irradiance. Atmospheric effects of the solar irradiance variations during 11-year solar activity cycle are investigated using different numerical models and observation data sets. It is shown that the direct and indirect (via ozone production) heating in the upper and middle stratosphere due to enhancement of the solar spectral irradiance leads to an acceleration of the polar night jets and suppression of the Brewer–Dobson circulation resulting in the ozone increase and warming of the lower tropical stratosphere. These stratospheric changes alter the tropospheric circulation leading to a statistically significant warming of the surface air over Russia, Europe and North America. The importance of the solar spectral irradiance variability for the attribution of the temperature changes in the upper stratosphere is also shown by the comparison of the simulated and observed temperature evolution during the last 25 years of the 20th century.

1 Introduction

Various aspects of Sun–Earth connections have been under consideration for many decades. Statistical evidences of such relation are given in extensive review by Hoyt and Schatten (1997), together with discussions of its robustness. It has been claimed (e.g., Reid 2000) that the response of the terrestrial climate to changes in solar activity in the course of the 20th century can be of the order of up to 0.5 K. Recently, the investigations of the solar variability contribution to the climate change gained a new impulse, because it is of great societal importance to know what part of the warming observed during the last century might be attributed to the solar activity variations (e.g., IPCC 2001). The physical mechanisms behind the above-mentioned

[1] Physical and Meteorological Observatory/World Radiation Center Davos, Switzerland
t.egorova@pmodwrc.ch, w.schmutz@pmodwrc.ch

[2] Institute for Atmospheric and Climate Science, ETH, Zurich, Switzerland
e.rozanov@pmodwrc.ch

S. Brönnimann et al. (eds.), *Climate Variability and Extremes during the Past 100 Years.*

correlations have not been clearly identified and the uncertainty of the solar forcing is still estimated by the experts as rather high (IPCC 2001). The first process to be considered is the variation of the total solar irradiance (TSI), which depends on the activity of the Sun. With precise measurements of the total solar irradiance over the last 20 years it has been shown that the total irradiance of the Sun varies by about 0.1% (~1.5 W/m^2) during one 11-year solar activity cycle (Fröhlich 2004). However, climate models showed that this forcing appears to be too small (North et al. 2004) to explain the response of the climate established from the statistical analysis of the observation records. From energy balance considerations the solar forcing of ~2 W/m^2 can change the surface temperature by about 0.2 K, if one assumes a mean value of climate sensitivity (IPCC 2001). This conclusion was recently confirmed by Stott et al. (2003). They applied sophisticated fingerprint analyses to the simulated and observed temperature long-term time series and showed that the model driven solely by TSI variability substantially underestimates the magnitude of the solar signal derived from the observations.

Thus, in order to amplify the simulated climate response to the solar variability some other mechanisms should be considered. Reid (2000) formulated a short list of the most promising candidates comprising of the variability of (i) spectral solar irradiance, (ii) energetic particles input to the terrestrial atmosphere, and, (iii) galactic cosmic rays. From the energy consideration these mechanisms cannot substantially change the global mean temperature, but they can redistribute the energy and modify the climate patterns, possibly causing significant local climate changes. The first mechanism, that is, the influence of the spectral solar irradiation variability on the global ozone and climate will be discussed in this paper. In Sect. 2 we consider the variability of the solar spectral irradiance and its direct effects on the temperature and ozone in the middle atmosphere. In Sect. 3 we discuss how the solar signal in the stratosphere and mesosphere can propagate toward the troposphere and affect the surface climate. In Sect. 4 we illustrate how this mechanism works in the chemistry-climate model SOCOL and compare the solar signal in the stratospheric ozone and temperature extracted from the simulation and available observations. In Sect. 5 we will show the influence of the solar activity on the temperature trends in the upper stratosphere.

2 Spectral Solar Irradiance Variability and its Direct Influence on the Temperature and Ozone in the Middle Atmosphere

Potential importance of the solar spectral irradiance (SSI) variability is based on the fact that it is much more variable than the TSI. Satellite monitoring revealed that the SSI variability, defined mostly by the bright features of the Sun (faculae and plagues), increases toward short-wave part of the spectrum (e.g., Rottman et al. 2004; Fox 2004). Figure 1 illustrates the relative variability of the SSI obtained from the SUSIM and SOLSTICE instruments onboard UARS satellite (data are available from http://lasp.colorado.edu/ and www.solar.nrl.navy.mil/) as well as

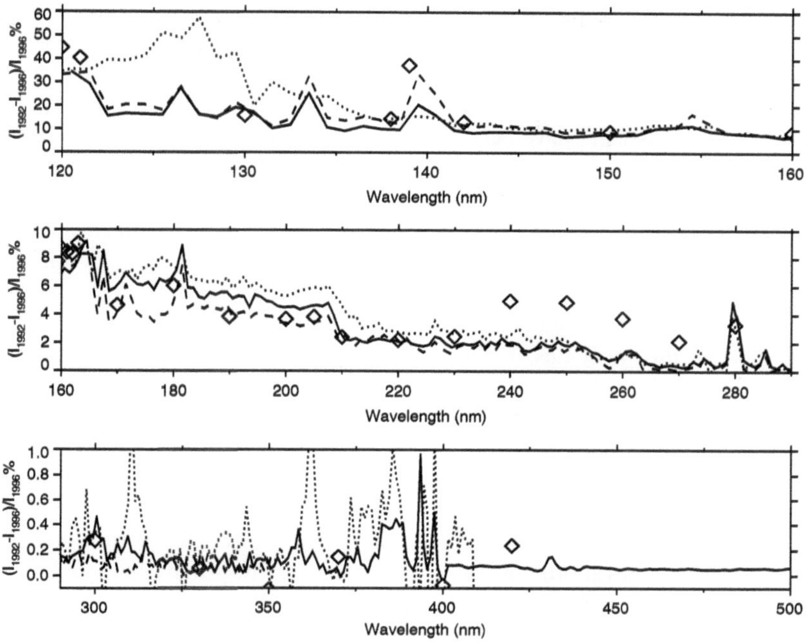

Fig. 1 Relative difference (%) between annual mean 1992 and 1996 SSI obtained from the satellite instruments SUSIM (dotted line) and SOLTICE (diamonds), and compiled by J. Lean (2000, solid line) and D. Marsh (Woods and Rottman 2002, dashed line).

from the reconstructions provided by J. Lean and D. Marsh (Lean 2000; Woods and Rottman 2002). The reconstruction data have been acquired from the SOLARIS homepage (http://strat-www.met.fu-berlin.de/~matthes/sparc/solaris.html). All data sets show small variability of about 0.3% in the spectral interval 400–290 nm, a steady increase of the variability up to 8% in the spectral interval 290–160 nm, almost constant variability of about 10–20% from 160 to 125 nm and a sharp increase of variability up to 40% near the Lyman-α line (~121.6 nm). The discussion of the satellite data quality can be found elsewhere (e.g., Floyd et al. 1998; Rottman et al. 2004; Fox 2004). It is worth noting that the most pronounced disagreement between the different data sets occurs in the spectral interval 170–210 nm. Substantial deviation of the SOLTICE data over 240–270 nm is interesting, but it is probably caused by the instrumental uncertainty.

While the visible radiation carries substantial energy (~43% of the TSI), its relative contribution to the TSI variability from the solar maximum to the solar minimum is only about 0.14%. Because the atmosphere is almost transparent for the visible radiation, this energy will be partly deposited at the surface or in the lower troposphere and partly reflected back to the space (e.g., by clouds). The total energy carried by the UV radiation is much smaller (~8% of TSI), but the relative contribution of the UV variability is as high as 67% (see also Krivova et al. 2006). In contrast to the visible radiation a substantial part of this energy is deposited in the middle atmosphere due to intensive absorption by the atmospheric ozone, oxygen and some other species (Brasseur and Solomon 1986).

The direct effects of the long-term SSI variability on the ozone and temperature of the middle atmosphere have been studied with 1D and 2D models (e.g., Brasseur et al. 1983; Garcia et al. 1984; Kiselev and Rozanov 1985; Huang and Brasseur 1993; Haigh 1994; Fleming et al. 1995). It was shown that an enhancement of the UV irradiance results in an enhanced oxygen photolysis, which, in turn, results in an intensification of the ozone production, and subsequent increase of the ozone mixing ratios in the middle stratosphere. The direct heating of the middle atmosphere due to the absorption of enhanced solar UV radiation and heating via elevated ozone concentration provides an additional source of energy for the solar maximum case and heats the stratosphere up mostly in the tropics and in the summer hemisphere, leading to the enhancement of the temperature gradient between winter hemisphere high-latitudes and tropical area.

In Fig. 2 we illustrate direct effects of the increased SSI using a simple 1D radiative-convective-photochemical model described by Rozanov et al. (2002). With the model we have carried out two simulations using the reconstruction of the SSI provided by J. Lean (2000) for 1992 (maximum of the solar activity) and 1996 (minimum of the solar activity). The applied solar forcing is illustrated in Fig. 1 (solid line). Figure 2a represents the temperature changes due to the SSI enhancement for different cases. The first case (dashed line) represents the temperature response when we assumed that relative changes of the solar irradiance are the same for all considered wavelengths. The obtained temperature signal is very small (~0.2 K) in the entire atmosphere. In the second case (dotted line) we imposed the reconstructed SSI changes, but did not allow ozone to respond. As can be seen from Fig. 2a, the difference in the temperature response between these two cases is considerable.

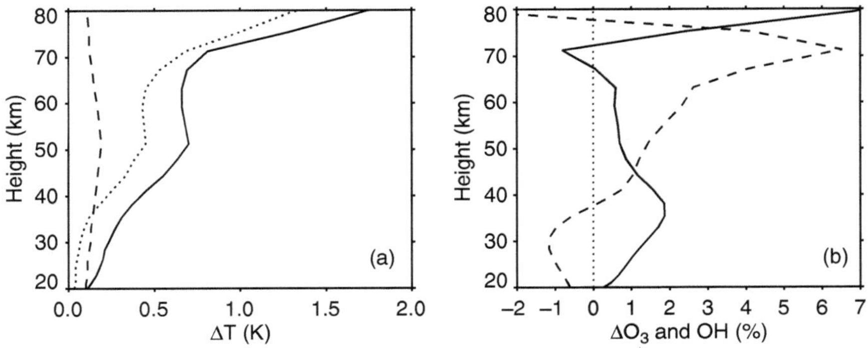

Fig. 2 The changes of the temperature (K), ozone (%) and hydroxyl (%) due to the enhancement of SSI from 1996 to1992. Panel (a): dashed line represents the case with homogeneous increase of SSI at all wavelength; dotted and solid lines represent the cases when the SSI from Lean (2000) reconstruction is used. Dotted line represent the case when the ozone was not allowed to respond. Panel (b): ozone changes (solid line), hydroxyl (dashed line).

In the second case the temperature signal is significantly higher in the stratosphere (~0.5 K at 50 km) and mesosphere (1.3 K at 80 km). It confirms the crucial role of the correct representation of the SSI variability in the model. The third case (solid line) is similar to the second case, but we applied the model with interactive photochemistry. Obtained ozone changes are responsible for an additional warming (~0.3 K) in the stratosphere and significant (~0.5 K) increase of the temperature response in the mesosphere. Near 70 km, however, this additional warming is smaller in comparison with other altitudes. The ozone changes presented in Fig. 2b can explain the difference in the temperature responses between the third and second case. The ozone mixing ratio increases up to 2% at 20–60 km and up to 7% above 75 km mainly as a result of the enhanced ozone production due to the intensification of the oxygen photolysis. The obtained increase of the ozone concentration magnifies the temperature response at these levels. The ozone decrease near 70 km layer occurs mainly due to elevated hydroxyl mixing ratio (illustrated with dashed line) caused by the enhanced H_2O photolysis. This ozone destruction results in smaller additional heating of this layer compared to the upper mesosphere and stratosphere.

3 Downward Propagation of the Solar Signal

Discussed in Sect. 2 direct heating enhances the temperature gradient between winter hemisphere high-latitudes and the tropical area. Enhanced equator-to-pole temperature gradient leads to an intensification of the Polar Night Jets (PNJ), which increases poleward and downward wave reflection suppressing the dynamical wave forcing in the stratosphere and the Brewer–Dobson circulation. The deceleration of the Brewer–Dobson circulation leads, in turn, to subsequent ozone enhancement and warming in the tropical lower stratosphere. Thus, the lower stratosphere warming can be considered as a key element of this mechanism. The change of the planetary wave pattern in the troposphere leads to an alteration of the Hadley cell position and strength and redistribution of the surface air temperature similar to the positive phase of the Arctic oscillation (Thompson and Wallace 1998). This mechanism has been suggested by several authors (Hines 1974; Haigh 1996; Kodera 1996; Hood et al. 1993; Shindell et al. 1999; Kodera and Kuroda 2002) on the basis of theoretical investigations and model simulations. Thus, the primary effects of the solar irradiance changes from solar minimum to solar maximum should be clearly seen in the composition, temperature and dynamics of the middle atmosphere. Moreover, the above-mentioned mechanism of dynamical coupling can facilitate downward propagation of this perturbation from the middle atmosphere to the troposphere. The tropospheric effects are expected to be the most pronounced during the cold season, when the PNJ exists in the stratosphere. Some evidences of the reliability of this mechanism can be obtained from the analysis of the solar signal extracted from the observation data. For example, Crooks and Gray (2005) have found statistically significant warming in the lower tropical stratosphere for the solar maximum case in the ERA-40 data. However, the solar signal is rather small and it is extremely difficult to extract it from noisy stratospheric zonal wind and surface temperature fields in the northern hemisphere. The reliability of the relationship between PNJ strength, lower stratospheric tropical temperature and surface climate can be demonstrated using the composite analysis by comparing the atmospheric quantities for different strength of the northern polar vortex. This analysis is not directly relevant to the solar signal, but provides a good illustration how this mechanism could work. The response of the observed temperature and ozone to the strength of the PNJ in the northern hemisphere has been analysed by Egorova et al. (2005). Using CPC data they confirmed that the above-mentioned theoretical mechanism works providing slower Brewer–Dobson circulation and warmer tropical stratosphere during the years when the PNJ is strong. Here we illustrate this mechanism using the zonal wind and temperature fields averaged over the northern cold season (November–March) from the ECMWF ERA-40 reanalysis data set. It is known that the years with strong PNJ are characterised by the deepening of the polar stratospheric depression, formation of the ridges over mid-latitudes and acceleration of the zonal wind. In the ERA-40 data the zonal wind difference between the composites depicted in Fig. 3 (left panel) reaches 10 m/s. Figure 3 (right panel) also shows the pattern of the temperature response, which consists of pronounced dipole-like structure with the cooling of up to 2 K in the middle-to-lower stratosphere and warming of up to 5 K in the upper stratosphere over the northern high latitudes. In the tropical lower stratosphere the expected warming is statistically significant and reaches 1.5 K. The temperature changes in other parts of the stratosphere are not statistically significant, while there are several significant spots of warming and cooling in the troposphere and near the surface.

The changes of the surface air temperature (SAT) and sea level pressure (SLP) are depicted in Fig. 4. The atmospheric state for the strong PNJ composite is characterised by the enhanced depression of up to 9 hPa over the polar area and formation of the positive pressure anomalies over the North Atlantic. As expected this pattern coincides with the behaviour of the troposphere during the positive phase of the Arctic and North Atlantic oscillations (e.g., Hurrell 1995; Thompson and Wallace 1998). The subsequent changes of the tropospheric circulation and storm tracks lead to the redistribution of the energy and warming of up to 3 K over Russia, Scandinavia and US (albeit the latter is not statistically significant) and cooling in

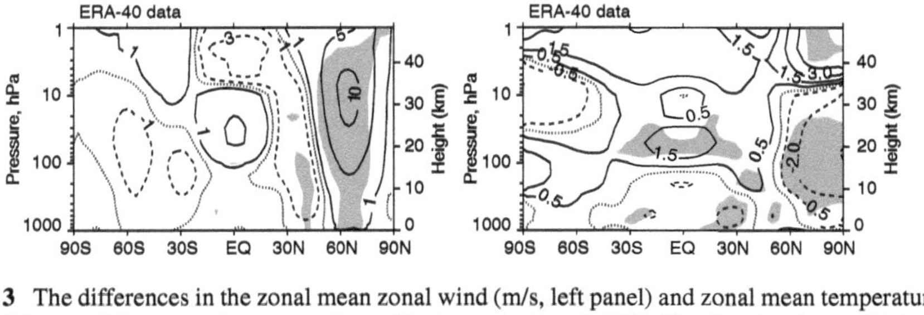

Fig. 3 The differences in the zonal mean zonal wind (m/s, left panel) and zonal mean temperature (K, right panel) between the composites with strong and weak PNJ. The data has been obtained from ECMWF ERA-40 archive. Shading marks the area where the difference is statistically significant at 95% confidence level.

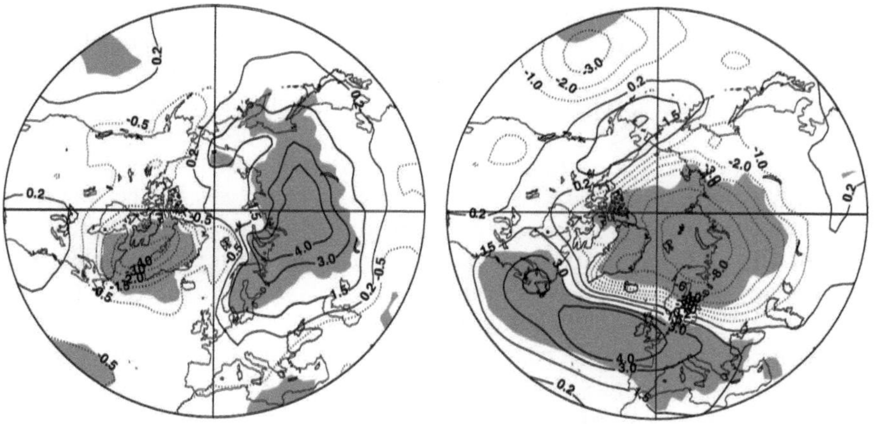

Fig. 4 The differences in the surface air temperature (K, left panel) and sea level pressure (hPa, right panel) between the composites with strong and weak PNJ. The data has been obtained from ECMWF ERA-40 archive. Shading marks the area where the difference is statistically significant at 95% confidence level.

the North Atlantic, Canada, Greenland and Central Asia (the latter is not shown). The explanation of the interannual variability of the northern PNJ is out of the scope of this paper. Here, we just mention that the alternation of the northern vortex can occur as a result of many natural phenomena at the surface (e.g., El-Nino) and in the stratosphere (e.g., volcanic eruptions). We just present this illustration to show that perturbations of the stratospheric PNJ really affect the temperature in the tropical lower stratosphere and in the lower troposphere over in the northern hemisphere. Therefore, if the zonal wind is sensitive to the variability of the solar irradiance these changes will be most likely detectable in these areas. Now, the question is how successfully can we simulate the changes in the atmosphere due to SSI variability?

4 Simulations of the Solar Signal in the Atmosphere and Comparison with Observations

There were many attempts to simulate the above-explained chain of physical and chemical processes involved in the Sun–Earth climate relationship with state-of-the-art general circulation and chemistry-climate models (e.g., Tourpali et al. 2003; Egorova et al. 2004; Matthes et al. 2004). However, virtually all attempts were performed in the time slice mode, that is, comparing perpetual solar maximum/ minimum conditions, which is not the case for the real world where the solar signal is transient. The comparison of the simulation results with the solar signal extracted from the observations is difficult in this case due to inconsistent experiment set-up and also due to different methods applied to extract the solar signal. To address this issue we have performed transient experiments with the chemistry-climate model (CCM) SOCOL (Rozanov et al. 2005a) covering 1975–2000 period of time and driven by realistic set of time evolving forcing, which includes sea-surface temperature and sea ice (SST/SI), greenhouse gases (GHG), ozone depleting substances (ODS), stratospheric aerosol (STA), and solar irradiance (SOL). The spectral solar irradiance in the CCM SOCOL affects solar fluxes, photolysis and heating rates. The set up of the experiment and the applied forcing have been described by Rozanov et al. (2005b) and Eyring et al. (2006). For the extraction of the solar signal from the simulated time series we have applied standard linear multiple-regression analysis. This approach is based on the representation of the simulated time series as a linear combination of the considered forcing mechanisms. For GHG and ODS forcing we have used time series of the applied CO_2 mixing ratio and total chlorine loading. As a proxy for volcanic forcing we used time series of the vertically integrated aerosol extinction at 550 nm in the tropical area. The solar irradiance variability is represented by the time series of the solar irradiance at 205 nm.

Figures 5 and 6 illustrate the solar signal in the zonal mean zonal wind, temperature and ozone averaged over the cold season in the northern hemisphere. Significant and rather homogeneous warming related mostly to the direct effects of SSI (see Fig. 2) is clearly visible in the upper stratosphere (above 40 km) and mesosphere in the tropics and the southern (summer) hemisphere. This warming is followed by the increase of the zonal wind in the both hemispheres. Not expected PNJ acceleration in the southern hemisphere appears due to strong acceleration and late breaking of the polar vortex in early November (not shown). In the northern hemisphere the

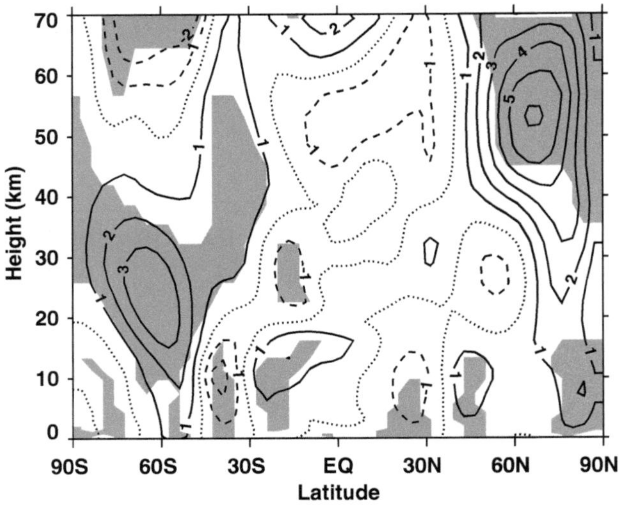

Fig. 5 Zonal mean solar signal in zonal wind (in m/s) averaged over the northern cold season. The shading marks regions with statistically significant solar signal at the 95% confidence level.

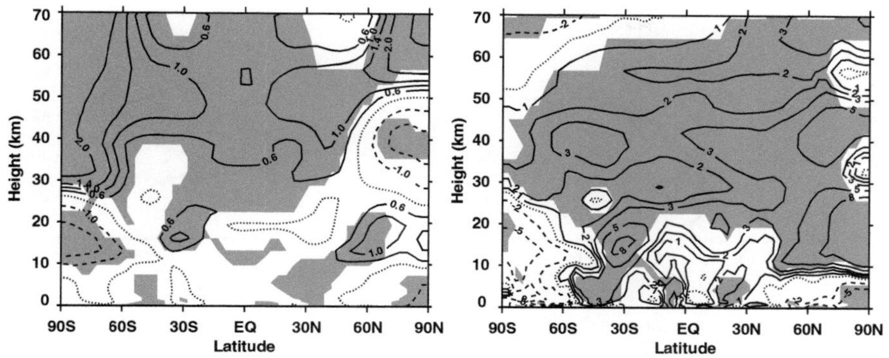

Fig. 6 Zonal mean solar signal in temperature (in K, left) and ozone mixing ratio (in %, right) averaged over the northern cold season. The shading marks regions with statistically significant solar signal at the 95% confidence level.

increase of the zonal wind occurs in the entire atmosphere, albeit is only significant above 40 km and below 20 km. Dipole-like temperature response in the southern lower atmosphere and in the northern middle atmosphere is connected to the zonal wind acceleration. The ozone response consists of the significant ozone increase almost in the entire stratosphere due to enhanced chemical production and changes of the Brewer–Dobson circulation caused by the zonal wind alteration. More or less homogeneous ozone increase of up to 4% occurs in the middle extra-polar stratosphere in agreement with the results of 1D model (see Fig. 2). The most extreme ozone change appears in the lower stratosphere around 30°S. This ozone increase most likely reflects the deceleration of the Brewer–Dobson circulation resulting in smaller influx of the ozone-poor air from the upper troposphere. This is supported by the appearance of the warming spot in the same area due to less intensive upward motions there and smaller decompression cooling. The obtained results confirm that our model is able to simulate theoretically anticipated chemical and dynamical changes in the atmosphere due to enhancement of the SSI presented in Sects. 2 and 3.

In Fig. 7 we compare the simulated annual mean solar signal in the tropical (averaged over 25°S–25°N) temperature and ozone with the solar signal extracted from the different observation data sets using similar linear multiple regression technique as well as with the results obtained from the time-slice experiments with CCM SOCOL (Egorova et al. 2004). The simulated solar signal in the tropical temperature is similar to the signal from the time-slice experiment (Egorova et al. 2004) in the entire atmosphere. However, the solar signal extracted from the transient experiment is more pronounced in the lower stratosphere and mesosphere. The difference in the mesosphere can be partially explained by the difference in the applied SSI perturbations. In Fig. 1 the SSI variability from SUSIM data (used by Egorova et al. 2004) is systematically higher than the SSI variability from the compilation of Lean (2000) used for the transient runs especially in the spectral

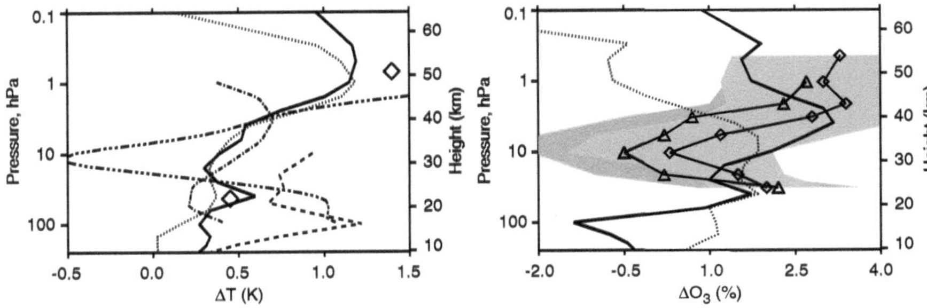

Fig. 7 Annual mean solar signal in the tropical (25°S–25°N averaged) temperature (K, left) and ozone (%, right) calculated with CCM SOCOL. Left panel: CCM SOCOL in transient mode (solid line); Egorova et al. (2004) (dotted line); SSU/MSU from Hood and Soukharev (2000) (dash-dotted line), CPC reanalysis from Hood, 2004 (dot-dot-dashed line); NCEP reanalysis from Labitzke et al. (2002) (dashed line), ERA-40 reanalysis from Crooks and Gray, 2005 (diamonds). Right panel: CCM SOCOL in transient mode (solid line); Egorova et al., 2004 (dotted line); SBUV from Soukharev and Hood (2006) (triangles); SAGE from Soukharev and Hood (2006) (diamonds). Shaded areas in the right panel illustrate the uncertainty of the solar signal extracted from the observation data.

range 170–210 nm, which is mostly responsible for the heating in the mesosphere. Higher temperature response in the lower stratosphere can be caused by the different experimental set-up or by the different method for the solar signal definition. The comparison with the observations is not straightforward, because the analysis of different data sets gives rather different results. The simulated temperature signal is close in the middle stratosphere to the analysis of SSU data by Hood and Soukharev (2000) and in the lower and upper stratosphere to the analysis of ERA-40 data by Crooks and Gray (2005). The simulated solar signal in ozone is rather similar to the Egorova et al. (2004) results only in the middle stratosphere. The difference in the upper stratosphere and mesosphere can be again explained by smaller SSI perturbations, which led to less intensive production of HO_x from water vapor photolysis. Possible contamination of the upper atmosphere during the long-term run under permanent solar maximum/minimum conditions or application of the linear multiple regression analysis for the extraction of the solar signal from the transient runs could also play some role. Taking into account the uncertainty of the observed solar signal we can conclude that the shape and magnitude of the solar signal in ozone from the transient run is in reasonable agreement with observations, however substantial disagreement around 30 km remains unexplained.

The simulated solar signal in the surface air temperature and sea-level pressure is illustrated in Fig. 8. In general agreement with the downward propagation mechanism considered in Sect. 3, the acceleration of the zonal wind in the stratosphere, shown in Fig. 5, affects the tropospheric circulation and surface temperature. The surface pressure changes have a dipole structure with a deepening of the polar depression of up to 3 hPa and formation of the positive pressure anomalies with the same magnitude in the Atlantic and Pacific sectors. Corresponding redistribution of

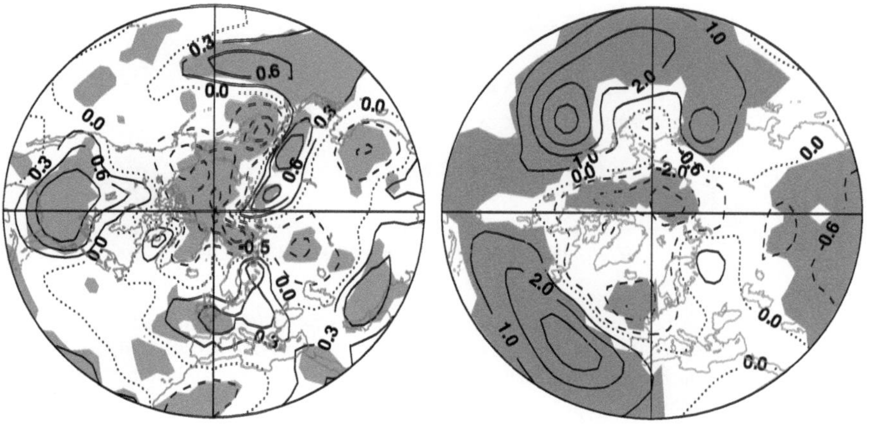

Fig. 8 The solar signal in the surface air temperature (K, left) and sea-level pressure (hPa, right) over the northern hemisphere during cold season. The shading marks regions with statistically significant solar signal at the 95% confidence level.

the energy transport leads to the warming over Eastern Russia, Europe, USA and Pacific Ocean of up to 0.8 K. The simulated warming resembles the warming pattern in Fig. 4; however, there is some difference in the location of the warming spot, its magnitude and statistical significance. The most substantial differences appear over USA and Pacific Ocean, where the solar signal in the surface air temperature is much higher and statistically significant. It should be noted, that the surface air temperature changes in Fig. 4 are not the result of the solar influence, but rather reflect the reaction of the surface climate to the increase of the PNJ strength, Therefore, the causes of this disagreement may result not only from the differences in the applied statistical techniques (composite analysis versus multiple linear regression analysis) but also from the different nature of the forcing. Nevertheless, the obtained results confirm that the SSI variability influences not only the middle atmosphere, but also the tropospheric climate.

5 Solar Variability and Temperature Trends in the Upper Stratosphere

The attribution of the climate changes requires a correct understanding of the temperature evolution in the upper stratosphere (IPCC 2001; Ramaswamy et al. 2001). It is commonly accepted that the main driving mechanisms of the observed cooling in the stratosphere during the satellite era have been the rising concentration of GHG and ozone depletion by man-made ODS production (Ramaswamy et al. 2001). However, it was pointed out by Randel (2004) that the evolution of the temperature in the upper stratosphere does not look like a straight line and cannot be

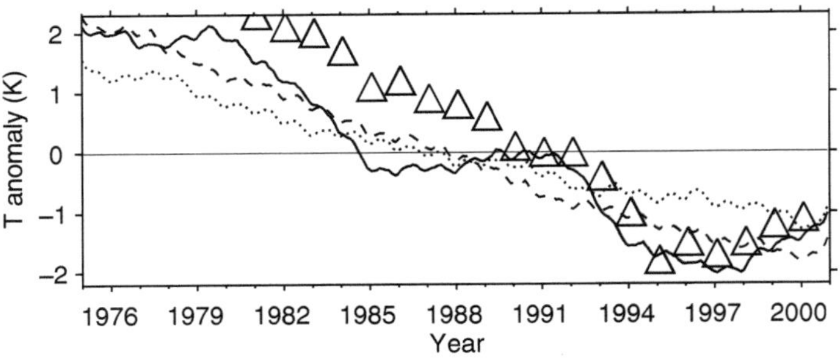

Fig. 9 Anomalies of the global mean temperature near the stratopause for the last 25 years of the 20th century from the CCM SOCOL runs driven by the time evolving SST/SI and GHG (dotted line); GHG+ODS (dashed line) and GHG+ODS+STA+SOL (solid line) and from the MSU data (triangles, Randel 2004).

explained only by the rather linear changes of GHG and ozone. Therefore, it is of great interest to estimate the influence of the solar irradiance variability on the temperature in the upper stratosphere in comparison with the GHG and ozone effects. To address this problem we have performed two additional 26-year-long runs with the CCM SOCOL. The first run was driven by the time evolving SST/SI and GHG, and the second run by SST/SI, GHG, and ODS. These runs do not include the variability of solar irradiance, therefore the comparison of the temperature in the upper stratosphere from the three CCM SOCOL runs with the observation data allows us to elucidate the influence of the solar irradiance. The anomalies of the temperature in the upper stratosphere relative to the mean temperature are presented in Fig. 9 for three model runs together with the observed anomalies presented by Randel (2004). The upper stratosphere in the model runs without solar irradiance variability steadily cools down with the time. The magnitude of the cooling depends on the applied forcing, it is about 50% higher for the case when the ozone depletion due to chlorine loading is included in the model. The temperature evolution in the case when the solar irradiance variability included is more complicated. The general cooling trend in this case is modulated by the localized warming during the solar activity maximum (1980, 1991, 2000) causing the no cooling or even small warming periods. This behaviour is also clearly visible in the observation data especially for the last solar maximum. Thus, the solar irradiance variability is very important for the correct detection and attribution of the stratospheric temperature trends.

6 Conclusions

Recent satellite observations show that the solar ultraviolet irradiance is much more variable than the total solar irradiance, therefore it has a potential to amplify the solar signal produced with models driven only by the TSI variability. The magnitude of the

SSI variability still has a large uncertainty in particular for the wavelength longer than 220 nm, but there is a hope that the future and ongoing measurements will help to establish SSI variability with higher accuracy. Presented here are simulations of the atmospheric effects of the solar irradiance variations during the 11-year solar activity cycle, which reveal that the direct and indirect (via ozone production) heating in the upper and middle stratosphere due to enhancement of the solar spectral irradiance leads to an acceleration of the polar night jets and suppression of the Brewer–Dobson circulation resulting in the ozone increase and warming of the lower tropical stratosphere. These stratospheric changes alter the tropospheric circulation leading to a statistically significant warming of the surface air over Russia, Europe and North America. The connection of the tropospheric climate with the intensity of the polar night jets was illustrated using the composite analysis of the ERA-40 data, which shows that an increase in the stratospheric zonal wind is linked to the surface air temperature and sea-level pressure changes, which resemble the simulated solar signal and confirm the reality of the downward propagation mechanism. Solar UV irradiance does not substantially amplify the solar signal in the global mean temperature, however in some areas (Russia, Europe and North America) SSI induced warming can reach 0.8 K. It means that additional warming due to SSI variability in these particular areas from 1900 to 1960, when the solar activity had undergone large changes can be as high as 1.5–2 K.

Despite some progress, the comparison of the simulated solar signal with the satellite observations is not very successful for the tropical temperature and ozone. One reason for this disagreement could be in the substantial uncertainty of the observational estimations. The absence of such processes as QBO and energetic particles in the model could also play a role and detailed studies of these processes are in progress.

The importance of the solar spectral irradiance variability for the attribution of the temperature changes in the upper stratosphere has been demonstrated by the comparison of the simulated and observed temperature evolution during the last 25 years of the 20th century. The comparison shows that spectrally resolved solar forcing is absolutely necessary to simulate the observed temperature evolution in the upper stratosphere.

Acknowledgements The work is supported by the Swiss Federal Institute of Technology. We thank T. Peter, M. Schranner, V. Zubov, M. Haberreiter, J. Stenflo, A. Ohmura, M. Wild, S. Brönnimann, A. Fischer, B. Luo, C. Schnadt, and I. Karol for their valuable contributions. ECMWF ERA-40 data used in this study have been obtained from the ECMWF data server.

References

Brasseur, G. and S. Solomon, 1986: *Aeronomy of the Middle Atmosphere*. D. Reidel, Dordrecht, The Netherlands.

Brasseur, G., P. De Baets, and A. De Rudder, 1983: Solar variability and minor constituents in the lower thermosphere and in the mesosphere. *Space Sci. Rev.*, **34**, 377–385.

Crooks, S. and L. Gray, 2005: Characterization of the 11-year solar signal using a multiple regression analysis of the ERA-40 dataset. *J. Climate*, **18**, 996–1015.
Egorova, T., E. Rozanov, and E. Manzini et al., 2004: Chemical and dynamical response to the 11-year variability of the solar irradiance simulated with a chemistry-climate model. *Geophys. Res. Lett.*, **31**, doi:10.1029/2003GL019294.
Egorova, T., E. Rozanov, and V. Zubov, et al., 2005: Chemistry-Climate Model SOCOL: a validation of the present-day climatology. *Atmos. Chem. Phys.*, **5**, 1557–1576.
Eyring, V., N. Butchart, and W. Waugh, et al., 2006: Assessment of the temperature, trace species, and ozone in chemistry-climate model simulations of the recent past. *J. Geophys. Res.*, **111**, doi:10.1029/2006JD007327.
Fleming, E., C. Chandra, and C. Jackman et al., 1995: The middle atmospheric response to short and long term solar UV variations: analysis of observations and 2D models results. *J. Atmos. Terr. Phys*. **57**, 333–365.
Floyd, L., P. Reiser, and P. Crane et al., 1998: Solar cycle 22 UV spectral irradiance variability: current measurements by SUSIM UARS. *Solar Phys.*, **177**, 79–87.
Fox, P., 2004: Solar activity and irradiance variations. In: Pap J and Fox P, eds., *Solar Variability and Its Effects on Climate, Geophys. Monograph*, vol 141. Washington, DC, 141.
Fröhlich, C., 2004: Solar Irradiance Variability. In: Pap J and Fox P., eds., *Solar Variability and Its Effects on Climate, Geophys Monograph*, vol 141. Washington, DC, 97.
Garcia, R., S. Solomon, and R. Roble et al., 1984: A numerical study of the response of the middle atmosphere to the 11 year cycle. *Planet Space Sci.*, **32**: 411–423.
Haigh, J. D., 1994: The role of stratospheric ozone in modulating the solar radiative forcing of climate. *Nature*, **370**, 544–546.
Haigh, J., 1996: The impact of solar variability on climate. *Science*, **272**, 981–984.
Hines, C., 1974: A possible mechanism for the production of Sun-weather correlations. *J. Atmos. Sci.*, **31**, 589–591.
Hood, L., 2004: Effects of solar UV variability on the stratosphere. In: Pap J and Fox P, eds., *Solar Variability and Its Effects on Climate, Geophys Monograph*, vol 141. Washington, DC, 283.
Hood, L. and B. Soukharev, 2000: The solar component of long-term stratospheric variability: observations, model comparison, and possible mechanisms. In: *Proceed. II SPARC General Assembly*, Mar del Plata, Argentina.
Hood, L., J. Jirikovich, and J. McCormack, 1993: Quasi-decadal variability of the stratosphere: influence of long-term solar ultraviolet variations. *J. Atmos. Sci.* **50**, 3941–3958.
Hoyt, D. and K. Schatten, 1997: The role of the Sun in climate change. Oxford University, New York.
Huang, T. and G. Brasseur, 1993: Effect of long-term solar variability in a two-dimensional interactive model of the middle atmosphere. *J. Geophys. Res*. **98**, 20,413–20,427.
Hurrell, J., 1995: Decadal trends in the North-Atlantic oscillation - regional temperatures and precipitation. *Science*, **269**, 676–679.
Intergovernmental Panel of Climate Change (IPCC), 2001: *Climate Change 2001: The Scientific Basis*, Cambridge University, New York.
Kiselev, A. and E. Rozanov, 1985: Radiative - photochemical modeling of the ozone mechanism of the Sun–Earth connections. In: VNIIGMI-WDC publication No. 452, 54.
Kodera, K., 1996: On the origin and nature of the interannual variability of the winter stratospheric circulation in the northern hemisphere. *J. Geophys. Res.* 100:14077–14087.
Kodera, K. and Y. Kuroda, 2002: Dynamical response to the solar cycle. *J. Geophys. Res.*, **107**: doi:10.1029/2002JD002224.
Krivova, N., S. Solanki, and L. Floyd, 2006: Reconstruction of solar UV irradiance in cycle 23. *Astron. Astr.*, **452**, 631–639.
Labitzke, K., J. Austin, and N. Butchart et al., 2002: The global signal of the 11-year solar cycle in the stratosphere: observations and models. *J. Atm. Sol- Terr. Phys.*, **64**, 203–210.
Lean, J., 2000: Evolution of the sun's spectral irradiance since the Maunder Minimum. *Geophys. Res. Lett.*, **27**, 2425–2428.
Matthes, K., U. Langematz, and L. Gray et al., 2004: Improved 11-year solar signal in the Freie Universität Berlin Climate Middle Atmosphere Model (FUB-CMAM). *J. Geophys. Res.*, **109**, doi:10.1029/2003JD004012.

North, G., Q. Wu, and M. Stevens, 2004: Detecting the 11-year Solar Cycle in the Surface Temperature Field. In: Pap J and Fox P., eds., *Solar Variability and Its Effects on Climate, Geophys Monograph*, vol 141. Washington, DC, 251.

Ramaswamy, V., M. L. Chanin, and J. Angell et al., 2001: Stratospheric temperature trends: observations and model simulations. Rev. Geophys. 39:71–122.

Randel, W., 2004: Detection, Attribution and Prediction: an integration of SPARC activities. Paper presented at 3rd General Assembly of SPARC, Victoria Conference Centre, Victoria (BC), 01–06 August 2004.

Reid, G., 2000: Solar variability and Earth's climate: introduction and overview. In: Solar Variability and Climate, vol 11, Friis-Christiansen E, Fröhlich C, Haigh J, Schüssler M, von Steiger R., eds., , Kluwer, The Netherlands, 1.

Rottman, G., L. Floyd, and R. Viereck, 2004: Measurement of the solar ultraviolet irradiance. In: Pap J and Fox P., eds., *Solar Variability and Its Effects on Climate, Geophys Monograph*, vol 141. Washington, DC, 111.

Rozanov, E. et al., 2002: Estimation of the ozone and temperature sensitivity to the variation of spectral solar flux. In: A. Wilson, ed., *From Solar Min to Max: Half a Solar Cycle with SOHO*, SOHO 11 Symposium, Davos, March 2002, vol SP-508, ESA, ESTEC Noordwijk, 181.

Rozanov, E., M. Schraner, and T. Egorova, et al., 2005a: Solar signal in atmospheric ozone, temperature and dynamics simulated with CCM SOCOL in transient mode. *Mem. S. A It* **76**:876–879.

Rozanov, E., M. Schraner, and C. Schnadt, et al., 2005b: Assessment of the ozone and temperature variability during 1979–1993 with the chemistry-climate model SOCOL. *Adv. Space Res.* **35**, 1375–1384.

Shindell, D., D. Rind, and N. Balachandran et al., 1999: Solar cycle variability, ozone, and climate. *Science*, **284**, 305–308.

Soukharev, B. and L. Hood, 2006: Solar cycle variation of stratospheric ozone: multiple regression analysis of longterm satellite data sets and comparisons with models. *J. Geophys. Res.* **111**, doi:10.1029/2006JD007107.

Stott, P., G. Jones, and J. Mitchell, 2003: Do models underestimate the solar contribution to recent climate change? *J. Climate*, **16**, 4079–4093.

Thompson, D. and J. Wallace, 1998: The arctic oscillation signature in the wintertime geopotential height and temperature fields. *Geophys. Res. Lett.*, **25**, 1297.

Tourpali, K., C. Schuurmans, and R. van Dorland et al., 2003: Stratospheric and tropospheric response to enhanced solar UV radiation: a model study. *Geophys Res Lett* 30:doi: 10.1029/2002GL016650.

Woods, T. and G. Rottman, 2002: Solar ultraviolet variability over time periods of aeronomic interest. In: Mendillo M, Nagy A, Waite J., eds., Comparative Aeronomy in the Solar System, Geophys. Monography, vol 130, Washington, DC, 221.

Stratospheric Ozone Variations Caused by Solar Proton Events Between 1963 and 2005

C. H. Jackman[1] and E. L. Fleming[2]

Abstract Some solar eruptions lead to solar proton events (SPEs) at the Earth, which typically last a few days. High energy solar protons associated with SPEs precipitate on the Earth's atmosphere and cause increases in odd hydrogen (HO_x) and odd nitrogen (NO_y) in the polar cap regions (>60° geomagnetic). The enhanced HO_x leads to short-lived ozone depletion (~days) due to the short lifetime of HO_x constituents. The enhanced NO_y leads to long-lived ozone changes because of the long lifetime of the NO_y family in the stratosphere and lower mesosphere. Very large SPEs occurred in 1972, 1989, 2000, 2001, and 2003 and were predicted to cause maximum total ozone depletions of 1–3%, which lasted for several months to years past the events. A long-term data set of solar proton fluxes used in these computations has been compiled for the time period 1963–2005. Several satellites, including the NASA Interplanetary Monitoring Platforms (1963–1993) and the NOAA Geostationary Operational Environmental Satellites (1994–2005), have been used to compile this data set.

1 Introduction

Explosions on the Sun sometimes result in large fluxes of high-energy solar protons at the Earth, especially near the maximum period of activity of a solar cycle. This disturbed time, wherein the solar proton flux is generally elevated for a few days, is known as a solar proton event (SPE). Solar protons are guided by the Earth's magnetic field and impact both the northern and southern polar cap regions (>60° geomagnetic latitude) (e.g., see Jackman and McPeters 2004). These protons can impact the neutral middle atmosphere (stratosphere and mesosphere) and produce

[1]Atmospheric Chemistry and Dynamics Branch, Code 613.3, NASA Goddard Space zigh Center, Greenbelt, MD 20771, USA
Charles.H.Jackman@nasa.gov

[2]Atmospheric Chemistry and Dynamics Branch, Code 613.3, NASA Goddard Space Flight Center, Greenbelt, MD 20771, USA
fleming@kahuna.nasa.gov

S. Brönnimann et al. (eds.), *Climate Variability and Extremes during the Past 100 Years.*

both HO_x (H, OH, HO_2) and NO_y (N, NO, NO_2, NO_3, N_2O_5, HNO_3, HO_2NO_2, $ClONO_2$, $BrONO_2$) constituents either directly or through a photochemical sequence (e.g., Swider and Keneshea 1973; Crutzen et al. 1975; Jackman et al. 1980; Solomon et al. 1981; McPeters 1986; Zadorozhny et al. 1992). Ozone is also impacted by the solar protons through direct photochemical destruction forced by the HO_x and NO_y enhancements (e.g., Weeks et al. 1972; Heath et al. 1977; Solomon et al. 1983; Jackman et al. 1990).

Although all sizes of SPEs can have an impact on the atmosphere, the extremely large SPEs cause the most pronounced changes. Several of these extremely large SPEs have occurred in the past 40 years. Huge fluxes of high-energy protons have impacted the Earth's atmosphere in 1972, 1989, 2000, 2001, and 2003. In this paper, the impact of SPEs over the 1963–2005 period will be discussed, concentrating particularly on the atmospheric effects during and after the huge SPEs.

The paper is divided into six primary sections, including the Introduction. We discuss the very important solar proton measurements and their production of odd hydrogen (HO_x) and odd nitrogen (NO_y) in Sect. 2. A comparison of the largest 15 SPEs in the past four solar cycles is also undertaken in Sect. 2. The GSFC two-dimensional model used to simulate the impact of the SPEs on the atmosphere is discussed in Sect. 3. The short-term impact of these SPEs on ozone during and for several days after particular events is given in Sect. 4. Longer term influences of the SPEs on the middle atmosphere are discussed in Sect. 5. Finally, the conclusions are given in Sect. 6.

2 Proton Fluxes: Odd Hydrogen (HO_x) and Odd Nitrogen (NO_y) Production

2.1 Proton Fluxes

Solar proton fluxes have been measured by a number of satellites in interplanetary space or in orbit around the Earth. The National Aeronautics and Space Administration (NASA) Interplanetary Monitoring Platform (IMP) series of satellites provided measurements of proton fluxes from 1963–1993. IMPs 1–7 were used for the fluxes from 1963–1973 (Jackman et al. 1990) and IMP 8 was used for the fluxes from 1974–1993 (Vitt and Jackman 1996). The National Oceanic and Atmospheric Administration (NOAA) Geostationary Operational Environmental Satellites (GOES) were used for proton fluxes from 1994–2005 (Jackman et al. 2005a). There are uncertainties associated with these proton flux data, especially given the large number of satellite instruments used to compile such a measurement record. We estimate the proton flux uncertainties to be up to 50%, given some straightforward comparisons of particular proton flux measuring instruments. It is beyond the scope of the present study to undertake a more detailed comparison, however, we do recommend that such a study be accomplished by experts in the field of solar particle observations.

Other precipitating particles are associated with SPEs, besides protons. Alpha particles comprise about 10% of the positively charged solar particles with other ions accounting for less than 1% of the remainder (e.g., Mewaldt et al. 2005). Measurements exist for other ions besides protons, but not enough for a continuous observation record from 1963–2005. Electrons also accompany SPEs (e.g., Mewaldt et al. 2005), however, their precipitation is not uniform over the polar caps. The energy deposition of electrons will primarily be in the auroral oval regions, which is more difficult to characterize, and will be confined mostly to the thermosphere and upper mesosphere. We only include solar protons in our computations and note that other charged particles could add modestly to the energy deposition in the middle atmosphere during SPEs.

Protons in their energy deposition process cause ionizations, dissociations, predissociations, and dissociative ionizations in collisions with atmospheric constituents. The protons thus produce secondary electrons, ions, excited molecules and atoms. The proton fluxes were used to compute daily average ion pair production profiles using an energy deposition scheme first discussed in Jackman et al. (1980). The scheme includes the deposition of energy by the protons and assumes 35 eV are required to produce one ion pair (Porter et al. 1976). Thereby, a data set of daily average ion pair production rates for the period 1963–2005 were created for use in model studies and is available at http://strat-www.met.fu-berlin.de/~matthes/sparc/inputdata.html.

2.2 *Odd Hydrogen (HO_x) Production*

Along with the ion pairs, the protons and their associated secondary electrons also produce odd hydrogen (HO_x). The production of HO_x relies on complicated ion chemistry that takes place after the initial formation of ion pairs (Swider and Keneshea 1973; Frederick 1976; Solomon et al. 1981). Solomon et al. (1981) computed HO_x production rates as a function of altitude and ion pair production. Each ion pair typically results in the production of around two HO_x constituents in the upper stratosphere and lower mesosphere. In the middle and upper mesosphere, an ion pair is computed to produce less than two HO_x constituents per ion pair. We include the HO_x production by SPEs in our model using a look-up table (see Jackman et al. 2005b) invoking the computations of Solomon et al. (1981). The HO_x constituents have lifetimes of only hours in the middle atmosphere, therefore, any further effects on other constituents from the HO_x group are apparent only during and shortly after an SPE.

2.3 *Odd Nitrogen (NO_y) Production*

Odd nitrogen is produced when the energetic charged particles (protons and associated secondary electrons) collide with and dissociate N_2. Following Porter et al. (1976) it is assumed that ~1.25 N atoms are produced per ion pair. The Porter et al. (1976) study also further divided the proton impact of N atom production between ground state (~45% or ~0.55 per ion pair) and excited state (~55% or ~0.7 per ion pair) nitrogen atoms. Ground state [$N(^4S)$] nitrogen atoms can create other NO_y constituents, such as NO, through

$$N(^4S) + O_2 \rightarrow NO + O \tag{1}$$

or can lead to NO_y destruction through

$$N(^4S) + NO \rightarrow N_2 + O. \tag{2}$$

Generally, excited states of atomic nitrogen, such as $N(^2D)$, result in the production of NO through

$$N(^2D) + O_2 \rightarrow NO + O \tag{3}$$

(e.g., Rusch et al. 1981; Rees 1989) and do not cause significant destruction of NO_y. Rusch et al. (1981) showed that there are huge differences in the final results of model computations of NO_y enhancements from SPEs that depend strongly on the branching ratios of the N atoms produced. We currently do not include any of the excited states of atomic nitrogen (e.g., $N(^2D)$, $N(^2P)$, and N^+) as computed constituents in our model. We use the following fairly accurate way to best represent the production of NO_y constituents by the SPEs: Assume that 45% of the N atoms produced per ion pair result in the production of $N(^4S)$ (~0.55 per ion pair) and that 55% of the N atoms produced per ion pair result in the production of NO (~0.7 per ion pair). There are uncertainties in these proton-caused NO_y production computations of about 20% (Jackman et al. 1979), which does not include the uncertainty in the measured precipitating proton flux.

The lifetime of odd nitrogen can vary dramatically depending on season and altitude. Odd nitrogen has a relatively short lifetime (~days) in the sunlit middle and upper mesosphere, however, lower mesospheric and stratospheric NO_y can last for weeks past an SPE. A large portion of the SPE-produced NO_y is conserved in a mostly dark polar middle atmosphere in the late fall and winter. This NO_y can then be transported to lower altitudes via the general downward flowing winds during this time of year and its lifetime can range from months to years, if transported all the way to the middle and lower stratosphere.

We have quantified middle atmospheric NO_y production previously (Jackman et al. 1990; Vitt and Jackman 1996; Jackman et al. 2005a) for years 1963 through 2003. We add NO_y computations in this study to these earlier calculations for years 2004 and 2005 and present the annual production from SPEs for the 43-year period 1963 through 2005 in Fig. 1. The annual-averaged sunspot number is also shown in Fig. 1 to illustrate the rough correlation between solar maximum periods and frequency of SPEs.

Substantial amounts of NO_y were produced near solar maximum in several years. The annual global NO_y production from solar protons is computed to be 3.7, 8.4, 6.7, 7.9, and 4.1×10^{33} molecules for the very active years 1972, 1989, 2000, 2001, and 2003, respectively. These annual production rates from SPEs are ~10–25%

of the largest global NO_y source (nitrous oxide oxidation, $N_2O + O(^1D)$) of about 3.3×10^{34} molecules/year (Vitt and Jackman 1996). The SPE sources of NO_y were very significant during these particular years for the middle atmosphere. Since the SPEs typically last only a few days, these impulses of NO_y from SPEs can impact the polar odd nitrogen amounts substantially during brief periods.

The 15 largest SPEs based on NO_y production in the past 40 years are given in Table 1. Surprisingly, eight of them occurred in the most recent solar maximum period.

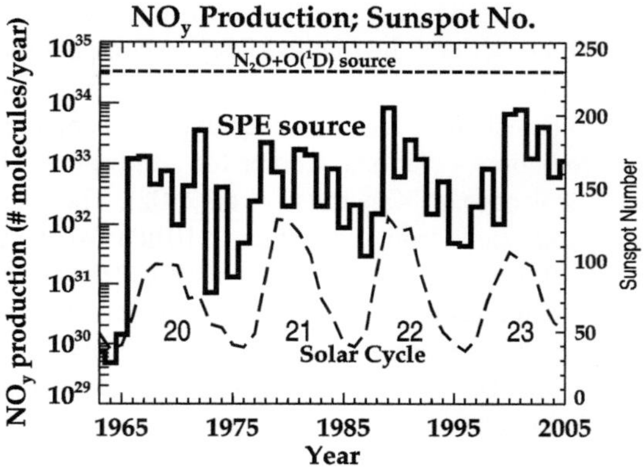

Fig. 1 Total global production of NO_y molecules per year in the polar stratosphere and mesosphere by SPEs (solid histogram – left ordinate) and the oxidation of nitrous oxide $[N_2O + O(^1D)]$ (dash-dot line – left ordinate) for years 1963–2005. The annually averaged sunspot number (dashed line – right ordinate) is also given.

Table 1 Largest 15 solar proton events in the past 40 years

Date of SPEs	Rank in size	NO_y production in the middle atmosphere (# of 10^{33} molecules)
October 19–27, 1989	1	6.7
August 2–10, 1972	2	3.6
July 14–16, 2000	3	3.5
October 28–31, 2003	4	3.4
November 5–7, 2001	5	3.2
November 9–11, 2000	6	2.3
September 24–30, 2001	7	2.0
August 13–26, 1989	8	1.8
November 23–25, 2001	9	1.7
September 2–7, 1966	10	1.2
January 15–23, 2005	11	1.1
September 29–October 3, 1989	12	1.0
January 28–February 1, 1967	13	0.99
March 23–29, 1991	14	0.89
September 7–17, 2005	15	0.88

3 GSFC Two-Dimensional Model Description and Simulations

The latest version of the Goddard Space Flight Center (GSFC) two-dimensional (2D) atmospheric model was used to predict atmospheric changes caused by the solar protons. The model has been in use since the late 1980s and has undergone extensive improvements over the years (Douglass et al. 1989; Jackman et al. 1990). The vertical range of the model, equally spaced in log pressure, is from the ground to approximately 90 km (0.0024 hPa) with approximately a 2 km grid spacing. Latitudes range from 85°S to 85°N with a 10°-grid spacing.

Fleming et al. (2002) described the methodology to compute the transport for the GSFC 2D model using the global winds and temperatures from meteorological data for particular years. This technique has now been applied using the National Centers for Environmental Prediction – National Center for Atmospheric Research (NCEP-NCAR) reanalysis-2 project (e.g., Kanamitsu et al. 2002). These data cover the time period 1958–present, and extend from the surface to 10 hPa. We have used the original NCEP analyses data (Gelman et al. 1986) for 10–1 hPa for 1979–present (climatological fields are used above 10 hPa prior to 1979). For the mesosphere for 1–0.002 hPa, we employ the temperature measurements made by the Microwave Limb Sounder (MLS) onboard UARS for September 1991 through June 1997 (Wu et al. 2003). The 2D model residual circulation and horizontal and vertical eddy diffusion quantities are then derived following the methodology described in Fleming et al. (2002, 2007).

The photochemical scheme includes all reactions that are thought to be important for ozone in the middle atmosphere. The reaction rates, including heterogeneous rates, are taken from Sander et al. (2003). A lookup table is employed in computing the photolytic source term, which is then used in computation of photodissociation rates for atmospheric constituents (Jackman et al. 1996). The GSFC 2D chemistry solver uses the Atmospheric Environmental Research (AER) 2D model scheme (Weisenstein et al. 2004), which computes a diurnal cycle every day. The ground boundary conditions for the source gases are taken from WMO (2003) for the particular simulated year. The model uses chemical families and computes 62 constituents (Jackman et al. 2005b).

We used the GSFC 2D model to compute two primary simulations, "base" and "perturbed," for the years 1960–2010. The transport for years 1960–2004 is driven by NCEP products for those particular years, whereas the transport for the individual years 2005–2010 is an average climatology of the 1958–2004 period. The "base" simulation includes no SPEs, whereas the "perturbed" simulation includes all SPEs from January 1, 1963 through December 31, 2005. The perturbation to the atmosphere was caused by the SPE-produced HO_x and NO_y enhancement.

4 Short-term Impact on Ozone

The ozone response due to very large SPEs is not subtle and has been observed due to numerous events to date (e.g., Jackman and McPeters 2004; López-Puertas et al. 2005; Seppälä et al. 2006). Ozone within the polar caps (60–90°S or 60–90°N geomagnetic) is generally depleted to some extent in the mesosphere and upper stratosphere (e.g., Jackman et al. 2005b) within hours of the start of the SPE. Decreases in mesospheric and upper stratospheric ozone are mostly caused by SPE-induced HO_x increases (see Solomon et al. 1981, 1983; Jackman and McPeters 1985; Jackman et al. 2005b; Verronen et al. 2006) and last only during and for a few hours after the SPEs. The Verronen et al. (2006) study implies that the solar proton-caused HO_x production could be up to 50% uncertain, which does include the uncertainty in the measured precipitating proton flux. SPE-caused NO_y enhancements can also drive upper stratospheric ozone depletion, but do not cause significant mesospheric ozone depletion (Jackman et al. 2001). Although these short-term SPE impacts on ozone merit study and have helped test atmospheric models (e.g., Jackman and McPeters 1987; Verronen et al. 2006), the longer-term SPE impacts on ozone are the more important component in polar stratospheric ozone variation and will be discussed in the next section.

5 Long-term Impact on Ozone

The longer-term impact of SPE-induced NO_y enhancements on ozone has been known for about 30 years. Heath et al. (1977) showed large stratospheric ozone reductions in Nimbus-4 BUV instrument data up to 19 days past the August 1972 events, which were probably caused by the NO_y enhancements. Several other papers (e.g., Reagan et al. 1981; Solomon and Crutzen 1981; Rusch et al. 1981; Jackman et al. 1990, 1995, 2000, 2005a; Randall et al. 2001) studied various aspects of NO_y influence on stratospheric ozone. The primary catalytic cycle for NO_y destruction of ozone is

$$NO + O_3 \rightarrow NO_2 + O_2 \tag{4}$$

$$NO_2 + O \rightarrow NO + O_2 \tag{5}$$

$$\text{Net: } O_3 + O \rightarrow O_2 + O_2 \tag{6}$$

The long lifetime of the NO_y constituents allows the influence on ozone to last for a number of months to years past the event. Long-term effects due to solar protons with durations of (a) about 2 months have been shown in measured NO_x (NO + NO_2) and ozone after the July 2000 SPE (Randall et al. 2001); and (b) about 5 months have been shown in measured NO_2 and ozone after the October 1989 SPEs (Jackman et al. 1995). Figure 2 shows the model predicted temporal behavior of profile ozone (lower plot) and NO_y (upper plot) for the polarmost northern hemisphere (NH) area (70°–90°N) for the time period 1963–2010. NO_y shows enhancements between 1% and about 10% in the lower stratosphere (below 10 hPa) for particular years. These very large SPEs in (a) August 1972; (b) August–September–October 1989; and (c) July and November 2000, September and November 2001, October 2003, and January and September 2005 cause the NO_y increases in years (a) 1972–1973; (b) 1989–1993; and (c) 2000–2006, respectively.

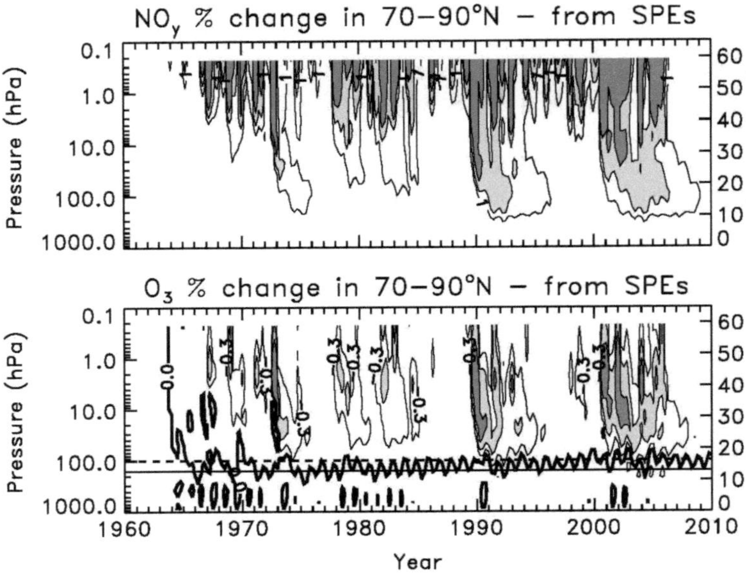

Fig. 2 Model computed percentage changes in NO_y and O_3 for the polar northern hemisphere area (70–90°N) for 1963–2010 resulting from SPEs in 1963–2005. Contour levels for NO_y (top plot) are +1%, +3%, and +10%. The "light gray" and "dark gray" highlighted areas for NO_y indicate increases from 3% to 10% and >10%, respectively. Contour levels for O_3 (bottom plot) are −3%, −1%, −0.3%, 0%, and +0.3%. The "light gray" and "dark gray" highlighted areas for O_3 indicate decreases from 1% to 3% and >3%, respectively. These changes were computed by comparing the "perturbed" to the "base" simulation. Horizontal lines at 160 hPa (solid) and 100 hPa (dashed) are given in the O_3 plot for assessing the temporal change of the 0% contour.

The increased NO_y led to a northern polar stratospheric ozone depletion for extended periods. SPE-caused depletions greater than 3% are highlighted in "dark gray" in Fig. 2 (lower plot). The polar southern hemisphere (SH) shows similar behavior (Fig. 3), however, there are differences caused by the seasonal differences for the occurrence of the SPEs (e.g., see discussion in Jackman et al. 2005b). SPEs that occur in the late fall/winter time period experience a lower amount of sunlight thus the loss process for NO_y via

$$NO + h\nu\,(<191\,\text{nm}) \rightarrow N + O \tag{7}$$

followed by

$$N + NO \rightarrow N_2 + O \tag{8}$$

is minimal. The vertical winds are generally downward at this time of year and NO_y is transported to lower altitudes, where photochemical loss is even less. SPEs in October 1989, November 2000, November 2001, October 2003, and January 2005 were thus most important in the NH. Likewise, the SPEs in August 1972, August

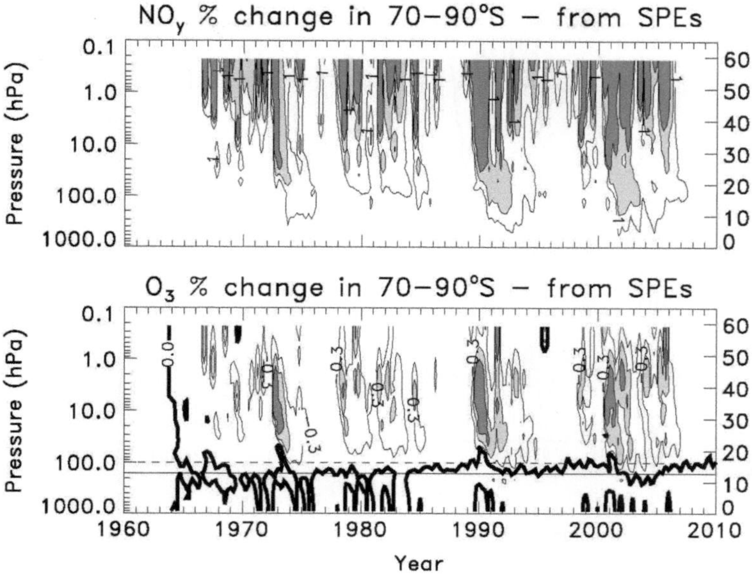

Fig. 3 Model computed percentage changes in NO_y (top) and O_3 (bottom) for the polar southern hemisphere area (70–90°S) for 1963–2010 resulting from SPEs in 1963–2005. Contour levels are the same as in Fig. 2. These changes were computed by comparing the "perturbed" to the "base" simulation. Horizontal lines at 160 hPa (solid) and 100 hPa (dashed) are given in the O_3 plot for assessing the temporal change of the 0% contour.

1989, and July 2000 were most important in the SH. Large SPEs, whether or not they occurred in late fall/winter, do cause perturbations in both hemispheres. For example, the October 1989 SPE was the largest in the past 40 years (see Table 1) and also caused a substantial impact in the SH (e.g., Jackman et al. 1995).

Enhanced levels of NO_y can also lead to ozone increases (Jackman et al. 2000). This is especially true in years of enhanced halogen loading. The ozone loss rate due to chlorine and bromine can be reduced through reactions such as

$$ClO + NO_2 + M \rightarrow ClONO_2 + M \tag{9}$$

and

$$BrO + NO_2 + M \rightarrow BrONO_2 + M \tag{10}$$

where chlorine and bromine reservoir constituents ($ClONO_2$ and $BrONO_2$) are produced at the expense of the ozone-reducing radicals (ClO and BrO).

Although such interference is relatively small in the NH and SH with computed ozone increases of just over +0.3% at most, the average altitude of the 0.0% contour line gradually rises upwards from about 160 hPa (~13 km) in 1980 to near 100 hPa

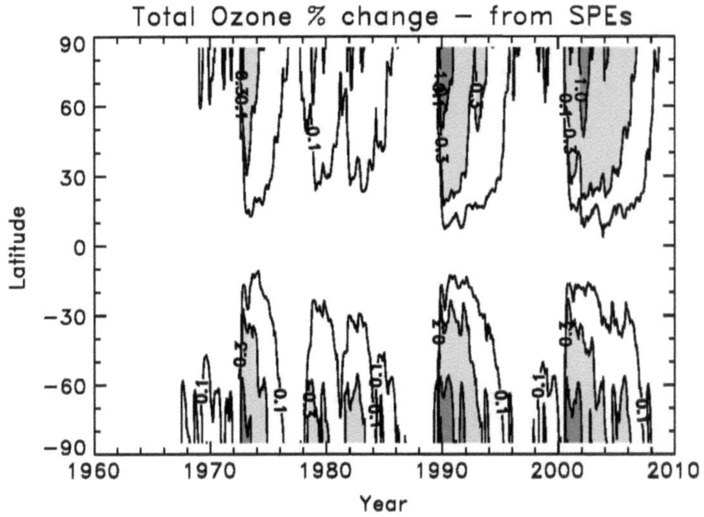

Fig. 4 Model computed percentage total ozone changes for 1963–2010 resulting from SPEs in 1963–2005. Contour intervals are −1%, −0.3%, and −0.1%. The "light gray" and "dark gray" highlighted areas indicate decreases from 0.3% to 1% and >1%, respectively. These changes were computed by comparing the "perturbed" to the "base" simulation.

(~16 km) in 2000 as the effective equivalent stratospheric chlorine increases from 1.8 to 3.2 ppbv over the same time period (see Fig. 1–23 of WMO 2003). There is significant seasonal variability (primarily driven by the annual solar change) and interannual variability (primarily associated with very large SPEs) in the 0.0% ozone contour line. Horizontal lines at 160 hPa (solid) and 100 hPa (dashed) are given in the bottom panels of Figs. 2 and 3 to aid in assessing the temporal change in the 0.0% contour line for ozone.

The impact on total ozone is shown in Fig. 4. Both hemispheres had extended periods of depleted ozone from 1–3% in 1972, 1989–1990, and 2000–2003. These changes are modest compared to the downward trends caused by halogen loading in the polar regions over the 1980–2000 time period [~3%/decade in the NH and ~4–9%/decade in the SH, WMO 2003], however, they need to be considered in understanding polar changes during particular years.

6 Conclusions

Several very large SPEs have occurred over the 43-year time period 1963–2005. These events, which occurred in 1972, 1989, 2000, 2001, and 2003, have led to significant polar enhancements in HO_x and NO_y. The HO_x enhancements led to short-term mesospheric and upper stratospheric ozone decreases, whereas the NO_y enhancements led to polar total ozone depletions of 1–3% lasting several months to years past the events. In more recent years, the NO_y enhancements were also found to lead to small ozone enhancements in the lowermost stratosphere because of interference with the ozone loss cycles for the chlorine and bromine constituents.

Acknowledgments This work was supported by the NASA Living with a Star Targeted Research and Technology Program and the NASA Atmospheric Composition Data and Analysis Program. We thank the Interplanetary Monitoring Platform and NOAA Geostationary Operational Environmental Satellite teams for providing the solar proton flux data. We thank two anonymous reviewers for valuable comments and suggested changes, which have led to an improved manuscript.

References

Crutzen, P. J., I. S. A. Isaksen, and G. C. Reid, 1975: Solar proton events: stratospheric sources of nitric oxide. *Science*, **189**, 457–458.

Douglass, A. R., C. H. Jackman, and R. S. Stolarski, 1989: Comparison of model results transporting the odd nitrogen family with results transporting separate odd nitrogen species. *J. Geophys. Res.*, **94**, 9862–9872.

Fleming, E. L., C. H. Jackman, J. E. Rosenfield, and D. B. Considine, 2002: Two-dimensional model simulations of the QBO in ozone and tracers in the tropical stratosphere. *J. Geophys. Res.*, **107**(D23), 4665, doi:10.1029/2001JD001146.

Fleming, E. L., C. H. Jackman, D. K. Weisenstein, and M. K. W. Ko, 2007: The impact of interannual variability on multi-decadal total ozone simulations. *J. Geophys. Res.* **112**, D10310, doi:10.1029/2006 JD007953.

Frederick, J. E., 1976: Solar corpuscular emission and neutral chemistry in the Earth's middle atmosphere. *J. Geophys. Res.*, **81**, 3179–3186.

Gelman, M. E., A. J. Miller, K. W. Jihnson, and R. N. Nagatani, 1986: Detection of long-term trends in global stratospheric temperature from NMC analyses derived from NOAA satellite data. *Adv. Space. Res.*, **6**, 17–26.

Heath, D. F., A. J. Krueger, and P. J. Crutzen, 1977: Solar proton event: influence on stratospheric ozone. *Science*, **197**, 886–889.

Jackman, C. H. and R. D. McPeters, 1985: The response of ozone to solar proton events during solar cycle 21: a theoretical interpretation. *J. Geophys. Res.*, **90**, 7955–7966.

Jackman, C. H., and R. D. McPeters, 1987: Solar proton events as tests for the fidelity of middle atmosphere models. *Physica Scripta*, **T18**, 309–316.

Jackman, C. H., and R. D. McPeters, 2004: The effect of solar proton events on ozone and other constituents. In: *Solar Variability and Its Effects on Climate*, J. M. Pap and P. Fox, eds., Geophys. Monograph 141, Washington, DC, 305–319.

Jackman, C. H., H. S. Porter, and J. E. Frederick, 1979: Upper limits on production rate of NO per ion pair. *Nature*, **280**, 170.

Jackman, C. H., J. E. Frederick, and R. S. Stolarski, 1980: Production of odd nitrogen in the stratosphere and mesosphere: an intercomparison of source strengths. *J. Geophys. Res.*, **85**, 7495–7505.

Jackman, C. H., A. R. Douglass, R. B. Rood, R. D. McPeters, and P. E. Meade, 1990: Effect of solar proton events on the middle atmosphere during the past two solar cycles as computed using a two-dimensional model. *J. Geophys. Res.*, **95**, 7417–7428.

Jackman, C. H., M. C. Cerniglia, J. E. Nielsen, D. J. Allen, J. M. Zawodny, R. D. McPeters, A. R. Douglass, J. E. Rosenfield, and R. B. Rood, 1995: Two-dimensional and three-dimensional model simulations, measurements, and interpretation of the influence of the October 1989 solar proton events on the middle atmosphere. *J. Geophys. Res.*, **100**, 11,641–11,660.

Jackman, C. H., E. L. Fleming, S. Chandra, D. B. Considine, and J. E. Rosenfield, 1996: Past, present and future modeled ozone trends with comparisons to observed trends. *J. Geophys. Res.*, **101**, 28,753–28,767.

Jackman, C. H., E. L. Fleming, and F. M. Vitt, 2000: Influence of extremely large solar proton events in a changing stratosphere. *J. Geophys. Res.*, **105**, 11659–11670.

Jackman, C. H., R. D. McPeters, G. J. Labow, E. L. Fleming, C. J. Praderas, and J. M. Russell, 2001: Northern hemisphere atmospheric effects due to the July 2000 solar proton event. *Geophys. Res. Lett.*, **28**, 2883–2886.

Jackman, C. H., M. T. DeLand, G. J. Labow, E. L. Fleming, D. K. Weisenstein, M. K. W. Ko, M. Sinnhuber, J. Anderson, and J. M. Russell, 2005a: The influence of the several very large solar events in years 2000–2003 on the neutral middle atmosphere. *Adv. Space Res.*, **35**, 445–450.

Jackman, C. H., M. T. DeLand, G. J. Labow, E. L. Fleming, D. K. Weisenstein, M. K. W. Ko, M. Sinnhuber, and J. M. Russell, 2005b: Neutral atmospheric influences of the solar proton events in October–November 2003. *J. Geophys. Res.*, **110**, A09S27, doi:10.1029/2004JA010888.

Kanamitsu, M., W. Ebisuzaki, J. Woollen, S.-K. Yang, J. J. Hnilo, M. Fiorino, and G. L. Potter, 2002: NCEP-DOE AMIP-II Reanalysis (R-2). *Bull. Am. Meteor. Soc.*, **83**, 1631–1643.

López-Puertas, M., B. Funke, S. Gil-López, T. von Clarmann, G. P. Stiller, M. Höpfner, S. Kellman, H. Fischer, and C. H. Jackman, 2005: Observation of NOx enhancement and ozone depletion in the northern and southern hemispheres after the October–November 2003 solar proton events. *J. Geophys. Res.*, **110**, A09S43, doi:10.1029/2005JA011050.

McPeters, R. D., 1986: A nitric oxide increase observed following the July 1982 solar proton event. *Geophys. Res. Lett.*, **13**, 667–670.

Mewaldt, R. A., C. M. S. Cohen, A. W. Labrador, R. A. Leske, G. M. Mason, M. I. Desai, M. D. Looper, J. E. Mazur, R. S. Selesnick, and D. K. Haggerty, 2005: Proton, helium, and electron spectra during the large solar particle events of October–November 2003. *J. Geophys. Res.*, **110**, A09S18, doi:10.1029/2005JA011038.

Porter, H. S., C. H. Jackman, and A. E. S. Green, 1976: Efficiencies for production of atomic nitrogen and oxygen by relativistic proton impact in air. *J. Chem. Phys.*, **65**, 154–167.

Randall, C. E., D. E. Siskind, and R. M. Bevilacqua, 2001: Stratospheric NO_x enhancements in the southern hemisphere vortex in winter/spring of 2000. *Geophys. Res. Lett.*, **28**, 2385–2388.

Reagan, J. B., R. E. Meyerott, R. W. Nightingale, R. C. Gunton, R. G. Johnson, J. E. Evans, W. L. Imhof, D. F. Heath, and A. J. Krueger, 1981: Effects of the August 1972 solar particle events on stratospheric ozone. *J. Geophys. Res.*, **86**, 1473–1494.

Rees, M. H., 1989: Physics and chemistry of the upper atmosphere, Cambridge University Press, Cambridge, 278–281.

Rusch, D. W., J.-C. Gerard, S. Solomon, P. J. Crutzen, and G. C. Reid, 1981: The effect of particle precipitation events on the neutral and ion chemistry of the middle atmosphere, 1, Odd nitrogen. *Planet. Space Sci.*, **29**, 767–774.

Sander, S. P. et al., 2003: Chemical kinetics and photochemical data for use in atmospheric studies. *JPL Publication 02–25.*

Seppälä, A., P. T. Verronen, V. F. Sofieva, J. Tamminen, E. Kyrölä, C. J. Rodger, and M. A. Clilverd, 2006: Destruction of the tertiary ozone maximum during a solar proton event. *Geophys. Res. Lett.*, **33**, L07804, doi:10.1029/2005GL025571.

Solomon, S. and P. J. Crutzen, 1981: Analysis of the August 1972 solar proton event including chlorine chemistry. *J. Geophys. Res.*, **86**, 1140–1146.

Solomon, S., D. W. Rusch, J.-C. Gerard, G. C. Reid, and P. J. Crutzen, 1981: The effect of particle precipitation events on the neutral and ion chemistry of the middle atmosphere, 2, Odd hydrogen. *Planet. Space Sci.*, **29**, 885–892.

Solomon, S., G. C. Reid, D. W. Rusch, and R. J. Thomas, 1983: Mesospheric ozone depletion during the solar proton event of July 13, 1982, 2, Comparison between theory and measurements. *Geophys. Res. Lett.*, **10**, 257–260.

Swider, W. and T. J. Keneshea, 1973: Decrease of ozone and atomic oxygen in the lower mesosphere during a PCA event. *Planet. Space Sci.*, **21**, 1969–1973.

Verronen, P. T., A. Seppala, E. Kyrola, J. Tamminen, H. M. Pickett, and E. Turunen, 2006: Production of odd hydrogen in the mesosphere during the January 2005 solar proton event. *Geophys. Res. Lett.*, **33**, L24811, doi:10.1029/2006GL28115.

Vitt, F. M. and C. H. Jackman, 1996: A comparison of sources of odd nitrogen production from 1974 through 1993 in the Earth's middle atmosphere as calculated using a two-dimensional model. *J. Geophys. Res.*, **101**, 6729–6739.

Weeks, L. H., R. S. CuiKay, and J. R. Corbin, 1972: Ozone measurements in the mesosphere during the solar proton event of 2 November 1969. *J. Atmos. Sci.*, **29**, 1138–1142.

Weisenstein, D. K., J. Eluszkiewicz, M. K. W. Ko, C. J. Scott, C. H. Jackman, E. L. Fleming, D. B. Considine, D. E. Kinnison, P. S. Connell, and D. A. Rotman, 2004: Separating chemistry and transport effects in 2-D models. *J. Geophys. Res.*, **109**, D18310, doi:10.1029/2004JD004744.

World Meteorological Organization (WMO), 2003: Scientific assessment of ozone depletion: 2002. *Rep. 47, Global Ozone Res. and Monit. Proj.*, Geneva.

Wu, D. L. et al., 2003: Mesospheric temperature from UARS MLS: retrieval and validation. *J. Atmos. Sol.-Terr. Phys.*, **65**, 245–267.

Zadorozhny, A. M., G. A. Tuchkov, V. N. Kikhtenko, J. Lastovicka, J. Boska, and A. Novak, 1992: Nitric oxide and lower ionosphere quantities during solar particle events of October 1989 after rocket and ground-based measurements. *J. Atmos. Terr. Phys.*, **54**, 183–192.

Sunspots, the QBO, and the Stratosphere in the North Polar Region: An Update*

K. Labitzke[1], M. Kunze[1], and S. Brönnimann[2]

Abstract The 11-year sunspot cycle (SSC) strongly affects the lower stratosphere. However, in order to detect the solar signal it is necessary to group the data according to the phase of the Quasi-Biennial Oscillation (QBO). Although this is valid throughout the year the effect of the SSC and the QBO on the stratosphere was largest during the northern winters (January/February). As the stratosphere can affect weather at the ground, the SSC effect on the lower stratosphere might provide a mechanism for solar-climate links. Here we analyse an extended, 65-year long data set of solar variability, QBO, and lower stratospheric dynamics. The results fully confirm earlier findings and suggest a significant effect of the SSC on the strength of the stratospheric polar vortex and on the mean meridional circulation.

1 Introduction

Even after 200 years of research, the relation between solar variability and Earth's climate remains a matter of debate in the scientific literature and a topic of foremost interest to the Earth science community. Effects of solar variability related to the 11-year sunspot cycle are most obvious in the stratosphere, though still not fully understood (van Loon and Labitzke 2000; Crooks and Gray 2005; Matthes et al. 2006). Hence, the effect of solar variability on climate at the ground might well proceed via the stratosphere.

Labitzke suggested in 1982 that the Sun influences the intensity of the north polar vortex in the stratosphere in winter, and that the Quasi-Biennial Oscillation (QBO) is needed to identify the solar signal. Based on these results, Labitzke found

[1] Institute for Meteorology, Free University of Berlin, Germany
Karin.Labitzke@mail.met.fu-berlin.de

[2] Institute for Atmospheric and Climate Science, ETH Zürich, Switzerland
stefan.bronnimann@env.ethz.ch

* Contribution is an updated version of the paper by Labitzke K et al. (2006) *Meteorologische Zeitschrift*, **15**, 355–363.

in 1987 that a signal of the 11-year sunspot cycle (SSC) emerged when the arctic stratospheric temperatures and geopotential heights were grouped into two categories determined by the direction of the equatorial wind in the stratosphere (QBO). The signal involves a change in strength of the northern polar vortex in the stratosphere. Through the mechanism of downward propagation (Baldwin and Dunkerton 2001) it might affect the sea-level pressure (SLP) field and lead to changes in the so-called Arctic Oscillation (AO) (Thompson and Wallace 1998).

This first study was based on 30 years of data (1957–1986), which is barely three solar cycles. Several publications criticized the short data record.

In the mean time, 20 more years of data became available, and the records have been extended into the past. Altogether, 65 years of data are now available, that is, 6.5 solar cycles, (see Fig. 3). In this paper we analyse the 65-year data set with respect to trends and to solar effects in the lower stratosphere (based on Labitzke et al. 2006).

2 Data and Methods

The work presented in this paper is mainly based on NCEP/NCAR reanalysis data (e.g., Kalnay et al. 1996) since 1948. In order to address the polar vortex in the stratosphere we use 30 hPa geopotential height at the North Pole. These data were extended back to 1942 using statistical reconstructions based on historical upper-level data as described in Labitzke et al. (2006). Reconstruction and validations methods as well as the historical data are identical to those used in Brönnimann et al. (2007a). The statistical approach is based on the assumption of a stationary relationship between predictors (mostly radiosonde data from the midlatitudes, but also from the Arctic) and the predictand. Hence, we are measuring the strength of the polar vortex via its imprint in the mid- to high-latitude upper troposphere. It is important to note, however, that the reconstructions are independent of both the QBO (no data south of 30°N were used) and solar variability and can therefore be used for the following statistical analyses. The skill of the reconstructions is shown to be sufficient for the analyses presented here (Labitzke et al. 2006). Data for the AO index were taken from Thompson and Wallace (1998).

The significance of our results depends on the number of solar cycles available. With these new data we have reached 6.5 solar cycles and we can now safely say that the results for the northern winters, especially in the west phase of the QBO, with a coefficient of correlation r reaching 0.7, are highly significant ($r = 0.5$ and 0.66 correspond to the 95% and 99% confidence levels, respectively).

The QBO is an oscillation in the atmosphere which is best observed in the stratospheric winds above the equator, where the zonal winds change between east and west with time. The period of the QBO varies in space and time, with an average value near 28 months at all levels, see reviews by Naujokat (1986) and Baldwin et al. (2001).

Information on the phase of the QBO was taken from the FU Berlin data back to 1953 (Labitzke et al. 2002). The series was extended back to 1942 based on pilot balloon wind observations from the tropical stratosphere. All zonal wind observations above 16 km were analysed in three latitude bins (20° S–5°S, 5°S–5°N, 5°N–20°N) and altitude bins of 2 km. According to lead-lag relationships determined from NCEP/NCAR data after 1957, the data in each bin were shifted to be in phase with the equatorial QBO at 40 hPa. The same data were used in Labitzke et al. (2006) to fit a periodic curve visually to the data and determine the phase of the QBO during winter.

Table 1 gives some more detailed information. For seven out of the 12 winters, the observations indicate a plausible QBO phase. In the other cases, the direct observations are less clear and the phase was constrained by the visual fit (see Labitzke et al. 2006), by other (non-winter) observations, or by information from the literature. The winter 1941/1942 was probably close to a phase change and hence must be considered uncertain, 1952/1953 also is close to a phase change, but here the westerly phase is relatively well documented. The other winters were probably not close to phase changes, though some uncertainty remains. The QBO phase attributed to these 12 winters is in excellent agreement with the QBO phase extracted from historical total ozone data for the same period (Brönnimann et al., 2007b).

The monthly mean values of the 10.7 cm solar flux are used as a proxy for variations through the SSC. The flux values are expressed in solar flux units: 1 s.f.u. = 10^{-22} W m^{-2} Hz^{-1}. This is an objectively measured radio wave intensity, highly and positively correlated with the 11-year SSC and particularly with the UV part of the solar spectrum (Hood 2003). For the earliest years, before regular measurements of the 10.7 cm solar flux became available in 1947, we derived the solar flux from a regression between the sunspot numbers and the flux (see Labitzke et al. 2006).

Table 1 Observed winds in the tropical stratosphere from November to February. Year refers to the year of the February, u_{med} is the median value of the zonal wind (m/s), n gives the number of observations, phase indicates the attributed phase

Year	u_{med}	n	Phase	Comment
1942	0.00	1	E	Possibly end of E phase; Schove (1969) indicates E in 1941. 2, see also visual fit to data in Labitzke et al. (2006)
1943	1.03	67	W	
1944	0.00	66	E	E in 1943 (Schove 1969), see also visual fit to data in Labitzke et al. (2006)
1945	0.00	1011	W	W in fall 1944 and spring 1945 (Labitzke et al. 2006)
1946	−5.52	159	E	E in 1946.2 (Schove 1969)
1947	7.07	131	W	
1948	3.54	785	E	Though the observed winds are W, this would mean that the W phase was extremely long, which is hardly plausible (see also visual fit to data in Labitzke et al. 2006); Schove (1969) indicates E in 1948.2
1949	0.00	712	W	W in fall 1948 and spring 1949; see also visual fit to data in Labitzke et al. (2006)
1950	−6.00	498	E	E in 1950.1 (Schove 1969)
1951	2.00	3483	W	
1952	−6.00	2099	E	E in 1952.4 (Schove 1969)
1953	1.16	3726	W	

3 Trends at 30 hPa Over the Arctic

The time series of monthly mean 30 hPa temperatures and heights over the North Pole for selected winter months, Fig. 1, gives a clear visualization of the large variability of the arctic stratosphere. In early winter, that is, December (Fig. 1a), the

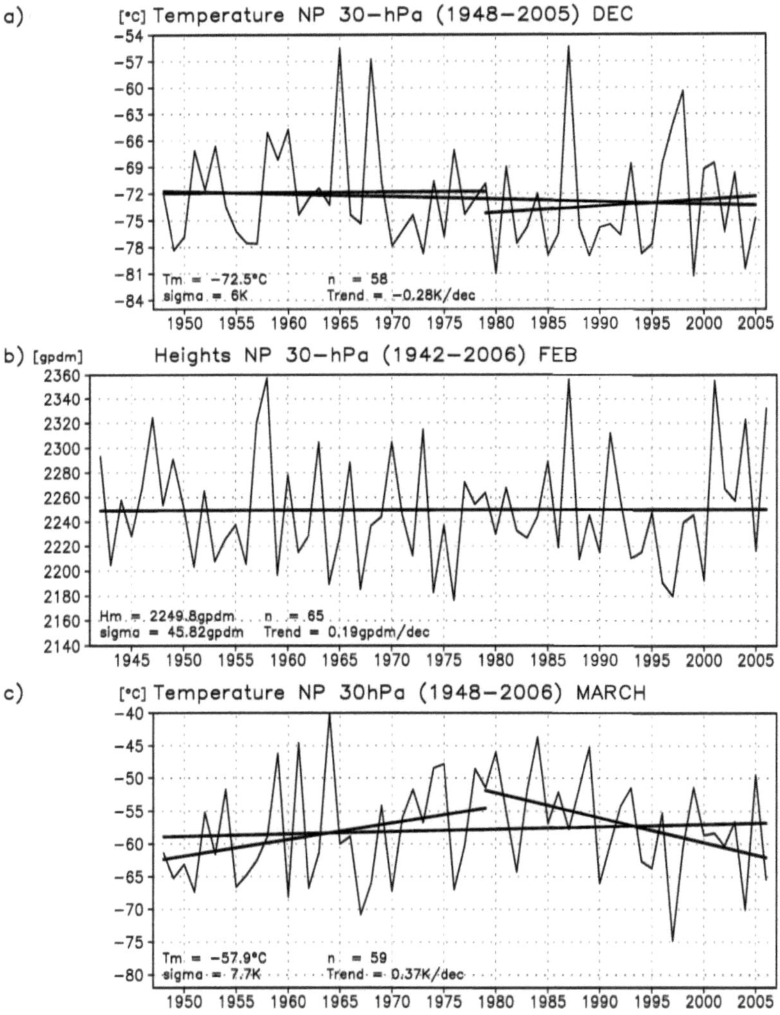

Fig. 1 (a) Time series of the monthly mean 30 hPa North Pole temperatures (°C) in December, 1948–2005. Trend lines are given for the whole period as well as for two sub-periods (1948–1979) and (1979–2005). The data are from NCEP/NCAR reanalysis. (b) Time series of the monthly mean 30 hPa North Pole heights (gpdm) in February, 1942–2006 – The data are from NCEP/NCAR and statistical reconstructions; (c) as (a), but for March, 1948–2006 (NCEP/NCAR reanalysis).

linear temperature trend for all 58 years is slightly negative, but slightly positive since 1979. The standard deviation is about 6 K.

For February (Fig. 1b), 65 years of 30 hPa height data are available – and the trend for all years is practically zero. Warm winters with a high 30 hPa level over the North Pole (like in February 1958, 1987 or 2001) alternate with very cold winters with a low 30 hPa level over the North Pole, (like in February 1976, 1997, or 1974, to name the three highest and lowest Februaries, respectively). There is no difference of the state of the arctic stratosphere between the early and the later Februaries.

In March (Fig. 1c), the temperature trend for all 59 years is slightly positive. Cutting the time series yields a strong positive trend in the early part of the period and a strong negative trend in the later period (starting in 1979). But this is not advisable. One cannot say that March has become colder (nor January or February, not shown) during the available period. Excluding the influence of the extreme in March 1997, which can be explained through atmospheric dynamics, the time series as a whole has no trend; and this realization is important in a discussion of the future development of the arctic ozone (Labitzke and van Loon 1999).

4 Influence of the 11-Year Sunspot Cycle on the Stratosphere in Late Winter

4.1 The Arctic Stratosphere Since 1942

Figure 2 shows in two scatter diagrams the 30 hPa heights over the North Pole in February when the modulation of the solar signal by the QBO is at its maximum. The correlations between the 30 hPa heights and the solar cycle are shown, with the winters in the east phase of the QBO in the left part of the figure, and the winters in the west phase of the QBO in the right part. The abscissas indicate the SSC. The correlations are clearly very different in the two groups, with negative correlations over the Arctic in the east phase of the QBO and large positive correlations there in the west phase. (The correlation for all years is 0.2, not shown.) The numbers in the scatter diagrams are the years of the individual Februaries. The total number of Februaries available is now 65, more than twice the number available in the beginning (Labitzke 1987; Labitzke and van Loon 1988) and comprises seven minima and six maxima, Fig. 3. As mentioned above, the 20 years after the first publication in 1987 fit very well into the scatter diagrams (Fig. 2) and confirm the earlier results. But also the NCEP/NCAR reanalysis from 1948 till 1957 (10 years) fit very well and the size of the correlations (particularly in the west phase of the QBO) did not change much. Most remarkable are the six Februaries from 1942 till 1947 (reconstructions).

The average height difference (ΔH Fig.2) between solar maxima and minima is very large in the west-phase winters, reaching 704 m which is more than one standard deviation of the interannual variability. Figure 3 presents the SSC based on the

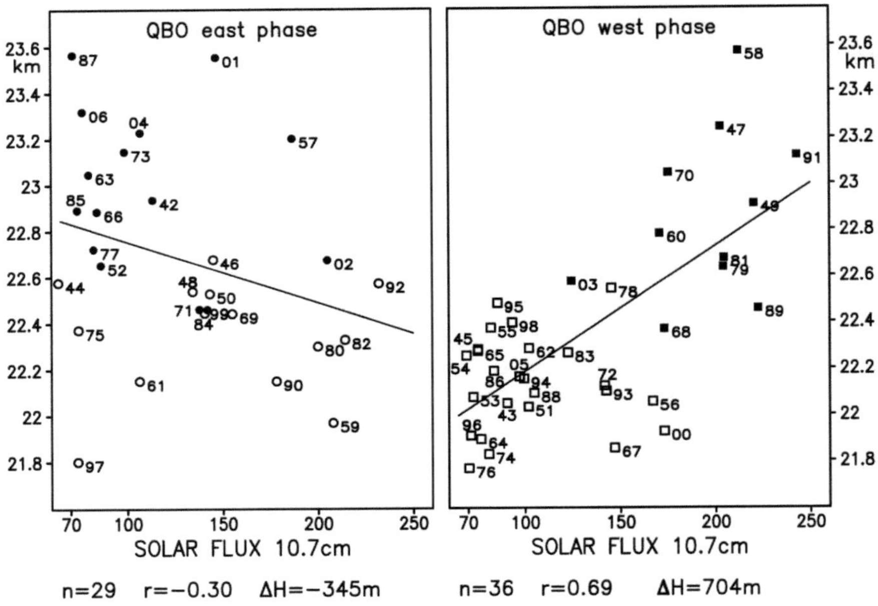

Fig. 2 Scatter diagrams of the monthly mean 30 hPa geopotential heights (geopot. km) in February at the North Pole (1942–2006), plotted against the 10.7 cm solar flux. Left: Circles: years in the east phase of the QBO (n = 29). Right: Squares: years in the west phase (n = 36). The numbers indicate the respective years; r = correlation coefficient; ΔH gives the mean difference of the heights (geopot. m) between solar maxima and minima. Filled squares and filled circles denote MMWs, that is, a winter with a reversal of the zonal wind over the Arctic at the 10 hPa level (data are from reconstructions 1942–1947 and NCEP/NCAR thereafter) (Labitzke and van Loon 1990, updated; Labitzke et al. 2006).

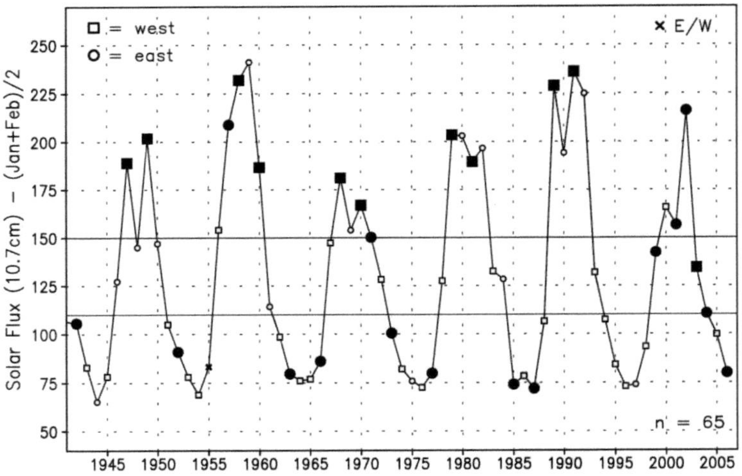

Fig. 3 Time series of the 10.7 cm solar flux, 1942–2006 ((January + February)/2). Squares denote winters in the west phase of the QBO, circles winters in the east phase. Large filled symbols characterize the occurrence of Major Midwinter Warmings (MMWs) in January or February. (Labitzke and van Loon 1990, updated); Labitzke et al. (2006).

10.7 cm solar flux for the period 1942–2006. It is indicated whether a winter (January/February) belonged to the west or east phase of the QBO. Further, filled symbols indicate if a MMW was observed. This can be well decided since 1950. The definition of a MMW is based on a reversal of the zonal wind over the Arctic at the 10 hPa level. As the very early data do not reach that high we must rely on comparisons with more recent events. The 30 hPa height values derived for 1947 and 1949 (west phase of the QBO) and 1942 (east phase) must be compared in Fig. 2 with neighbouring values, e.g., 1958, 1991, 1970, 1960 in the west group, or 1985, 1963 in the east phase group, which all represent well known MMWs. Therefore, we speculate that also during the winters 1942 (Brönnimann et al. 2004), 1947 and 1949 MMWs took place.

There is a very clear tendency for the MMWs in the west phase of the QBO to occur during solar maxima (solar flux above 150 units), Fig. 3: out of 11 cases 10 took place in solar max and none in solar min and this leads to the large positive correlations with the SSC over the Arctic, as discussed above.

For the MMWs in the east phase of the QBO the situation is less clear, but more MMWs took place during solar minima (solar flux below 110 units): 10 out of 15 MMWs cases, against four in solar maxima. This leads to the negative correlations in the east phase, see Figs. 2–5.

4.2 The Northern Hemisphere Stratosphere Since 1948

Figure 4 (left) shows for the northern hemisphere the correlations between the 10.7 cm solar flux and the detrended 30 hPa heights in February for the period 1948–2006, that is, 59 years and six solar cycles. On the right-hand side the height differences (geopotential meters) between solar maxima and minima are given. Upper panels: east phase of the QBO, lower panels: west phase of the QBO. The patterns of the correlations and the respective height differences are very different in the two phases of the QBO: the correlations are strongly positive (up to 0.68) over the Arctic in the west-phase winters, reflecting an intensification of the Brewer–Dobson Circulation (BDC) connected with MMWs and downwelling/warming over the Arctic in solar maxima (e.g., Kodera and Kuroda 2002; Salby and Callaghan 2004; 2006; Labitzke 2005; van Loon et al. 2007; Matthes et al. 2006). The correlations are negligible outside the Arctic because of dynamically forced upwelling/cooling.

In the east phase of the QBO the correlations are weakly negative over the Arctic (–0.35) but positive over the tropics and subtropics (larger than 0.6), connected here with an enhanced downwelling/warming, that is, a weakening of the BDC in solar maxima. This is consistent with our earlier results. It is of interest to compare these results based on 59 years of data (six solar maxima and six solar minima) with the results published earlier for the period 1968–2003 (Labitzke 2005), that is 36 years (four solar maxima and three solar minima): the patterns are very similar and the size of the height differences is practically the same. So, one can safely say that the earlier years since 1948 fit very well into the results obtained before with fewer data.

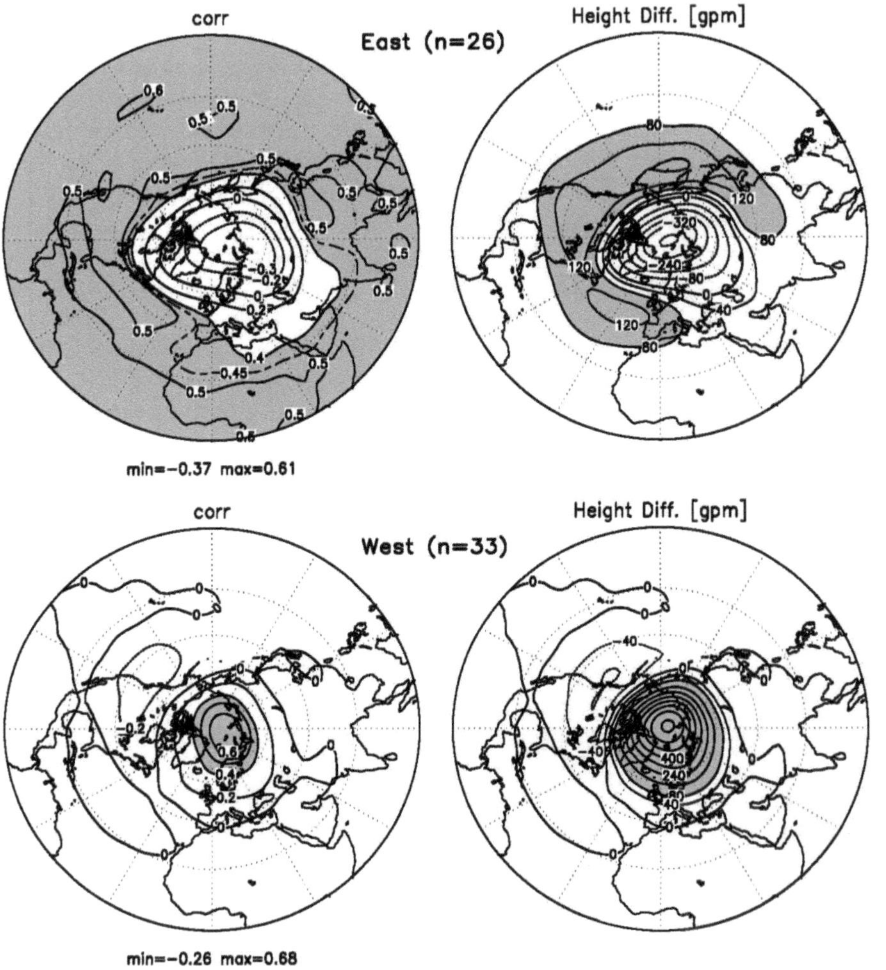

Fig. 4 Left: Correlations between the 10.7 cm solar flux (the 11-year solar cycle) and 30 hPa heights in February, shaded for emphasis where the correlations are above 0.4; upper panel: years in the east phase of the QBO; lower panel: years in the west phase of the QBO. Right: Respectively, height differences (geopot. m) between solar maxima and minima, shaded where the height differences are larger than 80 m. (NCEP/NCAR reanalysis, period: 1948–2006); (Labitzke 2002, updated).

Restricting the time period to the NCEP/NCAR reanalysis period (1948–2006), the vertical structure of the solar signal can be analysed. Figure 5 shows for February and for the period from 1948 till 2006 correlations and temperature differences based on zonal mean data, in vertical meridional sections ranging from 200 to 10 hPa and from 90°N to 90°S. The correlations for all Februaries and for the years in the respective phases of the QBO are given on the left-hand side, and the resulting temperature differences on the right-hand side. There is practically no signal of the SSC

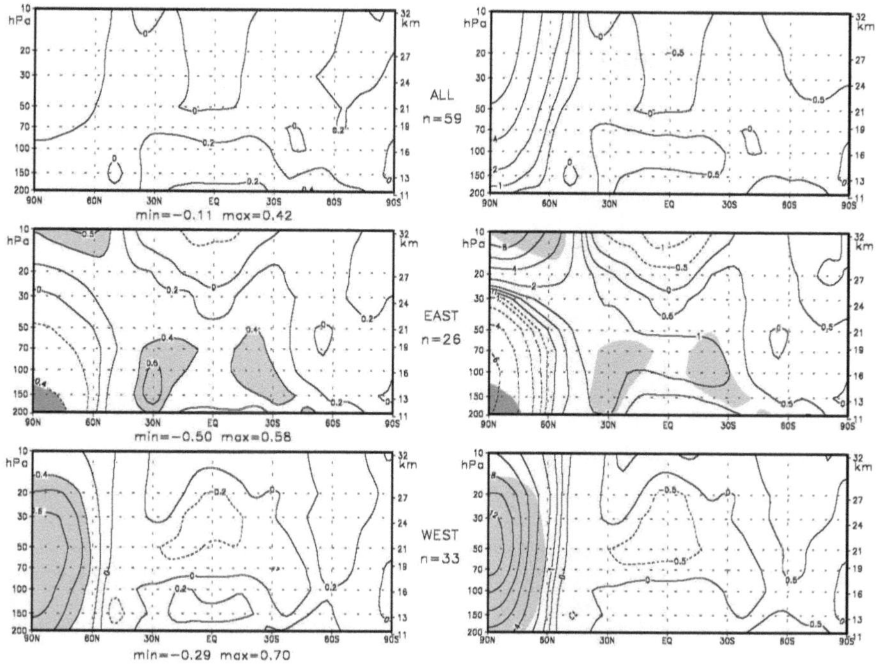

Fig. 5 Vertical meridional sections between 200 and 10 hPa (11 and 32 km) of (left): the correlations between the detrended zonally averaged, monthly mean temperatures for February and the 10.7 cm solar flux (shaded for emphasis where the correlations are larger than 0.4). Right: The respective temperature differences (K) between solar maxima and minima, shaded where the corresponding correlations on the left hand side are above 0.4. Upper panels: all years; middle panels: only years in the east phase of the QBO; lower panels: only years in the west phase of the QBO. (NCEP/NCAR reanalysis, 1948–2006). (Labitzke 2002, updated).

using all years. As discussed above the patterns of the correlations and of the differences are very different between the two phases of the QBO and the structure described above for the 30 hPa level is found to be consistent throughout the height range investigated here. The whole lower stratosphere is influenced by the SSC, with the different positive or negative correlations reaching down to the 200 hPa level over the Arctic, and with the opposite correlations reaching as far as 30°S. The size of the correlations and of the differences are almost the same as published, for example, by Labitzke (2005) for the period 1958–2003, with n = 46 years.

5 Summary

Clear effects of solar variability related to the SSC are seen in the stratosphere, most prominently in the northern polar region, but only if the QBO is taken into account. New results based on an extended, 65-year long data set fully confirm earlier findings and suggest a significant effect of the SSC on the strength of the stratospheric polar vortex and on the mean meridional circulation.

The strong signal in the stratosphere provokes the question whether any solar effect on climate at the Earth's surface might result from a downward propagation of an initial stratospheric response. This is under investigation.

Acknowledgements We thank the members of the Stratospheric Research Group, FUB for professional support, Harry van Loon for many years of close cooperation, and Roy Jenne (NCAR) for providing the historical pilot balloon data. Tracy Ewen, Andrea Grant, and Thomas Griesser (ETH Zürich) helped in the digitising and re-evaluation work of the historical upper-air data as well as in the reconstructions. The 10.7cm solar flux data are from the World Data Center A, Boulder, Colorado. SB was funded by the Swiss National Science Foundation.

References

Baldwin, M. P. and T. J. Dunkerton, 2001: Stratospheric harbingers of anomalous weather regimes. *Science*, **294**, 581–584.

Baldwin, M. P. et al., 2001: The quasi-biennial oscillation. *Rev. Geophys.* **39**, 179–229.

Brönnimann, S., J. Luterbacher, J. Staehelin, and T. Svendby, 2004: An extreme anomaly in stratospheric ozone over Europe in 1940–1942. *Geophys. Res. Lett.*, **31**, L08101, doi:10.1029/2004GL19611.

Brönnimann, S., T. Ewen, T. Griesser, and R. Jenne, 2007a: Multidecadal signal of solar variability in the upper troposphere during the 20th century. *Space Sci. Rev.*, **125**, 305–317.

Brönnimann, S., J. Annis, C. Vogler, and P. D. Jones, 2007b: Reconstructing the Quasi-Biennial Oscillation back to the early 1900s. *Geophys. Res. Lett.*, in press, doi:10.1029/2007GL O31354.

Crooks, S. A. and L. J. Gray, 2005: Characterization of the 11-year solar signal using a multiple regression analysis of the ERA-40 dataset. *J. Climate*, **18**, 996–1015.

Hood, L. L., 2003: Thermal response of the tropical tropopause region to solar ultraviolet variations. *Geophys. Res. Lett.*, **30**, 2215, doi:10.1029/2003 GL018364.

Kalnay, E. et al., 1996: The NCEP/NCAR 40-year re-analysis project. *Bull. Am. Meteorol Soc.*, **77**, 437–471.

Kodera, K. and Y. Kuroda, 2002: Dynamical response to the solar cycle. *J. Geophys. Res.*, **107**, 4749, doi:10.1029/2002JD002224.

Labitzke, K., 1982: On the interannual variability of the middle stratosphere during the northern winters. *J. Met. Soc. Jp.*, **60**, 124–139.

Labitzke, K., 1987: Sunspots, the QBO, and the stratospheric temperature in the north polar region. *Geophys. Res. Lett.*, **14**, 535–537.

Labitzke, K., 2002: The solar signal of the 11-year sunspot cycle in the stratosphere: differences between the northern and southern summers. *J. Met. Soc. Jp.*, **80**, 963–971.

Labitzke, K., 2005: On the solar cycle - QBO relationship: a summary. *J. Atm. Sol-Terr. Phys.*, **67**, 45–54.

Labitzke, K. et al., 2002: *The Berlin Stratospheric Data Series*; CD from Meteorological Institute, Free University Berlin.

Labitzke, K. and K. Kunze, 2005: Stratospheric temperatures over the Arctic: comparison of three data sets. *Meteorol Z*, **14**, 65–74.

Labitzke, K. and H. van Loon, 1988: Associations between the 11-year solar cycle, the QBO and the atmosphere. Part I: the troposphere and stratosphere in the northern hemisphere winter. *J. Atmos. Terr. Phys.*, **50**, 197–206.

Labitzke, K. and H. van Loon, 1990: Associations between the 11-year solar cycle, the Quasi-Biennial Oscillation and the atmosphere: a summary of recent work. *Phil. Trans. R. Soc. Lond A*, **330**, 577–589.

Labitzke, K. and H. van Loon, 1999: *The Stratosphere (Phenomena, History, and Relevance).* Springer, Berlin, Heidelberg, New York, 179 pp.

Labitzke, K., M. Kunze, and S. Brönnimann, 2006: Sunspots, the QBO, and the stratosphere in the north polar region – 20 years later. *Meteorol Z*, **15**, 355–363.

Matthes, K., Y. Kuroda, K. Kodera, and U. Langematz, 2006: The transfer of the solar signal from the stratosphere to the troposphere: northern winter. *J. Geophys. Res.*, **111**, D06108, doi:10.1029/2005JD006283.

Naujokat, B., 1986: An update of the observed Quasi-Biennial Oscillation of the stratospheric winds over the tropics. *J. Atmos. Sci.*, **43**, 1873–1877.

Salby, M. L. and P. Callaghan, 2004: Evidence of the solar cycle in the general circulation of the stratosphere. *J. Climate*, **17**, 34–46.

Salby, M. L. and P. F. Callaghan, 2006: Relationship of the Quasi-Biennial Oscillation to the stratospheric signature of the solar cycle. *J. Geophys. Res.*, **111**, DO6110, doi:10.1029/2005JD006012.

Schove, D. J., 1969: The bienniel oscillation, tree rings, and sunspots. *Weather*, **24**, 390–397.

Thompson, D. W. J. and J. M. Wallace, 1998: The Arctic Oscillation signature in the wintertime geopotential height and temperature fields. *Geophys. Res. Lett.*, **25**, 297–300.

van Loon, H. and K. Labitzke, 2000: The influence of the 11-year solar cycle on the stratosphere below 30km: a review. *Space Sci. Rev.*, **94**, 259–278.

van Loon, H., G. A. Meehl, and D. J. Shea, 2007: Coupled air-sea response to solar forcing in the Pacific region during northern winter. *J. Geophys. Res.*, **112**, doi:10.1029/2006JD007378.

Index

A

Aerosols, 9, 10, 18, 159, 196–201, 253
  direct effect, 10, 157, 201, 202
  indirect effect, 10, 157, 196, 201, 202
  volcanic, 18, 79, 172, 175, 212
Aircraft, 6, 7, 38, 40, 41, 51, 85, 105, 106, 108, 146, 273, 275, 276, 308, 312
Albedo, 3, 12, 132, 144, 169, 170, 173, 195, 196, 199, 202, 212, 215
Alps, 158, 191, 275
Antarctica, 6, 40, 43, 129, 134, 136, 137, 279, 313, 315
Arctic, 2–4, 29–31, 40–43, 47–61, 134, 136, 137, 143–145, 147, 149–151, 293–300, 321, 322, 348, 350, 351
Atlantic, 3, 9, 13, 32, 42, 61, 151, 175, 211, 212, 221, 222, 229, 230, 277–279, 322, 323
Atlantic Multidecadal Oscillation (AMO), 9, 13
Australia, 96, 107, 108, 279, 359

B

Brightening, global, 10, 155, 157–163

C

Carbon dioxide, 186, 273, 280, 283
CH_4, 16, 272, 280, 283, 284, 288
Clouds, 4, 7, 10, 11, 14, 66, 79, 132, 137, 169–173, 175, 195, 196, 198–200, 202, 203, 312
  lower-level, 11, 169–175, 177
  upper-level, 11, 169–177
CO_2, 186, 273, 280, 283

D

Data assimilation, 7, 103, 104, 108–111, 113, 115, 189
Decadal variability, 134, 178, 224, 240, 242, 264, 300
Detection and attribution, 9, 328
Dimming, global, 10, 155–157, 159, 161–165
Dobson, G.M.B., 129–131, 133–135, 138, 289, 305, 306, 310, 311, 317, 321, 322, 325, 353
Drought, 13, 14, 186, 188, 200, 209, 210, 212
Dust Bowl, 14

E

El Niño/Southern Oscillation (ENSO), 11, 13, 91, 211
Energy balance, 12, 156, 181, 183, 191, 212
ERA40, 6, 257
Europe, 16, 183, 185, 187, 188, 235–237, 239–243, 247, 251, 254, 255, 258, 259, 262–264
Evaporation, 112, 164, 186, 195, 196, 203
Extremes, 1, 2, 4–6, 12–16, 20, 235–238, 240, 243, 251, 252

F

Flooding, 2, 200, 210
Free troposphere, 275, 276

G

Greenhouse
  effect 29, 155, 161, 163–165, 283
  gases 9, 10, 16–18, 105, 120, 124, 144, 155, 161, 200, 203, 243, 272, 287, 288

H
Heatwave, 14, 186, 188, 235, 236, 239, 240, 242–246, 252, 254
Humidity, 13, 33, 34, 39, 40, 65, 66, 111, 112

I
Infrared radiation, 104, 226
Interannual variability, 4, 13, 52, 59, 108, 136, 137, 139, 180, 183, 186, 187, 251, 253, 255, 264, 296, 299, 300, 323, 342, 351
Intergovernmental Panel on Climate Change (IPCC), 1, 12, 86, 91, 104, 159, 183, 215, 216, 231, 235–237, 243, 244, 251–254, 259– 261, 264, 265, 272, 288, 317, 318, 327

L
Land-atmosphere interaction, 12, 13, 179–189, 215, 231
Land surface, 1, 4, 7, 11, 12, 16, 86, 88, 163, 179, 186–188, 203, 212, 221, 244, 253
Leaves, 13, 186, 215, 222, 224–227, 229–231
Long-wave radiation, 4, 10, 238, 241

M
Mediterranean, 14, 180, 182, 183, 185–187, 211, 214, 216
Methane (CH_4), 16, 272, 280, 283, 284, 288
Microwave Sounding Unit (MSU), 87, 88, 90, 91, 93, 108, 110, 119–127, 326, 328
Monsoon
  Asian, 212
  West African, 212, 215, 216

N
NCEP/NCAR, 6, 95, 107, 110, 148–150, 293, 295–300, 348–352, 354, 355
North America, 9, 13, 43, 180, 224, 226, 271, 275, 280, 317, 329
North Atlantic Oscillation (NAO), 13, 228–230, 277
Northern Annular Mode (NAM), 18, 19
NOx, 16, 17, 272, 273, 278, 284, 285, 290

O
Ocean-atmosphere interaction, 152
Ozone, 4, 6, 7, 16–19, 91, 124, 129–139

P
Pacific, 13, 58, 68, 106, 113, 152, 175, 200, 211, 237, 277, 326, 327
Phenology, 186, 187, 221, 222, 228, 230, 231
Plants, 180, 186–188, 222, 224–229
Polar vortex, 18–20, 137, 286, 293–300, 311, 322, 347, 348, 356
Precipitation, 5, 14, 15, 37, 88, 111, 112, 164, 175, 179

R
Radiative forcing, 9, 10, 158, 162–164, 169, 173, 195, 204, 216, 272, 280
Radiosonde, 6, 7, 38, 40–43, 85–96, 98, 99, 103, 105, 106, 109–111, 115, 119, 295, 348
Reanalysis, 6, 7, 58, 66, 85, 86, 88, 92–99, 103–105, 108–111
Remote sensing, 47–49, 52, 60, 87, 92, 110, 123, 143, 189, 231, 309

S
Sahel, 13, 14, 164, 180, 200, 210–213, 215
Sea ice, 2–4, 6, 7, 10, 18, 47–55, 57–61, 75, 77, 79, 105
Sea-surface temperature (SSTs), 6, 9, 13, 14, 18, 65–80, 105, 113, 175, 211, 212, 284, 288, 290, 324, 328
Short-wave radiation, 10, 19, 169, 173, 177, 240
SO_2, 274
Soil moisture, 16, 156, 163, 164, 179–183, 185, 186, 188, 189, 212, 243, 244, 246, 247
Solar irradiance, 9, 19, 20, 317, 318, 320, 321, 323, 324, 328, 329
Stratosphere, 1, 2, 4, 7, 10, 16–20, 34, 40, 42, 86, 89, 91, 92, 95
Storm, 29, 30, 103, 113, 322
Sulphate, 159, 202
Sulphur, 10, 159

T
Thermohaline circulation, 9
Total Ozone Mapping and monitoring system (TOMS), 17, 132–136
Transpiration, 180–183, 185, 186, 188

Trends, 2, 5–12, 14, 18, 20, 56, 59–61, 85–99
Tropical Pacific, 13, 175, 211

U

UV radiation, 17, 18, 319, 320

V

Volcanic
  aerosols, 18, 212
  eruptions, 2, 10–12, 19, 20, 65, 212, 285, 287, 290, 295, 313, 323

Advances in Global Change Research

1. P. Martens and J. Rotmans (eds.): *Climate Change: An Integrated Perspective.* 1999 ISBN 0-7923-5996-8
2. A. Gillespie and W.C.G. Burns (eds.): *Climate Change in the South Pacific: Impacts and Responses in Australia, New Zealand, and Small Island States.* 2000 ISBN 0-7923-6077-X
3. J.L. Innes, M. Beniston and M.M. Verstraete (eds.): *Biomass Burning and Its Inter-Relationships with the Climate Systems.* 2000 ISBN 0-7923-6107-5
4. M.M. Verstraete, M. Menenti and J. Peltoniemi (eds.): *Observing Land from Space: Science, Customers and Technology.* 2000 ISBN 0-7923-6503-8
5. T. Skodvin: *Structure and Agent in the Scientific Diplomacy of Climate Change.* An Empirical Case Study of Science-Policy Interaction in the Intergovernmental Panel on Climate Change. 2000 ISBN 0-7923-6637-9
6. S. McLaren and D. Kniveton: *Linking Climate Change to Land Surface Change.* 2000 ISBN 0-7923-6638-7
7. M. Beniston and M.M. Verstraete (eds.): *Remote Sensing and Climate Modeling: Synergies and Limitations.* 2001 ISBN 0-7923-6801-0
8. E. Jochem, J. Sathaye and D. Bouille (eds.): *Society, Behaviour, and Climate Change Mitigation.* 2000 ISBN 0-7923-6802-9
9. G. Visconti, M. Beniston, E.D. Iannorelli and D. Barba (eds.): *Global Change and Protected Areas.* 2001 ISBN 0-7923-6818-1
10. M. Beniston (ed.): Climatic Change: *Implications for the Hydrological Cycle and for Water Management.* 2002 ISBN 1-4020-0444-3
11. N.H. Ravindranath and J.A. Sathaye: *Climatic Change and Developing Countries.* 2002 ISBN 1-4020-0104-5; Pb 1-4020-0771-X
12. E.O. Odada and D.O. Olaga: *The East African Great Lakes: Limnology, Palaeolimnology and Biodiversity.* 2002 ISBN 1-4020-0772-8
13. F.S. Marzano and G. Visconti: *Remote Sensing of Atmosphere and Ocean from Space: Models, Instruments and Techniques.* 2002 ISBN 1-4020-0943-7
14. F.-K. Holtmeier: *Mountain Timberlines. Ecology, Patchiness, and Dynamics.* 2003 ISBN 1-4020-1356-6
15. H.F. Diaz (ed.): *Climate Variability and Change in High Elevation Regions: Past, Present & Future.* 2003 ISBN 1-4020-1386-8
16. H.F. Diaz and B.J. Morehouse (eds.): *Climate and Water: Transboundary Challenges in the Americas.* 2003 ISBN 1-4020-1529-1
17. A.V. Parisi, J. Sabburg and M.G. Kimlin: *Scattered and Filtered Solar UV Measurements.* 2004 ISBN 1-4020-1819-3
18. C. Granier, P. Artaxo and C.E. Reeves (eds.): *Emissions of Atmospheric Trace Compounds.* 2004 ISBN 1-4020-2166-6
19. M. Beniston: Climatic Change and its Impacts. *An Overview Focusing on Switzerland.* 2004 ISBN 1-4020-2345-6
20. J.D. Unruh, M.S. Krol and N. Kliot (eds.): *Environmental Change and its Implications for Population Migration.* 2004 ISBN 1-4020-2868-7
21. H.F. Diaz and R.S. Bradley (eds.): *The Hadley Circulation: Present, Past and Future.* 2004 ISBN 1-4020-2943-8
22. A. Haurie and L. Viguier (eds.): *The Coupling of Climate and Economic Dynamics.* Essays on Integrated Assessment. 2005 ISBN 1-4020-3424-5

Advances in Global Change Research

23. U.M Huber, H.K.M. Bugmann and M.A. Reasoner (eds.): *Global Change and Mountain Regions*. An Overview of Current Knowledge. 2005.
ISBN 1-4020-3506-3; Pb 1-4020-3507-1
24. A.A Chukhlantsev: *Microwave Radiometry of Vegetation Canopies*. 2006
ISBN 1-4020-4681-2
25. J. Mcbeath, J. Rosenbery: *Comparative Environmental Politics*. 2006
ISBN 1-4020-4762-2
26. M.E. Ibarrarán and R. Boyd: *Hacia el Futuro*. Energy, Economics, and the Environment in 21st Century Mexico. 2006 ISBN 1-4020-4770-3
27. N.J Roserberg: *A Biomass Future for the North American Great Plains*: Toward Sustainable Land Use and Mitigation of Greenhouse Warming. 2006.
ISBN 1-4020-5600-1
28. V. Levizzani, P. Bauer and F.J. Turk (eds.): *Measuring Precipitation from Space*. EURAINSAT and the Future. 2007 ISBN 978-1-4020-5834-9
29. N.H. Ravindranath, M. Ostwald: *Handbook for Greenhouse Gas Inventory. Carbon Mitigation and Roundwood Production Projects*. 2007 ISBN 978-1-4020-6546-0
31. E. Wiegandt: *Mountains*: Sources of Water, Sources of Knowledge. 2007
ISBN 978-1-4020-6747-1
32. C.F. Hutchinson, S.M. Herrmann: *The Future of Arid Lands-Revisited*. 2007.
ISBN: 978-1-4020-6688-7
33. S. Brönnimann, J. Luterbacher, T. Ewen, H.F. Diaz, R.S. Stolarski, U. Neu (eds.): *Climate Variability and Extremes during the Past 100 Years*. 2008.
ISBN: 978-1-4020-6765-5

MIX
Papier aus verantwortungsvollen Quellen
Paper from responsible sources
FSC® C105338

If you have any concerns about our products,
you can contact us on
ProductSafety@springernature.com

In case Publisher is established outside the EU,
the EU authorized representative is:
Springer Nature Customer Service Center GmbH
Europaplatz 3, 69115 Heidelberg, Germany

Printed by Libri Plureos GmbH
in Hamburg, Germany